Mastering Microsoft Defender for Office 365

Streamline Office 365 security with expert tips for setup, automation, and advanced threat hunting

Samuel Soto

Mastering Microsoft Defender for Office 365

Group Product Manager: Dhruv J. Kataria

Publishing Product Manager: Prachi Sawant

Book Project Manager: Ashwin Dinesh Kharwa

Senior Editor: Mudita S

Technical Editor: Rajat Sharma

Copy Editor: Safis Editing

Proofreader: Mudita S

Indexer: Hemangini Bari

Production Designer: Alishon Mendonca

DevRel Marketing Coordinator: Marylou De Mello

First published: September 2024

Production reference: 1140824

Published by Packt Publishing Ltd.

Grosvenor House

11 St Paul's Square

Birmingham

B3 1RB, UK.

ISBN 978-1-83546-828-9

www.packtpub.com

To my mother, Migdonia Serrano Quiñones, and the memory of my father, Samuel Soto Sosa, for their sacrifices and for exemplifying the importance and power of determination and humbleness. To my wife, Haruka, and my son, Kentarou, for their love and support during life's journey; behind every great man, there is a loving family supporting him. Finally, to you, the reader, for your desire to learn and become a better security professional.

– Samuel Soto

Contributors

About the author

Samuel Soto, a seasoned cybersecurity expert, has forged a 25-year career across both the public and private sectors worldwide. Since joining Microsoft in 2020, he has been regularly engaged in thwarting complex security challenges. His portfolio includes high-profile engagements in cyber threat intelligence, specifically dealing with nation-state adversaries and organized cybercrime gangs. Samuel's experience and leadership in digital transformations, coupled with an entrepreneurial spirit, has seamlessly bridged the technology-business gap, allowing him to make significant strides during critical recovery and transformation efforts for many governments' environments and Fortune-100 companies.

I want to thank the people who have been close to me and supported me, especially my wife, Haruka, my son, Kentarou, and my parents, Migdonia Serrano Quiñones and Samuel Soto Sosa.

About the reviewers

Chris Tierney is a principal incident response security researcher on the Microsoft **Detection and Response Team (DART)** and has been helping customers respond to their security incidents during their time of need. He has worked in security for a decade and, broadly, in IT for the last 18 years. He has lived and worked in Japan since 2014 across the defense and public sector industries. Recently, he graduated from the SANS Institute with a master's in information security engineering. Chris's true professional passions are mentoring new and upcoming members of the security community to break into the field, cultivate their passions, and achieve their goals.

Paul Sudduth is a cybersecurity enthusiast with over two decades of experience in safeguarding critical systems, in both the military and corporate sectors. He holds a bachelor's degree in cybersecurity and information assurance and possesses an array of prestigious certifications, including GIAC, GCIH, and GSEC. His background encompasses contributions to numerous high-profile cybersecurity projects and initiatives. He is a member of the SANS Advisory Board.

Table of Contents

2

Basic Components of Defender for Office 365   19

3

Basic Checks and Balances   45

4

Part 2 - Day-to-Day Operations

5

6

7

Strengthening Email Security   147

8

Catching What Passed the Initial Controls   169

9

Incidents and Security Operations   197

Part 3 – Making the Tool Work for Your Organization

10

Magnifying the Unseen – Threat Intelligence and Reports   247

11

12

Preface

Over the course of my 25+ years in the information technology and security fields, I have had the opportunity to support many organizations with varying security budgets. One common observation I've made is the tendency for security strategies to rely on a single set of isolated security technologies. This approach often overlooks the most vulnerable areas, including email clients such as Outlook, document processing software such as Word, and even messaging applications such as Teams.

Unfortunately, even organizations with substantial security budgets lack a focused security strategy that addresses real-world attacks. As a result, many of these organizations experience breaches, some of which go unnoticed until their entire environment is compromised. It's a recurring theme in the news, with major companies from the Global 500 reporting significant breaches, facing hefty fines, and losing customer trust.

Sometimes, a breach can be as simple as an employee accidentally clicking on a link in an email while trying to assist someone, leading to the infection of internal corporate systems. The aftermath of such incidents involves not only lost business but also extensive damage control, including costly investigations and recovery processes that can run into millions of dollars.

The absence of an integrated and focused security strategy results in a lack of proactive and effective protection. Consequently, organizations are forced to resort to incident response and containment measures once a breach occurs. Recognizing this issue, Microsoft, a leader in software, cloud, and security, has developed Defender for Office 365. This set of security tools is specifically designed to safeguard the everyday tools that end users interact with.

In this book, I aim to provide you with not only a comprehensive understanding of the threats you need protection from but also insights into how Defender for Office 365 operates, as well as how you can leverage it to ensure your organization doesn't fall victim to poor productivity tool security.

Who this book is for

If you're looking to enhance the security of productivity software such as Office 365, you've come to the right book. Whether you're a novice or an experienced IT or security professional, this book offers practical insights to support and administer Defender for Office 365 deployments. It starts by covering fundamental security concepts and Office 365 deployments to ensure inclusiveness for all readers. The discussion then delves into the various components of Defender for Office 365, guiding users in how to determine the optimal configuration for their environment and providing methods to track their success.

What this book covers

Chapter 1, The Security Wild West, introduces you to common security concepts and how Microsoft Defender security products protect your organization. Zero trust is covered, along with how to get executive support.

Chapter 2, Basic Components of Defender for Office 365, explores the basic components of Defender for Office 365, how these work against common security threats, and the impact of misconfiguration on the end user.

Chapter 3, Basic Checks and Balances, examines the security frameworks and approaches used by many organizations and how to identify what works for your organization. Guidance is provided on how to qualify these requirements into trackable metrics to help strategize your deployment.

Chapter 4, Basics of Configuration, walks you through a basic deployment and how to ensure it aligns with an organization's security requirements, while minimizing end the user impact.

Chapter 5, Common Troubleshooting, covers the common approaches to troubleshooting issues in Defender for Office 365 and tips on saving time during maintenance. Effective approaches are introduced to handle rare and complex issues.

Chapter 6, Message Quarantine Procedures, discusses how to manage message quarantines and strike a good balance between effective quarantines and minimal end user impact.

Chapter 7, Strengthening Email Security, dives into advanced configuration, including measures to minimize malicious messages coming from your environment.

Chapter 8, Catching What Passed the Initial Controls, covers more advanced protections to handle malicious messages that evade the initial controls deployed for advanced attacks or internal threats. Guidance is provided on the proper analysis and control of message routing by using mail flow and message tracing.

Chapter 9, Incidents and Security Operations, explores effective security operations to decrease missed threats. Automation is introduced to improve efficiency, increase visibility, decrease wasted man-hours, and decrease alert fatigue among the security team members.

Chapter 10, Magnifying the Unseen – Threat Intelligence and Reports, examines threat intelligence and the many options available to enrich signals and alerts, helping you to further improve security operations and threat hunting. Reports are also discussed to track the effectiveness of security efforts and threat intelligence quality.

Chapter 11, Integration and Artificial Intelligence, discusses approaches to leveraging information from third-party tools to improve security operations, including approaches to integration. Artificial intelligence is introduced, including how to use Copilot for Security to further improve security operations.

Chapter 12, *User Awareness and Education*, provides guidance on how to execute effective security training, as well as how to use the features available in Defender for Office 365 to execute training that mimics real-world attacks.

To get the most out of this book

You will need a basic understanding of how to use a web browser, as well as evaluation licenses for Office 365 and Defender for Office 365 if your organization does not have any licenses yet.

Software/hardware covered in the book	Operating system requirements
Microsoft Outlook	Windows or macOS
Microsoft Teams	Windows or macOS
Microsoft Word	Windows or macOS
Microsoft Excel	Windows or macOS

Conventions used

There are a number of text conventions used throughout this book.

`Code in text`: Indicates code words in text, database table names, folder names, filenames, file extensions, pathnames, dummy URLs, user input, and Twitter handles. Here is an example: "Open a PowerShell window and connect to your EOP tenant by using `Connect-ExchangeOnline`."

A block of code is set as follows:

```
Hostname: _dmarc
TXT value: v=DMARC1; p=<reject | quarantine | none>;
pct=<0-100>; rua=mailto:<DMARCAggregateReportURI>;
ruf=mailto:<DMARCForensicReportURI>
```

When we wish to draw your attention to a particular part of a code block, the relevant lines or items are set in bold:

```
Set-HostedContentFilterPolicy -Identity "MyASFPolicy" -[ASFSetting]
Test
```

Bold: Indicates a new term, an important word, or words that you see on screen. For instance, words in menus or dialog boxes appear in **bold**. Here is an example: "They can also revoke the **Remember MFA on the device** option."

> **Tips or important notes**
> Appear like this.

Get in touch

Feedback from our readers is always welcome.

General feedback: If you have questions about any aspect of this book, email us at `customercare@packtpub.com` and mention the book title in the subject of your message.

Errata: Although we have taken every care to ensure the accuracy of our content, mistakes do happen. If you have found a mistake in this book, we would be grateful if you would report this to us. Please visit `www.packtpub.com/support/errata` and fill in the form.

Piracy: If you come across any illegal copies of our works in any form on the internet, we would be grateful if you would provide us with the location address or website name. Please contact us at `copyright@packt.com` with a link to the material.

If you are interested in becoming an author: If there is a topic that you have expertise in and you are interested in either writing or contributing to a book, please visit `authors.packtpub.com`.

Share Your Thoughts

Once you've read *Mastering Microsoft Defender for Office 365*, we'd love to hear your thoughts! Scan the QR code below to go straight to the Amazon review page for this book and share your feedback.

`https://packt.link/r/1835468284`

Your review is important to us and the tech community and will help us make sure we're delivering excellent quality content.

Download a free PDF copy of this book

Thanks for purchasing this book!

Do you like to read on the go but are unable to carry your print books everywhere?

Is your eBook purchase not compatible with the device of your choice?

Don't worry, now with every Packt book you get a DRM-free PDF version of that book at no cost.

Read anywhere, any place, on any device. Search, copy, and paste code from your favorite technical books directly into your application.

The perks don't stop there, you can get exclusive access to discounts, newsletters, and great free content in your inbox daily

Follow these simple steps to get the benefits:

1. Scan the QR code or visit the link below

https://packt.link/free-ebook/978-1-83546-828-9

2. Submit your proof of purchase

3. That's it! We'll send your free PDF and other benefits to your email directly

Part 1 – Introduction and Basic Configuration

In this part, we will introduce you to the cyber security threat landscape, security strategies, the Defender ecosystem, and the security deficiencies that Defender for Office 365 should correct. We will also explain the different components of Defender for Office 365 and how they work together to protect an organization. We will discuss what happens when a component is misconfigured. Next, we will help you identify the type of security your organization needs and align your Defender for Office 365 deployment accordingly. Finally, we will teach you how to perform a basic deployment of Defender for Office 365 and explain the impact it will have on end users.

This part contains the following chapters:

- *Chapter 1, The Security Wild West*
- *Chapter 2, Basic Components of Defender for Office 365*
- *Chapter 3, Basic Checks and Balances*
- *Chapter 4, Basics of Configuration*

1

The Security Wild West

Welcome to this detailed guide on Microsoft Defender for Office 365. The dangers of constantly evolving security threats have never been more evident. Understanding the security tools that impact most end user activities is the key to lowering your organization's security risks. In this book, we will explore the complexities of deploying Microsoft Defender for Office 365. This is one of Microsoft's premier security tools for protecting Office 365, a productivity and communication suite used by most organizations. In the following chapters, our focus will be to provide rationale for the reader to understand the importance of this security tool, along with guidance on configuring, using, and leveraging its advanced features, such as integration and proactive threat hunting.

Our security expedition will begin in *Chapter 1*, where we will navigate today's dangerous cyber threat landscape. We will establish a foundation for security by providing insights into the diverse facets of cyber threats, including advanced persistent threats, and into the impact of emerging technologies on security. We will explore the common attack vectors associated with Office 365, offering a contextual understanding of the daily threats organizations face. This understanding is crucial for comprehending the full scope of the capabilities of Defender for Office 365, and for understanding how the various components protect and integrate into a typical Office 365 deployment.

To ensure a complete understanding and provide guidance for a more diverse set of organizations, we will cover the reasons why migrating from **Exchange Online Protection** (**EOP**) to Defender for Office 365 is not just a change in product, but also a significant advancement in fortifying a productivity environment. The concept of Zero Trust will also be visited to include how this popular architecture design approach is supported by Defender for Office 365, and to discuss how we can benefit from it in our environment. The chapter ends with a discussion on some ways to kick-start the conversation with an organization's executives on what **return on investment** (**ROI**) can be expected from implementing Defender for Office 365. By the time you reach the end of this book, you will possess a thorough understanding of Defender for Office 365, how to gain executive backing for its implementation, and how to leverage its functionalities to leapfrog security efforts in your organization. I welcome you on your quest to become a master of Microsoft Defender for Office 365.

This chapter will cover the following topics:

- The security threat landscape and how it impacts your organization
- Typical approaches to deploying and attacking productivity tools
- The security tools that Microsoft offers and how they support security strategies
- The typical protection approach for productivity tools
- Discussing the ROI of Defender for Office 365 with your C-suite executives

Let our journey begin!

The cyber threat landscape – how do others get attacked?

Security is always a game of cat and mouse. Adversaries are constantly learning new tricks and developing new attacks, both on the technical and social engineering sides. For example, let's consider credit cards. In the mid-nineties, fake number generators were a major problem. To defend against this, credit card companies introduced ways to verify numbers in real time, so attackers had to find a way to capture real numbers online via fake websites. Credit card companies again smartened up and set up further protections to prevent captured credit cards from being used. As such, adversaries are now using credit card skimmers to physically copy and clone cards. This constant back and forth is also occurring on a different scale and impacting not only individuals, but organizations and governments too. Remember, the technological advancements we are currently experiencing are not used exclusively by law-abiding individuals. The advancements are giving adversaries multiple new platforms to change their approaches.

Cyber threats and their evolution

Over the years, cyber threats have changed, transitioning from simple viruses and worms to more intricate types of attacks, such as ransomware, phishing, and **advanced persistent threats** (**APTs**). Ransomware is a notable threat that involves encrypting data and demanding a ransom for its release (*Newman & Burgess, 2023*). In contrast, phishing attacks focus on manipulating individuals, luring them into revealing sensitive information through deceitful emails or fake websites (*Wong, 2023*).

The emergence of APTs adds a layer of intricacy to the ever-evolving landscape of cybersecurity threats. These groups' calculated attacks, frequently funded and directed by governments, have a long-term perspective, aiming to either steal sensitive data or disrupt operations. A key characteristic of most APTs is their persistent nature, with some groups operating covertly for many months at a time due to substantial resources backing their operations (*Yasar & Rosencrance, 2023*). A famous example is the SolarWinds attack, which saw the Russian-based group APT29, or Cozy Bear, going to great lengths to hide their presence for nine months.

The role of emerging technologies

The emergence of **Artificial Intelligence (AI)**, **Machine Learning (ML)**, and the **Internet of Things (IoT)** has made the cybersecurity landscape increasingly complex. Despite the benefits of these technologies, it's vital to acknowledge the potential vulnerabilities they introduce, making them susceptible to exploitation by cybercriminals. Attackers find IoT devices particularly enticing due to their lack of security controls out of the box, as well as many instances of organizations not securing IoT devices and thereby exposing them to the internet.

While many organizations have leveraged AI and ML to improve security tools, such as next-generation antivirus agents, the potential for misuse still exists. Cybercriminals have identified multiple opportunities to use these technologies to automate attacks, evade detection, and mimic human behavior, all to deceive unsuspecting victims (*Tamer Charife & Michael Mossad, (n.d.)*). This problem has been particularly notable with the rise of deepfake videos, which have been used by nation-states to cause confusion among opposing nations' citizens, and to influence public opinion in the adversary's favor.

The human factor

Even with the constant progress of technology, we cannot ignore the fact that the human element remains a crucial weakness. Social engineering attacks, known for their ability to manipulate individuals and extract sensitive information, have a high success rate. This highlights the significance of continuous education and awareness, as it enables individuals to more effectively identify and react to these threats (*SecurityScoreCard.com, 2024*).

Having discussed the most common vectors seen in general, it helps to have a look at the most common vectors as they refer to Office 365.

Common attack vectors related to Office 365

Long gone are the days of faxes and punch cards. Using productivity tools has become crucial for running a successful business in today's world. The Office 365 suite has proven to be invaluable in maintaining efficient operations not only due to the flexibility in content creation, but also the ability to continue working anywhere, even if you need to change devices, just by using a web browser. However, the widespread use of these technologies has made them a bigger target for cybercriminals, as there is a higher chance of success for them. The following are just some of the many attacks common to Office 365 and multiple other productivity suites.

Email-based attacks

Among the many attack vectors, it is crucial to highlight that email continues to be one of the most common and extensively abused methods. Cybercriminals often resort to phishing or spear-phishing attacks, where they craft deceptive emails that mimic email messages with the aim of tricking individuals into revealing sensitive information, clicking on a link, or even downloading malicious attachments. The popularity of Office 365's Outlook has caused it to become a key component of many attacks. Adversaries include malicious messages that mimic the look of messages in Outlook originating from corporations or even Microsoft itself (*GeeksforGeeks, 2023*).

Credential theft

Credential theft is another common attack vector. It involves attackers employing techniques such as trying commonly used or publicly leaked passwords to gain unauthorized access to an account, either by password sprays or brute force. These types of attacks leverage the lack of identity management security practices in many environments, including not using Conditional Access policies or modern **multi-factor authentication** (**MFA**) approaches. These security practices may be omitted to prevent user notification fatigue, which attackers take advantage of. Once an attacker successfully accesses a user account and gains access to the Office 365 environment, they can take advantage of the user's privileges for enumeration of the environment and lateral movement (*Groenewald, 2022*). Such an attack was observed in the 2022 Uber breach, which saw an account being compromised due to MFA notification fatigue.

Third-party integrations

Office 365's integration with third-party apps provides attackers with an opportunity to exploit integration misconfigurations. A common attack vector is the use of legacy authentication methods, such as those observed in many organizations' Exchange Online deployments. This weakness was so commonly used that Microsoft decided to disable legacy authentication in any out-of-the-box configurations after October 1, 2022. Weak application passwords are another common misconfiguration that Microsoft security guidance recommends not to use, as they can be leveraged to circumvent MFA.

Cloud-based attacks

By exploiting synchronization tokens, **Attacker-in-the-Cloud** (**AitC**) attacks take advantage of the authentication and data synchronization processes used by cloud services. Unlike traditional **Attacker-in-the-Middle** (**AitM**) attacks that eavesdrop on data in transit, AitC attacks manipulate authentication tokens, giving attackers unrestricted access to a user's cloud-stored data without compromising their login credentials directly.

The subtlety of AitC attacks makes them particularly treacherous. By being ignorant of the authentication token tampering, the user remains unaware of the attacker's presence as they confidently access and store data in the cloud. As a result, an intruder can silently observe, manipulate, or pilfer data without detection.

Malicious insiders

Lastly, we cannot forget malicious insiders, which are present in all organizations and present one of the most damaging vectors. In an organization, certain individuals take advantage of their access to the Office 365 environment for malicious activities, such as stealing sensitive information, disrupting operations, or facilitating external attacks (*Anastasov, 2023*).

To gain a better understanding of the role these attack vectors play in productivity tools, it is good to have a look at what a typical deployment looks like.

Office productivity tool deployments – how do others deploy?

Microsoft Office 365 offers a suite of cloud-based productivity tools and services designed to empower an organization's digital landscape. Let's explore the key aspects of Microsoft 365, including its components and typical architecture, from an organization's viewpoint, covering hybrid and cloud-only environments.

Components of Microsoft 365

Microsoft 365 encompasses far more than just email. It includes a complete suite of collaboration tools that, aside from the traditional office productivity applications, facilitate content creation and sharing. The tools you will most likely encounter in most organizations include the following:

- **Exchange Online**: This service enables efficient communication for organizations through email hosting and management. Some of its features include shared calendars and contacts (*Microsoft, 2023*).
- **SharePoint Online**: SharePoint is a collaborative platform designed for document management and content sharing. It allows teams to create, share, and manage content and applications (*Microsoft, 2023*).
- **Teams**: Microsoft Teams acts as a collaboration hub, encompassing chat, video conferencing, file sharing, and integration with other Microsoft 365 apps (*Microsoft, 2023*).
- **OneDrive for Business**: Users can securely store, share, and access files from anywhere with this personal cloud storage service (*Microsoft, 2023*).

- **Office apps**: Also known as Microsoft 365, Office apps are accessible via the browser and on different devices. Microsoft 365 includes familiar Office apps such as Word, Excel, PowerPoint, and more (*Microsoft, 2023*).

- **Microsoft Entra ID** (**Azure Active Directory**): Entra ID, formerly Azure Active Directory, offers identity and access management for Microsoft 365 services, improving security and user administration (*Microsoft, 2023*).

Microsoft 365 is not only about the components it contains, but also about how these components are deployed and integrated into an organization's environment. It helps to view the different possible architectures.

Microsoft 365 cloud architecture

Organizations have the flexibility to deploy productivity tools in various ways to meet their specific needs. The two most common approaches are cloud-only and hybrid:

- **Cloud-only environment**: With a cloud-only configuration, all Microsoft 365 services and data are housed in the cloud. This simplifies management, reduces infrastructure costs, and offers scalability. However, it may demand a reliable internet connection.

- **Hybrid environment**: Many organizations choose a hybrid approach, mixing on-premises infrastructure with cloud services. This enables them to use their current investments while transitioning to the cloud. An example is integrating on-premises Active Directory with Entra ID.

Having a thorough understanding of Microsoft 365 components is just one piece of the puzzle. Microsoft's reputation for offering comprehensive, integrated environments that boost office productivity means that security has not been forgotten. A nicely integrated suite of security tools is offered that aims to help maintain an organization's security posture with minimal impact on productivity.

Microsoft Defender – a primer

Microsoft Defender creates a comprehensive ecosystem of security technologies, which intertwine seamlessly to deliver a complete and holistic approach to organizational security. By integrating multiple components, this platform enhances cybersecurity postures across identities, endpoints, cloud applications, and digital estates. This fosters a collaborative defense that adapts to organizational requirements while protecting against a constant barrage of new security threats.

Overview of the Microsoft Defender ecosystem

The ecosystem comprises several robust components, each contributing to different aspects of cybersecurity:

Figure 1.1 – The Microsoft Defender line of products and their focus

- **Microsoft 365 Defender XDR**: This central hub and security portal, formerly called *Microsoft Threat Protection*, acts as a unified front for all the alerts and signals provided by the entire Defender security suite. The acronym **XDR** stands for **Extended Detection and Response**. This comprehensive security solution offers end-to-end insights across endpoints, email, data, identities, and applications. By automating security operations, incident response efforts become more efficient.

- **Microsoft Defender for Endpoint**: This **endpoint detection and response** (EDR) solution included out of the box in all modern Windows operating systems via its unified agent strengthens security measures at the device level. This component surpasses ordinary next-generation antivirus solutions by providing **advanced threat protection** (ATP) for endpoints. It does this by using artificial intelligence to identify, examine, and address sophisticated threats, vulnerabilities, and breaches.

- **Microsoft Defender for Office 365**: With a focus on safeguarding the enterprise's communication and collaboration footprint, this module provides protection against threats such as phishing, malware, and more. It covers email, OneDrive, SharePoint, and other Office 365 services.

- **Microsoft Defender for Identity**: With a primary focus on safeguarding organizational identities, this agent offers increased visibility into the intricate processes and traffic associated with identity management in a domain controller. Leveraging on-premises Active Directory traffic in-depth analysis at the domain controller level to include certificate services and federation can help with identifying and investigating advanced threats, compromised identities, and malicious insider actions targeting the organization.

- **Microsoft Entra Privileged Identity Management**: This component improves upon the out-of-the-box protections and identity governance controls provided by Entra to include just-in-time privileged access, automated access reviews, enhanced audit history, and other features.

- **Microsoft Defender for Cloud Apps**: This security solution acts as a protective bubble, safeguarding **software-as-a-service (SaaS)** applications and monitoring the control plane of cloud platforms. It offers insights into cloud application usage, shadow IT, misconfiguration, and potential cybersecurity risks.

- **Microsoft Defender for Cloud**: As of the time of writing, this component is only accessible from the Azure portal, but alerts and findings are shared with Microsoft Defender XDR and visible from the Microsoft Defender 365 portal. Formerly named *Azure Security Center*, this tool provides a comprehensive and unified security posture management solution. It provides auditing according to common industry recommendations and correction via automation across all of Microsoft's cloud environments, including Azure, Microsoft 365, and hybrid environments, as well as third-party cloud environments such as AWS and GCP.

- **Microsoft Sentinel**: This tool resides in the Azure platform, but can be accessed from both the Azure portal and the Defender portal. This cloud-native solution combines **Security Information and Event Management (SIEM)** and **Security Orchestration Automated Response (SOAR)** capabilities to deliver advanced security analytics and threat intelligence throughout the organization, allowing for the identification of sophisticated attacks that involve multiple elements. By leveraging the capabilities of AI, Sentinel empowers organizations to address security threats swiftly and effectively by filtering and acting on the most relevant security data.

- **Microsoft Purview**: This is an all-inclusive data governance service that aids organizations in uncovering, comprehending, categorizing, and safeguarding their data. It allows for safeguarding across diverse sources, whether on-premises, in multi-cloud environments, or within SaaS applications. Purview's goal is to overcome the obstacles of data discovery, cataloging, and maintaining data sensitivity and compliance.

- **Microsoft Defender EASM**: Microsoft's offering for external attack surface management provides continuous discovery and tracking of an organization's external digital footprint to discover openings and vulnerabilities that could be exploited.

- **Microsoft Defender Threat Intelligence**: This product augments the threat intelligence already offered by Microsoft Defender products from incident investigations, detections in the field, and analysis from Microsoft security experts. This product provides in-depth information on current threats, actors, and other points of interest.

Implementing these tools can tremendously boost your security visibility and posture, especially when these are properly tuned, but it is not enough. An effective security strategy goes beyond just deploying tools. It involves designing an environment with security in mind. The Zero Trust approach can provide a way for many organizations to transition to a stronger security foundation from the beginning.

Holistic approach to security and Zero Trust

Microsoft's Defender suite draws special attention to the principle of integrated security by delivering an interconnected security strategy across an organization's entire digital estate. By sharing signals across its ecosystem, Defender effectively uses its extensive visibility, machine learning, and automation to proactively address, identify, analyze, and counter threats across all essential domains (*Microsoft, 2023*).

Suppose a phishing campaign targets an organization. In such a case, Microsoft Defender for Office 365 would swiftly flag the suspicious email and collaborate with Microsoft 365 Defender to automate investigations and expedite incident response (*Microsoft, 2023*). The combination of increased visibility and interoperability seamlessly aligns with the Zero Trust approach to security architecture.

The term Zero Trust centers on a straightforward yet powerful concept: trust no one and always verify access attempts. Unlike traditional perimeter-based security models, Zero Trust acknowledges the potential for threats to emerge from both external and internal sources within the organization. This approach shifts the emphasis from basic perimeter defense to a comprehensive defense strategy that carefully examines every access request, regardless of its source (*Irei & Shea, 2022*).

The essence of Zero Trust lies in three fundamental principles:

- **Verify identity**: Verify that all users and devices trying to access resources in an environment have been properly authenticated and authorized

- **Least privilege access**: Provide users with only the access they require to fulfill their job duties, and no extra permissions

- **Continuous monitoring**: Continuously monitor network activities, swiftly identifying and resolving any irregularities that may occur

The Microsoft Defender suite goes beyond traditional security measures by incorporating the Zero Trust model into its design, ensuring that its tools and services align with the principles of this security approach:

- **Endpoint security**: Microsoft Defender for Endpoint diligently monitors endpoints at all times, ensuring strict adherence to security policies before granting access. If anomalies are detected, the solution can swiftly address threats, maintaining the constant application of the verify principle of Zero Trust (*ThreatLocker, 2023*).

- **Identity management**: With the help of AI-driven technology, Microsoft Defender for Identity and Entra ID's user risk, sign-in risk, and conditional access policies can proactively detect and stop identity-based threats. By meticulously confirming the identity of each user and device, it seamlessly adheres to the Zero Trust principle, refusing to grant blind trust.

- **Information protection**: Microsoft Defender for Office 365 ensures real-time monitoring of data access. By communicating with Entra ID, verifying user identities, and restricting access to sensitive data, the least privilege principle is upheld (*Irei & Shea, 2022*).

- **Integrated threat intelligence**: The suite's threat intelligence capabilities provide constant surveillance of the threat landscape, adjusting to emerging risks and ensuring that security measures and tools keep pace with threats. This proactive approach aligns with the Zero Trust model's emphasis on remaining constantly vigilant (*ThreatLocker, 2023*).

To enhance their resilience against cyber threats, organizations are advised to use the full range of tools available in the Microsoft Defender suite. By leveraging the complete ecosystem, they gain advantages such as a cohesive security stance, easy information exchange, and efficient incident handling. Continuous monitoring ensures that an organization's defenses are constantly vigilant. Now that we have seen how all these tools and approaches work at a high level, let's dig deeper into what makes Defender for Office 365 special compared to more typical productivity security tools.

Protecting your productivity tools

Before we discuss what Microsoft Defender for Office 365 can do for your organization, it's crucial to grasp the underlying security solution that forms the basis for email protection: **EOP**. EOP, a cloud-based email filtering service developed by Microsoft, offers advanced protection against spam and malware. Its effectiveness has made it a key component in protecting many organizations' mailboxes by blocking malicious files, spam, and phishing attempts. With its integration with Microsoft Exchange Online and Office 365, EOP offers enhanced security and reliability features to protect organizational communications.

EOP components

With EOP, organizations can enjoy a wide range of features that are specifically designed to safeguard their email communications:

- **Anti-malware protection**: With its multi-layered anti-malware engine, EOP thoroughly examines and filters email content, ensuring any known malicious software is detected. This practice serves as a barrier, preventing harmful content from infiltrating an organization's inboxes (*Davis et al., 2023*).

- **Spam filtering**: Through its advanced algorithms, EOP efficiently detects and removes unwanted emails, ensuring users' inboxes remain clutter-free and secure.

- **Connection filtering**: EOP effectively blocks emails from malicious IP addresses by utilizing real-time block lists and establishing a safe sender list, ensuring seamless delivery of genuine emails (*Davis et al., 2023*).

- **Policy tips**: When integrated with **Data Loss Prevention** (**DLP**), EOP offers users policy tips to alert users about potential policy violations before sending an email, providing a way to minimize unintentional sharing of sensitive data (*Davis et al., 2023*).

- **Transport rules**: By configuring transport rules, administrators can enforce specific actions based on predetermined conditions, ensuring that emails meeting certain criteria are managed accordingly.

- **Safe attachments**: This feature scans email attachments in a special environment before they are delivered to the recipient, ensuring that no hidden malicious content can execute unauthorized processes within an organization's network (*Davis et al., 2023*).

No comprehensive enterprise security tool is free. Understanding the license structure is the key to securing good ROI.

Licensing for EOP

Microsoft offers EOP as part of both Exchange Online and Office 365 subscription packages. However, organizations that maintain their own mail servers can also get EOP as a standalone service to take advantage of its security measures. Here is the breakdown of the licensing structure:

- **EOP standalone**: Tailored to meet the needs of organizations utilizing on-premises Exchange servers. With this subscription, you get access to comprehensive protection that includes anti-malware and anti-spam filtering.

- **Office 365 E1**: This comprehensive package combines EOP with other productivity tools such as OneDrive, SharePoint, and Teams, giving you everything you need in one bundle.

- **Office 365 E3 and E5**: These advanced enterprise packages go beyond EOP and offer additional security features such as ATP and data governance tools.

The licensing costs may differ based on region, number of users, and any additional services or customizations requested by the organization. Precise pricing details can be obtained by directly consulting with Microsoft or its partners.

Defender for Office 365 – why not just stay with EOP?

Although EOP provides a strong security foundation, the increasing sophistication of cyber threats has expanded the range of attack vectors beyond email. Microsoft Defender for Office 365 aims to bolster the security of email and collaboration tools, providing a robust level of protection against evolving threats. By utilizing Defender for Office 365 instead of solely relying on EOP, users can enjoy heightened security measures and greater peace of mind (*MSFTTracyP & Davis, 2023*). Defender for Office 365 builds on many of the features offered by EOP, as shown in the following figure:

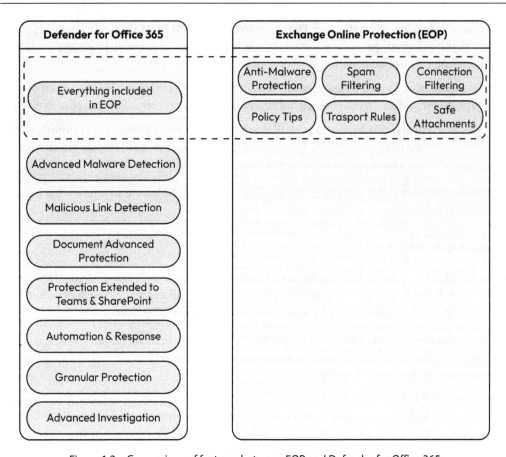

Figure 1.2 – Comparison of features between EOP and Defender for Office 365

Now that we have a general overview of the inclusions during the migration from EOP to Defender for Office 365, let's explore the nuances of these additional attributes.

Advanced security capabilities

EOP provides essential protection against threats, such as spam, malware, and phishing. However, Microsoft Defender for Office 365 strengthens this safeguard by incorporating extra measures such as Safe Attachment and Safe Links, which provide more advanced protection against harmful attachments and links. These features analyze links and attachments in real time to protect against potential harm.

Comprehensive protection across tools

Microsoft Defender for Office 365 is a versatile security service that considers more than just email threats. It ensures the safety of links and attachments across collaboration tools within Microsoft 365. As collaboration tools are more deeply incorporated into daily operations, having a holistic security solution becomes crucial.

Automation and response

One major benefit of Microsoft Defender for Office 365 is its **Automated Investigation and Response** (**AIR**) capability. This feature helps security teams by automating threat monitoring and remediation.

Customization and flexibility

While EOP offers fundamental protection, Microsoft Defender for Office 365 provides enhanced flexibility through custom policies. Organizations can customize security settings to match their specific requirements.

Advanced investigation tools

Not only does Defender for Office 365 offer protection, but it also includes tools for extensive threat investigation. Features such as Threat Explorer empower security teams to analyze the intricacies of a security event, providing critical insights for mitigating threats and preventing future incidents.

Comprehensive plans

There are different plans available with Microsoft Defender for Office 365. These are tailored to suit different organizational needs. For example, Defender for Office 365 Plan 1 includes all EOP features and provides real-time detections. On the other hand, Plan 2 provides additional advanced features, such as post-breach capabilities and attack simulation training.

You have learned about Defender for Office 365 features and benefits, but how do you communicate them to your executives? You should consider their expectations for the ROI and the time frame for achieving it.

Understanding the ROI

When organizations consider implementing Defender for Office 365, executives will inquire about the ROI. Financial loss, damage to brand reputation, and loss of customer trust can all be consequences of cyberattacks. Since calculating ROI for security investments is beyond the scope of this book, we will focus on essential points to facilitate organization-wide discussions.

The direct and indirect costs of cyber threats

Understanding the typical cost of a cyberattack is crucial for organizations to discuss the benefits of implementing Defender for Office 365. The cost and impact can differ significantly, so to estimate what an attack would cost your organization, we need to examine the following direct and indirect costs based on industry, size, laws, and location:

- **Direct costs**: Some immediate financial outlays comprise paid ransoms, system restoration, investigation, and regulatory and legal fees. A study conducted by IBM found that the average direct cost of a data breach to a company is $3.86 million. The cost of each lost or stolen record is estimated to be approximately $146. Direct breach costs are influenced by the cause of breach, number of records lost, and industry. The direct costs of healthcare breaches can reach an average of $7.13 million per breach, as stated by IBM (*IBM, 2024*).

- **Indirect costs**: The consequences of cyberattacks include damage to brand reputation, erosion of customer trust, and potential business loss from system downtime. According to Microsoft's 2020 Global Threat Report, the average indirect cost of a data breach for surveyed global organizations is $8.64 million. This estimate covers costs from business disruption, revenue loss, and brand reputation damage after a breach. According to Microsoft's report, breached organizations experienced a notable decline in customer retention, requiring an average of over 14 months to regain customer trust. The biggest factor in indirect breach costs is the loss of loyal customers. The report revealed that small and medium-sized businesses have higher indirect breach costs than larger enterprises (*Microsoft, 2020*).

When you talk to executives, you need to clarify the problem you want to solve, because they won't allocate funds for new tools based on intuition. You should also quantify the negative impact of the problem in financial terms to highlight its urgency.

The impact of Defender for Office 365 on ROI

To persuade your executives to adopt Defender for Office 365, you need to demonstrate how it addresses the gap you have identified in your current security posture. You can structure your argument around three main benefits: threat prevention, operational efficiency, and compliance enhancement. Use relevant data and examples to support each point, and tailor them to your organization's specific needs and goals. These three main benefits can be thought out in the following manner.

- **Threat mitigation**: ATP features, such as anti-phishing, anti-malware, and safe attachments, are included in Microsoft Defender for Office 365. By utilizing these features, the solution can deflect a significant number of cyberattacks and avoid associated costs.

- **Operational efficiency**: Traditional security solutions' level of false positives often overwhelms IT teams. Microsoft Defender's AI-driven algorithms minimize false positives, allowing IT staff to concentrate on genuine threats. This improves productivity and reduces expenses caused by wasted man-hours.

- **Compliance and governance**: Failure to comply with data protection regulations can lead to significant financial penalties. By using Microsoft Defender for Office 365, organizations can adhere to these regulations and potentially avoid expensive penalties.

The positives mentioned are extra points that you can account for when discussing with executives why Defender for Office 365 deployment should be funded. Remember that some executives might already be desensitized to negative news. Thus, providing points that touch on improving operations and lowering costs might put the deployment on the fast track for funding approval.

Summary

The benefits of using Microsoft Defender for Office 365 are many and varied. Although financial savings are obvious, the intangible benefits are just as important. Investing in strong security solutions not only protects organizations from financial loss, but also strengthens their brand value in today's digital world.

Cyber threats, ranging from phishing campaigns to intricate ransomware attacks, challenge organizations daily. These threats apply to all organizations, big or small, with Office 365 being particularly targeted because of its widespread usage. The vulnerability doesn't stop at emails; it encompasses file sharing and other collaboration tools as well. That said, Microsoft 365 provides more than just productivity. It also integrates with the Defender tools. Microsoft Defender surpasses the status of being a mere set of security tools. Its features, ranging from real-time threat detection to sophisticated investigation tools, are a testament to its holistic protective approach. Along with its obvious protective features, Defender for Office 365 also brings intangible benefits. Those include peace of mind, together with enhanced organizational reputation and fostered stakeholder trust, which leads to an amplified ROI. Microsoft Defender showcases the dynamic nature of cybersecurity, emphasizing the necessity of proactive defense and adaptability. The upcoming chapter will provide a granular look at each component, explaining how they contribute to bolstering an organization's security strategy and how we need to think about their configuration before attempting a deployment.

References

- Wong, D. (2023). *The Evolution of Phishing Attacks*. Cybersecurity Today. Retrieved from `https://cybersecurity.att.com/blogs/security-essentials/the-evolution-of-phishing-attacks`

- Yasar, K., & Rosencrance, L. (2023). *advanced persistent threat (APT)*. Security. `https://www.techtarget.com/searchsecurity/definition/advanced-persistent-threat-APT`

- SecurityScoreCard.com (2024). *The Human Factor in Cybersecurity*. `https://security-scorecard.com/blog/the-human-factor-in-cybersecurity/`

- Newman, L. H., & Burgess, M. (2023, July 12). *Ransomware Attacks Are on the Rise, Again*. WIRED. `https://www.wired.com/story/ransomware-attacks-rise-2023/`

- Tamer Charife & Michael Mossad (n.d.) *AI in cybersecurity: A double-edged sword*. Retrieved from `https://www2.deloitte.com/xe/en/pages/about-deloitte/articles/securing-the-future/ai-in-cybersecurity.html`

- Groenewald, S. (2022). *This is How Hackers Are Stealing Your Microsoft 365 Credentials*. Micro Pro IT Support. `https://micropro.com/blog/this-is-how-hackers-are-stealing-your-microsoft-365-credentials/`

- Anastasov, K. (2023). *Insider Threats in Office 365: Detection and Prevention Strategies*. Medium. `https://medium.com/cybersecurity-science/insider-threats-in-office-365-detection-and-prevention-strategies-1744d112e145`

- GeeksforGeeks. (2023, May 30). *Types of Email Attacks*. GeeksforGeeks. `https://www.geeksforgeeks.org/types-of-email-attacks/`

- Microsoft. (2023). *Microsoft 365 and Office 365 service descriptions.* Retrieved from `https://learn.microsoft.com/en-us/office365/servicedescriptions/office-365-service-descriptions-technet-library`

- Microsoft. (2023). *Microsoft Digital Defense Report.* Retrieved from `https://www.microsoft.com/content/dam/microsoft/final/en-us/microsoft-brand/documents/MDDR_FINAL_2023_1004.pdf`

- Microsoft. (2023). *Defending new vectors: Threat actors attempt SQL Server to cloud lateral movement.* Retrieved from `https://www.microsoft.com/en-us/security/blog/2023/10/03/defending-new-vectors-threat-actors-attempt-sql-server-to-cloud-lateral-movement/`

- ThreatLocker. (2023, September 29). *The evolution of Endpoint Security.* ThreatLocker. `https://www.threatlocker.com/blog/the-evolution-of-endpoint-security`

- Irei, A., & Shea, S. (2022, October 20). *What is the zero-trust security model?* Security. `https://www.techtarget.com/searchsecurity/definition/zero-trust-model-zero-trust-network`

- Davis, C., Chakrabarti, R., & Simpson, D. (2023, October 24). *Exchange Online Protection (EOP) overview.* Retrieved from Microsoft Learn: `https://learn.microsoft.com/en-us/microsoft-365/security/office-365-security/eop-about?view=o365-worldwide`

- MSFT TracyP & Davis, C. (2023, October 25). *Why do I need Microsoft Defender for Office 365?.* Retrieved from Microsoft Learn: `https://learn.microsoft.com/en-us/microsoft-365/security/office-365-security/mdo-about?view=o365-worldwide`

- IBM. (2024). *Cost of a Data Breach Report 2024.* Retrieved from `https://www.ibm.com/security/digital-assets/cost-data-breach-report/#/`

- Microsoft. (2020). *Microsoft Digital Defense Report.* Retrieved from `https://download.microsoft.com/download/f/8/1/f816b8b6-bee3-41e5-b6cc-e925a5688f61/Microsoft_Digital_Defense_Report_2020_September.pdf`

2

Basic Components of Defender for Office 365

The successful implementation of a security solution necessitates a comprehensive grasp of its purpose and functionality. This comprehension can aid in further delineating the sequence and execution of the deployment, as well as gaining a more profound insight into the solution's impact on the organization. Throughout this chapter, we will take a detailed journey through the various aspects and functionalities of Defender for Office 365, shedding light on its critical role in modern cybersecurity. Please note that this chapter will serve as a review of how the components protect, as configuration and deployment will be covered in detail in later chapters, so it is recommended for the reader to focus more on how the component fits within their organization's security strategy versus how it is configured.

We will begin by examining the robust mechanisms employed by Defender for Office 365 to block malicious files and attachments, which serve as a frontline defense against many cyber threats. Following this, we will delve into how it effectively shields your organization from malicious links and phishing attempts, as these are among the most common and damaging cyber-attacks.

Moreover, we will discuss how Defender for Office 365 empowers users by equipping them with the knowledge and tools to recognize and avoid potential threats. We will also conduct an in-depth analysis of its monitoring and investigation capabilities, highlighting how it enables you to understand and respond to security incidents within your environment.

We will explore the seamless integration of Defender for Office 365 with other Defender security products, enhancing your overall security posture. Lastly, we will review famous historical cyberattacks, analyzing how the deployment of Defender for Office 365 could have mitigated or even prevented these incidents. This analysis will provide valuable insights and lessons for strengthening your defense strategies.

This chapter will cover the following topics:

- Ways to stop malicious files and attachments in their tracks
- Ways to prevent your users from being victims of malicious links and phishing

- How to allow your users to support your organization's security goals
- Ways to perform security investigations in your environment
- Ways to leverage Defender's integration capabilities
- A look at some famous attacks and how our solution could have been used

Let's continue our journey!

Blocking malicious files and attachments

Microsoft Defender for Office 365 offers comprehensive protection against malicious files across the entire Microsoft 365 collaboration product lineup. The protection encompasses email, SharePoint, OneDrive, Teams, and even offers document-level security through integration with Defender for Endpoint. In this section, we delve into these components, exploring their intricacies and examining the consequences of misconfiguration.

Safe Attachments

Safe Attachments in Microsoft Defender for Office 365 offers an extra layer of protection from malware in email attachments. By opening the attachments in a virtual environment, commonly referred to as a *sandbox*, before they reach the recipients, it ensures their safety. The process, known as *detonation*, enables the Safe Attachments component to examine the attachments' behavior and impact with no risk of harm to the recipients' devices or systems. Depending on the recipient's Safe Attachments policy, the attachment can be blocked, replaced, or monitored if it is deemed malicious (*Microsoft, 2023*).

The Safe Attachments component can scan a wide range of file formats, including popular archive types including .zip and .rar. It can also scan files that are hidden within other files, such as a Word document housing an Excel spreadsheet. The Safe Attachments component can process attachments of up to 25 MB per message.

Safe Attachments policies, configurable in the Microsoft 365 Defender portal or Exchange Online PowerShell, govern the operation of the Safe Attachments component. Although there is no default Safe Attachments policy, there is a security policy called Built-in protection preset that automatically offers Safe Attachments protection to all recipients, unless covered by other custom or preset policies. The misconfiguration of this component can have various impacts on the security and productivity of the organization and its users. Some of the possible impacts are as follows:

- Inadvertently setting **block** or **redirect** on the wrong users can lead to delayed delivery and a decrease in user control over attachments. For example, if a valid attachment is mistakenly flagged as malicious, it will be prevented from reaching the intended recipient, substituted with a text document, or rerouted to an alternative email address. This can lead to frustration and confusion, which will affect trust and collaboration between users and external parties.

- Opting for **off** or **monitor** could expose recipients to harmful attachments that can damage their device or data. For instance, if an unchecked malicious attachment is sent, it can reach the recipient unnoticed or unaltered. The recipient could unknowingly open the attachment, triggering the execution of the malware and posing a threat to their system or network.

An example of a malicious attachment stopped via the Safe Attachments feature is shown in the following screenshot.

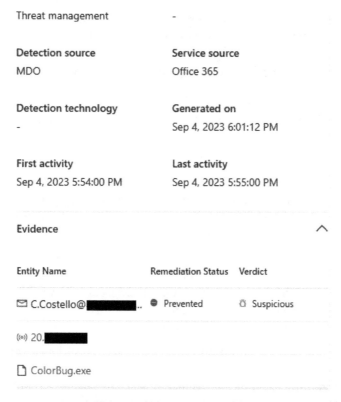

Figure 2.1 – Malicious attachments stopped

Protection extends beyond email attachments to encompass productivity files stored in cloud storage and devices.

Safe Documents

By extending the Safe Attachments configuration, the Safe Documents component adds an extra layer of security to Office documents opened in Protected View or Application Guard for Office, safeguarding against malware. The cloud backend of Microsoft Defender for Endpoint is utilized to scan documents and files (Defender for Endpoint does not need to be installed on the device), providing an extra layer of security before granting users permission to exit Protected View or Application Guard for Office. The process, known as *verification*, enables the Safe Documents component to examine the behavior and effects of documents and files without interfering with users' devices or systems. If the document or file is determined to be malicious, it may be blocked, quarantined, or monitored based on the user's Safe Documents policy settings (*Microsoft, 2023*).

The Safe Documents component can scan various types of files, including common Office formats such as `.docx`, `.xlsx`, and `.pptx`, as well as archive formats such as `.zip` and `.rar`. It can also scan files that are hidden within other files, such as a Word document with an embedded Excel spreadsheet. The Safe Documents component is capable of processing documents with file sizes of up to 60 MB.

Safe Attachments for SharePoint, OneDrive, and Teams

The enhanced safeguard provided by Safe Attachments can be expanded to SharePoint, OneDrive, and Teams. This approach safeguards organizations against additional vectors resulting from users or malicious individuals introducing and disseminating harmful files to an environment through methods other than email (*Microsoft, 2023*). This protection works as follows:

- **File scanning**: Just like the Safe Attachments feature for email attachments, files are scanned and opened in a virtual environment to observe any potential consequences (known as *detonation*). This feature is key for modern threats that have not been previously identified in malware signatures and that do not behave in a way that can be easily detected by typical machine learning models.

- **Password-protected files**: During detonation, the system verifies password-protected files by cross-referencing them with a database of commonly used passwords or patterns employed by malicious actors. This feature is quite useful in speeding up analysis as many actors will try to encrypt and password protect files to prevent analysis in hopes of frustrating the end user to the point of them ignoring the security controls.

- **Detecting and blocking malicious files**: Files identified as malicious within team sites and document libraries are immediately blocked from opening. This is a control against a modern attack vector in which bad actors first compromise a low-visibility account and use internal chat systems to spread malware in hopes of taking over high-privilege accounts.

- **File locking**: The direct integration with SharePoint, OneDrive, and Teams file stores ensures that malicious files are effectively locked away. Despite being listed and accessible, the blocked file defies users' attempts to open, copy, relocate, or share it; only deletion is permitted. This feature protects the user's system from modern highly evasive malware used by advanced state-backed actors, as many of these files tend to use new techniques and signatures not previously detected and so can stay dormant, evading detection and being shared by multiple users. Once the file is detected as malware, any copies in the environment are locked to prevent execution.

- **Downloading blocked files**: Users are given the choice to download a blocked file as the default setting, but SharePoint Online admins have the power to restrict this action to enhance protection against malware propagation. This feature is part of allowing security teams flexibility on how their security controls are implemented. In some mature environments, security teams might have a group of experts that can be allowed to download the file in a controlled environment for further analysis to determine if the finding is a false positive along with other actions such as forensics.

The presence of misconfiguration or lax security settings for the Safe Attachments feature in SharePoint, OneDrive, and Teams components could create vulnerabilities that enable the internal spread of threats if a single account is compromised. The collaborative nature of SharePoint and Teams makes this particularly risky, as it could have a widespread impact on the organization. As such, always remember that there is a fine balance that is different in each organization. The higher the security, the more restrictions and less flexibility users will experience. Always compare the possible security impact versus the impact on the business before lowering security controls.

Protecting from malicious links and phishing

Securing an organization is akin to a cat-and-mouse game, where cyber criminals adapt and evolve their tactics as defenses against malicious email attachments grow stronger. One tactic used by criminals is to include phishing or malicious links in emails and documents, hoping that someone will click and unknowingly grant them access to the organization. Microsoft has enhanced Defender for Office 365 by incorporating Safe Links and anti-phishing policies, ensuring the safety of user data and corporate information systems. Let's take a closer look at these components and explore how they safeguard an organization.

Safe Links technology

Safe Links enables real-time verification of URLs in emails, Office documents, and Teams (*Microsoft, 2023a*). Here's how it operates:

- **Real-time scanning**: When a user clicks on a link, Safe Links diligently checks it against a constantly updated database of malicious URLs. If a match is detected, the user might be alerted or prevented from entering the site. This is a key feature of security products with ample usage such as Defender products, as any malicious public links detected anywhere in the world are added to a database for other users to be automatically protected.

- **URL rewriting**: URLs are automatically rewritten to go through Microsoft's scanning engine, providing an added level of protection as they are checked each time a user tries to access them. This feature plays a significant role in protecting against phishing attacks, where hackers commonly mask dangerous links. It is also an important step to prevent users from spreading malicious links by mistake and ensure that anyone clicking the link is automatically routed to the Defender URL real-time scanning service.

- **User education and alerts**: Clicking on malicious links triggers an immediate redirection to a warning page, which serves as an educational tool to inform users about the associated risks and promote safe browsing habits. This is part of what is known as security in depth, as multiple controls are placed on potentially impactful actions in case the user skipped previous controls by mistake, along with serving as an education tool to make users more cautious.

Frustration among users can stem from misconfiguration in safe links, resulting in multiple areas of concern, such as the following:

- Users may be exposed to potential threats if malicious links aren't rewritten and bypass security checks

- Strict policies that unnecessarily block legitimate URLs, hindering productivity and causing frustration

- If users choose to visit unsafe sites, they may bypass warnings, potentially resulting in security breaches

An example of a malicious link stopped via the Safe Links feature is shown in the following screenshot:

Figure 2.2 – Malicious URL stopped

Safe Links is particularly effective in thwarting **zero-hour attacks,** which involve cybercriminals exploiting vulnerabilities that were previously undisclosed. When users click on links, the real-time verification can alert them if the URL redirects to a suspicious site, even if it wasn't flagged as dangerous in the initial email. With a significant portion of cyber-attacks, namely phishing attacks, originating from emails, this dynamic protection becomes an essential asset. Let's also not forget that many organizations might take months to detect a breach, which makes this a key feature in a wider proactive approach.

Anti-phishing policies

In addition to Safe Links, Microsoft Defender for Office 365 implements anti-phishing policies to protect against identity spoofing and fraudulent email attempts. These policies are driven by cutting-edge machine learning models and advanced analysis techniques, operating in the following manner (*Microsoft, 2023*):

- **User and domain impersonation protection**: These policies effectively detect and flag emails that try to mimic well-known usernames and domain names, diminishing the success rate of spear-phishing attempts that rely on impersonating trusted entities. Examples of the types of attack this protects against are when an attacker sends an urgent email pretending to be an executive to a help desk to get administrators to click on a malicious link, or even when the attacker sends malicious messages to external entities via compromised parts of a victim's infrastructure as seen in business email compromise attacks. In this typical type of attack, a simple character will be changed, for example, instead of CONTOSO.COM, the attacker could replace one of the letters with a number like CONTOSO.COM or even use Cyrillic character substitution. The domain used will be an existing domain and might even include a certificate to fool the victim.

- **Spoof intelligence**: Implementing algorithms that can identify sender spoofing allows for the quarantine of potentially harmful emails, safeguarding users from deceptive content that masquerades as coming from a legitimate source. Much like user and domain impersonation, the attacker is trying to fool the victim into making the malicious message look like it is coming from a valid domain, but in this case, this change is at the message header level and there might not even be an existing domain. Some attackers might perform this attack as it is lower cost and easier.

- **Customizable protection settings**: Organizations can adjust settings based on their specific risk profiles, enabling strict controls in areas such as financial departments. This option is quite powerful for allowing customizable security while minimizing the impact on the business. Remember that the more security is implemented, the less flexibility the end user will be allowed, but different groups have different privileges and impact on the environment which means they will require different levels of security.

A misconfigured anti-phishing policy can lead to the following issues:

- **False positives/negatives**: Legitimate emails can be mistakenly identified as phishing (*false positives*) or phishing attempts can go undetected (*false negatives*), both with serious consequences. Imagine the impact of mistakenly blocking the CEO's emails multiple times and you can understand why proper configuration is important.

- **User disruption**: Legitimate emails being quarantined can cause disruptions and workflow interruptions, leading to missed communications. Imagine your team missing some important project information due to emails being quarantined – depending on the message, the impact could be very costly to the business.

- **Security risks**: Inadequately configured anti-phishing policies can leave organizations at risk of phishing attacks, potentially leading to data breaches and financial losses. Imagine experiencing a major breach and tracking it down to a malicious email that is commonly stopped by Defender, but was allowed through due to improper whitelisting in a policy. The security team would have a lot of explaining to do regarding this configuration to multiple executives; not a nice look.

The impact of these policies on **Business Email Compromise** (**BEC**) scams is clear, serving as a testament to the policies' effectiveness. Using machine learning, Microsoft's anti-phishing measures have successfully detected and flagged unusual patterns in email content, header information, and sender behavior. This has greatly reduced the risk of employees falling victim to deceptive requests and unknowingly transferring funds into criminals' accounts.

An example of a BEC attack stopped is shown in the following screenshot:

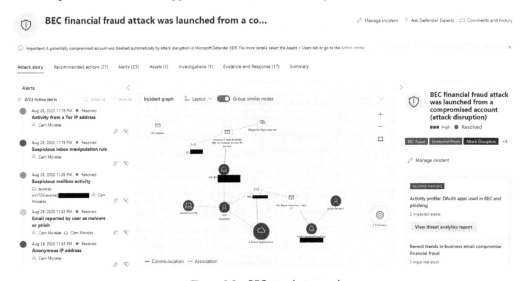

Figure 2.3 – BEC attack stopped

Effective security extends beyond fortifying against attacks and should also engage users as active participants.

Empowering your users

Despite the robust detection capabilities of Defender for Office 365, the tactics and methods employed by attackers constantly change. Hence, it is essential to give users a platform to report any suspicious behavior that might have been overlooked. This feedback is invaluable in refining security tools and bolstering your defense. This capability is offered through the Report Message add-in feature.

The Report Message add-in

The Report Message add-in, an integral element of Microsoft Defender for Office 365, empowers users to take charge of their email security. This feature, accessible through Outlook, enables users to report suspicious messages to their organization's IT department and Microsoft. Such messages may include deceitful phishing attempts, annoying spam, or any communication that raises suspicions of a security threat (*Microsoft, 2023*). This component's functionality focuses on the following areas:

- **Ease of reporting**: With the add-in installed, reporting an email is as easy as clicking a button within their Outlook client or on the web. The report button is one of the first controls that allows the security team to detect malicious messages that have evaded controls and users should be properly trained on its location and use.

- **Integration with security systems**: Once a message is reported, it's analyzed by Microsoft's security systems. By analyzing the data, the email filtering service can improve its efficiency and effectiveness, enabling it to better identify and block malicious emails in the future.

- **Feedback loop**: The information collected from user reports enhances Microsoft's machine learning models, resulting in a more agile and effective defense against email-based threats.

- **Administrative action**: Upon receiving the reported messages, administrators have the authority to examine the message headers or content in detail, conducting a comprehensive investigation. They may also blacklist URLs or domains if required.

The efficacy of this component could be compromised if it is not properly configured. The following are some examples of the risks:

- **Overlooked threats**: If the add-in isn't configured correctly for all users, there is a risk that some employees may overlook or fail to report malicious emails, which could result in avoidable security incidents or data breaches.

- **False positives**: If misconfigured, legitimate emails may be flagged as suspicious, possibly resulting in important communications being missed or delayed. Additionally, this issue may desensitize users to legitimate threats, leading to a decrease in overall alertness.

- **Increased vulnerability**: If the automated investigation and response capabilities linked to the Report Message add-in are not properly set up, the organization becomes more susceptible to attacks as the time to respond to and address threats lengthens.

An example of an email reported by a user thanks to the Report Message add-in is shown in the following screenshot:

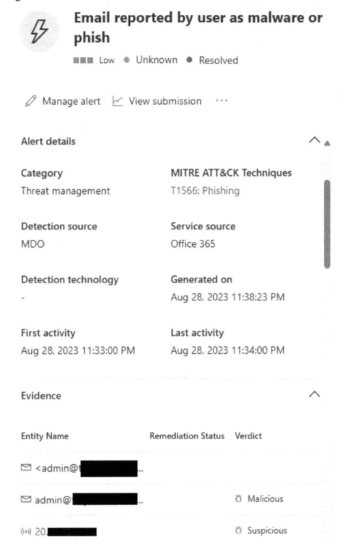

Figure 2.4 – Email reported as suspicious by user

The true importance of this component is revealed in the face of zero-day malware attacks. Microsoft uses user-reported suspicious files or links to improve its security features by analyzing content in a secure environment to detect new threats. When many users report numerous messages, it offers valuable insights that appear in the Security Dashboard and other reports. This information can serve as an indicator for your organization's security team of the need to fine-tune their policies.

Protecting against compromised internal accounts or devices

A comprehensive security program must address threats from external and internal sources. The potential harm caused by internal threats, whether from compromised accounts or malicious insiders, is equally significant. The focus of the advanced protection for internal mail is to address the need for protection from this perspective.

How advanced protection for internal mail works

Within Microsoft Defender for Office 365, advanced protection is used to safeguard internal mail by scanning for suspicious activity originating from and directed within the organization. This component utilizes the previously mentioned features: Safe Attachments, Safe Links, anti-phishing, and the Report Message add-in. It also incorporates Defender for Office 365 threat analytics and reporting, as well as automated investigation. These functionalities help analyze attachments, links, and message content to identify any potentially malicious content or behavior (*Microsoft, 2023*). Here's an overview of the functioning of the advanced protection for internal mail feature in Defender for Office 365:

- **Email filtering**: Defender for Office 365 employs a filtering stack that rigorously assesses internal emails in the same manner as external emails. It examines email content for dubious links and attachments, which are indicators of phishing and malware dissemination endeavors.

- **Safe Attachments**: This feature operates by isolating email attachments in a virtual environment, separate from the user's system, to conduct analysis and testing for potential malware, mitigating any risk to network integrity.

- **Safe Links**: Like Safe Attachments, Safe Links thoroughly inspect URLs found in emails. It evaluates whether the links direct to recognized malicious websites, effectively thwarting users from clicking on harmful links.

- **AI-enabled detection**: By leveraging artificial intelligence and data derived from Microsoft's comprehensive threat intelligence and analysis initiatives, the service generates patterns for recognizing malicious content, including both familiar and novel malware, enabling the autonomous identification of harmful and questionable material.

- **Threat investigation and response**: Defender for Office 365 streamlines the investigation process, empowering security teams to efficiently track, prioritize, and mitigate threats. This encompasses state-of-the-art hunting capabilities that can identify nuanced indicators of security breaches.

Attackers often resort to internal attacks, specifically lateral movement attacks, to gain credentials with elevated privileges. The significance of this component cannot be overstated; it is crucial to assess how attacks prevalent are against your organization both before and after its implementation for a clearer understanding of its impact.

Examples of attacks mitigated

Let's look at some of the most common attacks that this component aims to mitigate:

- **Insider threats**: Employees, whether intentionally or inadvertently, could transmit confidential information to unauthorized individuals. Advanced protection ensures constant surveillance and regulation of internal communications, effectively thwarting any potential breaches.

- **Account takeover attacks**: If an employee's email account is compromised, it can be used as a conduit for spreading malware or launching phishing attacks on other employees within the organization. Through the monitoring of internal emails, advanced protection can identify unusual activity, often linked to compromised accounts, and stop the propagation of malware or the success of phishing attacks.

Visibility is an asset; however, it alone is insufficient. We must also comprehend the criteria for identifying genuine attacks and the corresponding remedial measures. Let us examine investigations.

Knowing and investigating what is happening in your environment

The features discussed so far can provide a strong foundation and protect an organization from the most common attacks, but that alone may not suffice. Organizations striving for a mature security program tailored to their needs, industry, and geography require tools that offer visibility and feedback. These tools should not only assess the effectiveness of implemented controls, but also monitor malicious activities in the wild. Defender for Office 365 encompasses various elements that contribute to its effectiveness, such as real-time reports, threat trackers, campaign views, advanced threat investigation, and automated investigation and response (*Microsoft, 2023*). Let's dive into this section and discover the advantages you can enjoy by utilizing these components.

Real-time reports

By utilizing real-time reports in Microsoft Defender for Office 365, organizations can stay always updated on their email security posture. By providing real-time insights into email threats, these reports enable administrators to take a proactive approach to monitoring the security landscape. These reports are accessible via the Microsoft 365 Defender portal (`https://security.microsoft.com`), specifically on the **Email and Collaboration Reports** page under the **Reports** section, and provide a comprehensive overview of the email volume, identified threats, and how well the security policies are working.

One of the crucial reports is the **Mail Latency Report**, which provides a comprehensive overview of mail delivery times and detonation latencies. It's crucial for comprehending the effectiveness of your email system, offering percentile charts (50th, 90th, and 99th) that depict the delivery durations of messages within your organization.

An additional important report is the **Threat Protection Status Report**, which gathers data on malicious content detected and intercepted by Defender for Office 365. This report serves as a fundamental resource for cybersecurity teams, as it consolidates intelligence on the threats that have been thwarted from infiltrating your network.

Users with **Defender for Office 365 Plan 2** can utilize the **Threat Explorer** tool. This is an enhanced reporting tool that provides real-time visibility into malware and phishing emails, showcasing active attack campaigns. It is especially beneficial for incident response, as it facilitates the initiation of automated investigations and the execution of remedial measures directly from the tool.

Those with **Defender for Office 365 Plan 1** have access to **Real-time detections**. While it doesn't offer some of the advanced features found in Threat Explorer, it still provides a real-time view of email threats, allowing for immediate visibility into malware detections. Users can filter these detections by various attributes and expand their search up to a 30-day range.

Both tools, **Threat Explorer** and **Real-time detections**, offer comprehensive insights across different filters such as **All Email, Malware, and Phishing**. They have undergone updates to enhance accessibility and streamline workflow, guaranteeing users can efficiently navigate and utilize the tools. Furthermore, the reports offer the opportunity to conduct thorough investigations by providing email summaries, recipient information, and the ability to export data for additional analysis.

Reports can offer a mechanism for promptly detecting issues and attacks, as shown in the following screenshot:

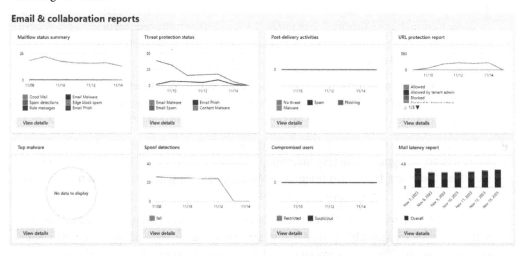

Figure 2.5 – Email and collaboration-related reports

Threat trackers

By using threat trackers, organizations can effectively trace the sources and routes of email threats. By offering comprehensive details about the origin of a threat, how it is delivered, and any accompanying malicious URLs or attachments, security teams can understand how an attack occurred and identify potential points of vulnerability in the environment. Here are more details on how it works:

- **Intelligence briefing**: Threat trackers serve as a comprehensive security intelligence briefing, delivering real-time updates on cybersecurity threats that may have an adverse effect on an organization, including emerging malware campaigns and phishing patterns. This assists organizations in proactively mitigating threats by staying well-informed.

- **Categories of trackers**: Threat trackers encompass diverse categories, each designed for a specific function:

    - **Noteworthy trackers**: Emphasize the critical and novel threats that have been identified and require urgent attention

    - **Trending trackers**: Display patterns in malware and phishing incidents within an organization's Office 365 environment, generally spanning the previous week

    - **Tracked queries**: These enable security teams to establish and regularly evaluate activity using predetermined queries that can be tailored to suit the organization's requirements

- **Threat information**: The **Threat tracker** page in the Microsoft 365 Defender portal displays in-depth information on threats. For example, if Microsoft's security features detect a novel form of ransomware or a phishing scheme, it will be flagged by the **Noteworthy** tracker, prompting the security team to assess and execute any requisite measures.

- **Operational efficiency**: Through the provision of this intelligence, threat trackers enhance the operational efficiency of security teams. They can prioritize threats by utilizing the information provided, concentrating their efforts on the most substantial risks at any given moment.

- **Integration with response tools**: Threat trackers, while not remediation tools on their own, are seamlessly integrated with response tools in Defender for Office 365, particularly for users with Plan 2 subscriptions. This integration enables a smooth transition from tracking and analyzing threats to implementing measures against them.

- **Accessibility and workflow**: The interface for threat trackers has been enhanced to conform with contemporary accessibility standards, guaranteeing that all team members can proficiently utilize the tool. Additionally, the workflow has been optimized to enhance expedited and effortless access to vital information, thereby making the threat response process more streamlined.

Threat trackers can offer a rapid overview of areas of concern in an environment, as shown in the following screenshot:

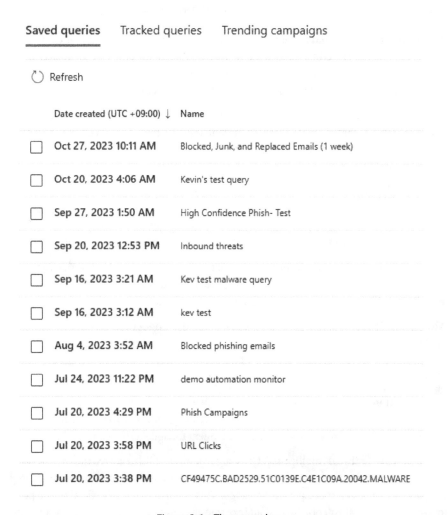

Threat tracker

Saved queries   Tracked queries   Trending campaigns

◯ Refresh

Date created (UTC +09:00) ↓	Name
☐ Oct 27, 2023 10:11 AM	Blocked, Junk, and Replaced Emails (1 week)
☐ Oct 20, 2023 4:06 AM	Kevin's test query
☐ Sep 27, 2023 1:50 AM	High Confidence Phish- Test
☐ Sep 20, 2023 12:53 PM	Inbound threats
☐ Sep 16, 2023 3:21 AM	Kev test malware query
☐ Sep 16, 2023 3:12 AM	kev test
☐ Aug 4, 2023 3:52 AM	Blocked phishing emails
☐ Jul 24, 2023 11:22 PM	demo automation monitor
☐ Jul 20, 2023 4:29 PM	Phish Campaigns
☐ Jul 20, 2023 3:58 PM	URL Clicks
☐ Jul 20, 2023 3:38 PM	CF49475C.BAD2529.51C0139E.C4E1C09A.20042.MALWARE

Figure 2.6 – Threat tracker

The value of this level of threat intelligence alone justifies the cost of licensing. As your organization's security maturity level advances, employing this intelligence will shift your security team's approach from reactive to proactive.

Campaign views

Campaign views offer a comprehensive perspective on email-based attacks targeting an organization. This component provides security teams with a comprehensive understanding of the magnitude and extent of coordinated attacks frequently started by sophisticated threat actors. By consolidating data from various emails, campaign views expose common attack tactics and patterns. Their functionality includes the following:

- **Aggregation of threat data**: Campaign views compile and present data on email threats that have specifically targeted an organization. This comprises a range of attack vectors, including phishing, malware, and other nefarious activities. By compiling this data, campaign views offer a more comprehensive understanding of threat campaigns rather than individual occurrences.

- **Visualization of attack campaigns**: The feature visually represents the attack patterns, displaying the evolution of a particular campaign over time. This can encompass details such as the quantity of malicious emails sent, the strategies and methods used by attackers, and a chronological sequence of the attack.

- **Contextual information**: Campaign views not only display events but also offer contextual insight. This includes information on the sender, the infrastructure used (including IP addresses and domains), and the campaign's targets within the organization. The profundity of contextual understanding is paramount in assessing the nature and potential consequences of an attack.

- **Integrated response actions**: Security teams can take response actions from within campaign views. If a campaign is determined to be malicious, measures such as URL or email address blocking, as well as email removal from users' inboxes, can be directly initiated from the Campaign View.

- **Aiding incident response**: Through the provision of a condensed summary, campaign views facilitate incident response by enabling security teams to rapidly gauge the scale and severity of an attack. This enhances the speed of decision-making and response times, which play a pivotal role in minimizing the consequences of cyber-attacks.

- **Customization and tracking**: Security teams can personalize the view to emphasize specific aspects of a campaign that hold the utmost relevance to their organization. They can additionally monitor the real-time progress of an attack, a vital component for agile response and mitigation strategies.

- **Correlation with threat intelligence**: Campaign views establish a correlation between individual threats and broader threat intelligence, enabling the identification of whether a campaign is connected to a larger, more pervasive attack. This is particularly beneficial in the preparation for and defense against coordinated attacks that may impact a whole sector or industry.

Campaign views enable you to assess the performance of your environment in comparison to the prevalent attacks observed in the wild, as demonstrated in the following screenshot:

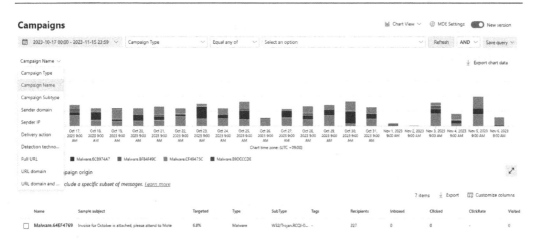

Figure 2.7 – Campaign views

Using campaign views can significantly enhance the refinement of your security efforts to align with your industry and environment. Combining this information with automation can lead to exceedingly effective security operations.

Automated investigation and response

Automation plays a crucial role in enhancing the efficacy of incident response. AIR amplifies the capabilities of security operations teams by automating the investigation and response to known threats. This functionality aids in the reduction of time spent on email threat investigation and remediation, while also minimizing alert noise, decreasing response times, and reducing manual effort in handling security incidents.

When AIR is activated, commonly triggered by an alert, an automated investigation begins to gather data and details that are pertinent to the alert. As an example, in the case of an alert concerning suspicious email activity, AIR will commence analysis of the email and any associated entities. The investigation's scope may widen if new and relevant alerts surface, ensuring the system's responsiveness to evolving threats.

As the investigation progresses, AIR develops findings and suggests remedial actions. These recommendations are presented to the security team, who can assess the outcomes of the automated analysis and decide whether to approve or decline the proposed actions. This step is crucial as it enables human supervision to ensure that the automated system's decisions are in line with the organization's security policies and risk tolerance.

The remediation actions that AIR might propose entail quarantining malicious files, stopping suspicious processes, or isolating devices that may be compromised. Depending on the organization's preferences and settings, certain actions may be automated by AIR, while others may necessitate explicit approval from the security team.

AIR provides great reports on actions taken, as shown in the following screenshot:

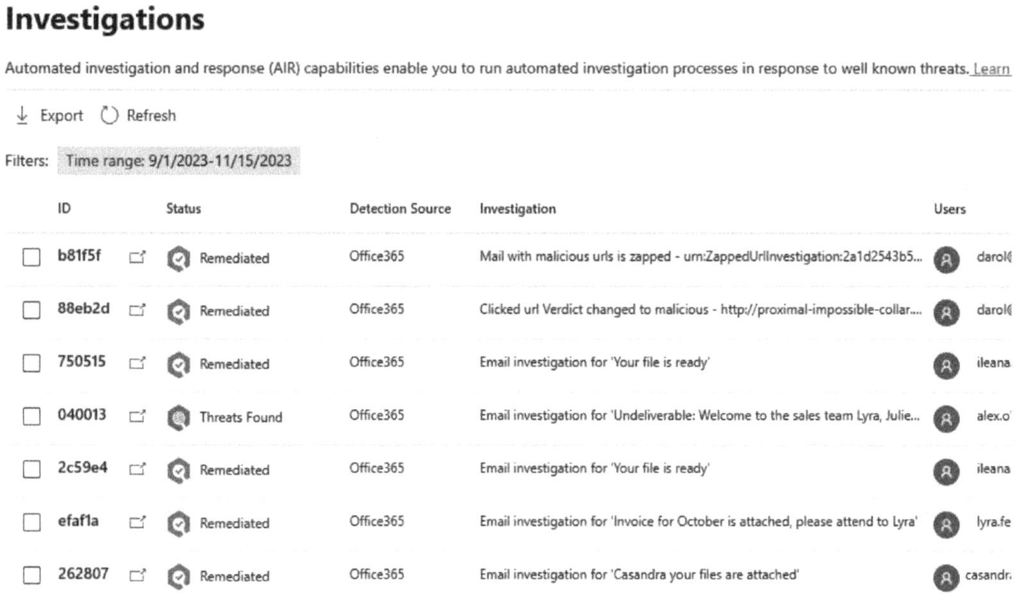

Investigations

Automated investigation and response (AIR) capabilities enable you to run automated investigation processes in response to well known threats. Learn

↓ Export   ○ Refresh

Filters: Time range: 9/1/2023-11/15/2023

	ID		Status	Detection Source	Investigation		Users
☐	b81f5f	☐	Remediated	Office365	Mail with malicious urls is zapped - urn:ZappedUrlInvestigation:2a1d2543b5...	⊗	darol(
☐	88eb2d	☐	Remediated	Office365	Clicked url Verdict changed to malicious - http://proximal-impossible-collar...	⊗	darol(
☐	750515	☐	Remediated	Office365	Email investigation for 'Your file is ready'	⊗	ileana
☐	040013	☐	Threats Found	Office365	Email investigation for 'Undeliverable: Welcome to the sales team Lyra, Julie...	⊗	alex.o
☐	2c59e4	☐	Remediated	Office365	Email investigation for 'Your file is ready'	⊗	ileana
☐	efaf1a	☐	Remediated	Office365	Email investigation for 'Invoice for October is attached, please attend to Lyra'	⊗	lyra.fe
☐	262807	☐	Remediated	Office365	Email investigation for 'Casandra your files are attached'	⊗	casandr.

Figure 2.8 – Automated investigation and response pane

One of the key benefits of AIR is its uninterrupted operation, functioning as a virtual analyst who tirelessly works towards resolving alerts. By correlating signals from different incidents and alerts across Microsoft Defender products, AIR establishes judgments on the maliciousness, suspiciousness, or harmlessness of evidence.

The inclusion of AIR in the Defender for Office 365 package depends on the activation of audit logging. This feature guarantees that all actions and decisions made by AIR are documented, offering a transparent trail for security teams to review and audit.

Integration with other Defender security products

The main selling point of the Microsoft Defender suite is its seamless integration with various tools. Instead of relying on a single tool, this suite combines multiple tools that excel in their respective domains. By working together, these tools provide the security team with a comprehensive view of the environment. This approach allows for greater precision and a diverse range of solutions to support organizations at any stage of their security journey. It is essential to discuss how the components of Defender for Office 365 integrate with other Microsoft security tools to fully understand its capabilities.

Microsoft Defender for Endpoint

Microsoft Defender for Endpoint is a comprehensive enterprise endpoint security platform engineered to aid in the prevention, detection, investigation, and response to sophisticated threats. Its functionality centers on multiple core processes:

- **Threat and vulnerability management**: This proactive component identifies vulnerabilities and misconfigurations on endpoints, allowing IT teams to prioritize and fix these issues to fortify the organization's security posture.

- **Attack surface reduction**: Defender for Endpoint implements a range of protective measures to minimize the vulnerability of potential cyberattack targets, such as setting up configurations and rules to thwart malicious scripts, ransomware, and other fileless threats.

- **Next-generation protection**: It leverages advanced heuristics, machine learning, and behavior analysis to ensure instantaneous protection against an extensive variety of threats, including emerging and unknown ones. It ensures the recognition of obscure malware through the analysis of suspicious behavior.

- **Endpoint Detection and Response (EDR)**: Should a breach occur, EDR capabilities facilitate the recognition and handling of advanced threats. It offers tools for ongoing monitoring and response actions, such as device isolation and collection of investigation packages.

- **Automated investigation and remediation**: Through the automation of alert investigations and prompt breach resolution, Defender for Endpoint substantially alleviates the workload on security operations teams and enhances response times.

- **Microsoft Defender Experts**: This is an additional service where Microsoft's security professionals offer precise attack notifications and expert advice to handle specific threats.

Microsoft Defender for Endpoint additionally gathers and analyzes an extensive volume of signal data, which, when combined with Microsoft's cloud-based analytics, bolsters detection and response capabilities. For example, if a suspicious activity is detected on an endpoint, Defender for Endpoint doesn't just alert the security team; it also provides a detailed story of the attack progression and suggests remediation steps.

When Microsoft Defender for Endpoint detects a possible threat on a device, it communicates this information to Defender for Office 365. This collaboration enables a swift response, such as directing Defender for Office 365, to eliminate a malicious file discovered on an endpoint from all email messages. Hence, the offending file is effectively barred not only on the device but also across the complete Microsoft 365 security suite.

Furthermore, the integration enables the examination of attack entry points, encompassing email, endpoints, and applications. This is particularly advantageous for security analysts who need to meticulously trace the trajectory of an attack and gain insight into its origins and dissemination within the network.

In addition, the shared threat intelligence between Defender for Endpoint and Defender for Office 365 facilitates more effective containment of attacks. Upon detecting a threat, one system can relay this information to the other system, enabling it to obstruct the threat and prevent its dissemination.

Microsoft Defender for Cloud Apps

Microsoft Defender for Cloud Apps functions as a **Cloud Access Security Broker** (**CASB**), providing oversight and authority over both authorized and unauthorized cloud applications in an organization's ecosystem. Here's a comprehensive breakdown of how Microsoft Defender for Cloud Apps operates:

- **Integration with Microsoft 365 Defender**: Defender for Cloud Apps is seamlessly integrated into the Microsoft 365 Defender portal, offering a unified interface for overseeing and controlling security across diverse domains like identities, data, devices, applications, and infrastructure. This integration ensures that alerts from Defender for Cloud Apps are visible in the incidents and alerts queue of the Microsoft 365 Defender portal, simplifying the workflow for security teams.

- **Traffic monitoring**: By leveraging traffic information collected about cloud apps and services accessed from managed devices, Defender for Cloud Apps simplifies the discovery of Shadow IT through its native integration with Microsoft Defender for Endpoint. This information is analyzed to provide insights into user and device activities without the need for supplementary deployment or convoluted integration steps.

- **Session policies**: Defender for Cloud Apps employs session policies to exert precise control over cloud app usage. It can oversee activities, prohibiting specific actions and enforcing heightened authentication measures by routing traffic through its reverse proxy. This mechanism additionally enables Defender for Cloud Apps to scan files for malware during upload or download sessions and enforce content policies, such as prohibiting the transfer of sensitive information through copy-pasting, thereby mitigating the risk of data leakage.

- **Cloud Discovery**: It establishes connections with a range of SaaS applications and performs scans for sensitive data, analyzing user interactions and identifying storage locations. This extent of exploration is critical for evaluating the security position of deployed cloud applications and enforcing necessary protocols.

- **Threat Protection**: As a component of Microsoft's **Extended Detection and Response** (**XDR**) solution, Defender for Cloud Apps integrates signals across the Microsoft ecosystem to deliver continuous threat protection. This aids in the identification and mitigation of threats spanning multiple vectors.

- **Data protection**: It offers core CASB capabilities including visibility into cloud app usage, data protection, and features for managing **SaaS Security Posture Management** (**SSPM**). The latter enhances an organization's security stance in connected applications by conducting scans for sensitive data and managing its circulation.

- **Compliance enforcement**: By means of scanning and identifying sensitive data in cloud apps, Defender for Cloud Apps empowers organizations to ensure compliance with data protection regulations. It also enables the enforcement of controls such as encryption and access policies to secure data throughout the cloud environment.

- **Policy application**: Administrators can implement specific access policies, evaluate the security status of cloud applications, and resolve any issues. This could involve the application of policies for data sharing, access privileges, and other security-related configurations.

Microsoft Defender for Cloud Apps is purposefully designed to seamlessly integrate with Defender for Office 365, presenting a consolidated approach to security management for organizations. The integration leverages the app connector API to establish a connection with a Microsoft 365 account, enabling enhanced oversight and management of Microsoft 365 usage, as well as access to SaaS security posture management features through Microsoft Secure Score.

When Defender for Cloud Apps is connected with Defender for Office 365, it taps directly into Microsoft 365's audit logs. This signifies that it can accept audited events from supported services, including Exchange, Teams, and Power BI, resulting in a comprehensive overview of activities across these platforms. The integration also encompasses Microsoft Entra ID (previously Azure Active Directory), fortifying the security posture through the surveillance and administration of identities.

Microsoft Purview Data Loss Prevention

Microsoft Purview DLP is a comprehensive service specifically developed to assist organizations in preventing the accidental or deliberate dissemination of sensitive data beyond their established limits. This is vital in an era where data breaches can cause substantial financial and reputational consequences.

When discussing examples, think about a healthcare organization that is mandated to follow HIPAA regulations. Purview DLP can automatically detect and safeguard **patient health information (PHI)** across various platforms, such as emails, documents, and data repositories. Should an employee attempt to send the PHI outside the secure network, the DLP system has the capability of obstructing the transfer and alerting administrators.

Here is a comprehensive breakdown of how Microsoft Purview DLP functions:

- **Identification of sensitive data**: The first step of Purview DLP involves the identification of sensitive data across multiple platforms. It is capable of functioning with Microsoft 365 services, Office applications, Windows and macOS endpoints, and even non-Microsoft cloud apps, as well as on-premises file shares and SharePoint.

- **Deep content analysis**: The system employs advanced content analysis techniques, as opposed to basic text scanning, for sensitive information detection. This can entail the examination of keywords, regex expressions, internal functions, and the proximity of data to other susceptible elements.

- **Policy application**: Once sensitive data is detected, DLP policies are enforced. These policies essentially encompass rules that delineate the definition of sensitive data and dictate the appropriate course of action upon encountering such data. Actions can involve obstructing the sharing of data, notifying administrators, or transferring the files to a more secure destination.

- **Automated protection**: The DLP solution automatically implements these policies, thereby ensuring real-time data protection without the necessity for ongoing human monitoring.

- **User behavior analytics**: In DLP, purview analyzes user behavior to detect any suspicious activities. If a user tries to transfer sensitive data in a manner that violates established policies, the system has the capability to detect and implement suitable action.

- **Regulatory compliance maintenance**: Purview DLP assists organizations in meeting a multitude of regulatory obligations by actively monitoring and securing sensitive data, thereby mitigating the risk of penalties and other compliance-related challenges.

- **Implementation process**: To implement DLP successfully, you need to plan strategically, test thoroughly, and fine-tune meticulously to match your organization's unique environment. You can use the Purview portal or Microsoft 365 Defender to monitor alerts and activities.

- **Unified capabilities**: Microsoft Purview DLP provides a consolidated approach to data loss prevention across endpoints, applications, and services, streamlining data security management.

When integrated with Defender for Office 365, Microsoft Purview DLP enhances security processes by applying data loss prevention policies directly to the flow of information within an organization's email and collaboration platforms. This integration enables the implementation of advanced monitoring and protection measures, such as automatic encryption of emails or prevention of sharing sensitive content. The system is finely tuned to discern various manifestations of sensitive information, and its versatility permits the formation of bespoke policies, rendering it suitable for compliance with regulations such as HIPAA. Consequently, organizations can attain a unified and regulated security stance, guaranteeing proper handling of sensitive data in line with established data governance policies.

How Defender for Office 365 could have averted famous attacks

Let's examine notable attacks that involved attack vectors targeted by Defender for Office 365. Although it would be unwise to claim that a single tool could have prevented these attacks, it is crucial to assess its potential positive influence on the outcome of the attack.

Democratic National Committee email hack

In 2016, the **Democratic National Committee** (**DNC**) noticed some unusual activity in its email systems. They hired a cybersecurity firm, CrowdStrike, for incident response and investigation support. This investigation led to the discovery of two state-sponsored hacking groups, Cozy Bear (APT29) and Fancy Bear (APT28), involved in the breach.

Attack vector

To breach the DNC's network, the attackers mainly relied on spear-phishing emails. These emails mimicked legitimate ones and often contained malicious attachments or links. The attackers could install malware on the computers of the DNC staff members who opened these attachments or links, enabling the attackers to get login credentials and access the DNC's email systems.

Data exfiltration

After breaching the DNC's network, the attackers stole and leaked confidential emails and documents. The public exposure of the data through outlets such as WikiLeaks sparked intense political and media attention.

How could Microsoft Defender for Office 365 have helped?

The DNC email hack might have been prevented or mitigated by Microsoft Defender for Office 365, which can do the following things:

- **Phishing email detection**: The attackers tried to fool the recipients by faking the sender's email address, making it look like the phishing emails came from trusted sources within the DNC or other legitimate organizations. The goal of this technique was to convince the recipients that the emails were genuine. Microsoft Defender for Office 365 could have detected and stopped the spear-phishing emails used in the attack by using machine learning and threat intelligence to spot and flag any suspicious emails that show signs of phishing attempts.

- **Attachment scanning**: Phishing emails sometimes contain harmful attachments, such as PDF documents or Word files infected with malware. If the recipient opens these attachments, they may expose their computer to malware or trigger a breach. Microsoft Defender for Office 365 can scan and analyze the malicious attachments in spear-phishing emails with its anti-malware and anti-phishing features. It may isolate or block any attachments that are malicious or phishing attempts, preventing them from reaching the intended recipients.

- **Link protection**: Some phishing emails contained fake login pages that looked like authentic websites, such as email login portals. The primary goal of these pages was to steal the email credentials of the people who received the emails. If the emails had malicious links, users could have avoided the spear-phishing risks by not clicking on them. However, Microsoft Defender for Office 365's Safe Links feature can help users stay safe from such threats. It scans and blocks any dangerous links and prevents users from visiting harmful websites by mistake.

- **User training and awareness**: Some phishing emails use urgency or social engineering tactics to pressure recipients into acting quickly. They may use phrases such as *Urgent: Account Compromised* or *Immediate Action Required* to create a sense of urgency. Microsoft Defender for Office 365 offers training tools that can help users learn the best practices for email security and become more aware of phishing threats.

Sony Pictures Entertainment hack

In late 2014, Sony Pictures Entertainment, an American film and television company, suffered a severe and widely reported cyberattack. The attackers exposed the company's confidential data, including unreleased movies, internal emails, and employees' personal information.

Attribution

The U.S. government eventually blamed North Korea for the cyberattack that targeted Sony Pictures in 2014. The attack was supposedly motivated by North Korea's anger over the film *The Interview*, which made fun of a fictional attempt to kill its leader.

Stolen data

Hackers stole and publicly exposed a huge volume of confidential data from Sony Pictures Entertainment, such as upcoming movies, internal communications, staff personal details, executive pay information, and more. The data breach caused severe harm to the company's reputation and finances.

How could Microsoft Defender for Office 365 have helped?

To prevent or reduce the damage of the Sony Pictures Entertainment hack, Microsoft Defender for Office 365 could have done the following:

- **Phishing email detection**: The breach began when Sony Pictures Entertainment employees received spear-phishing emails. These emails looked authentic and matched the recipient's job role or responsibilities within the company. For instance, some emails impersonated fellow employees, executives, or external business partners. Microsoft Defender for Office 365 has advanced phishing detection capabilities that leverage machine learning and threat intelligence could identify and block these phishing emails, which can trick users into revealing login credentials.

- **Attachment scanning**: The attackers used deceptive emails to trick recipients into opening harmful attachments, such as corrupted Word or PDF files. These attachments contained malware that ran on the recipients' computers and gave the attackers access to the company's network. Microsoft Defender for Office 365 could have protected the users from these attachments by scanning and blocking them before they could be opened or run.

- **Malicious document detection**: The malware was delivered and executed by macros embedded in some of the malicious documents. Macros are scripts that can perform tasks automatically within documents. The attackers exploited vulnerabilities and installed malware using macros. These malicious documents would have been detected, blocked from download, and prevented from running macros by Microsoft Defender for Office's 365 Safe Documents feature. This would have stopped the malware from spreading.

- **Security alerts and monitoring**: After gaining access to the network, the attackers stole large amounts of sensitive data, such as unreleased films, internal emails, employee personal information, executive salary details, and more. They later leaked the data to the public. Microsoft Defender for Office 365 alerts and monitors security administrators in real time about unusual activities, which could help detect breaches and unauthorized access early. Combined with other tools such as Defender for Cloud Apps, it could have alerted the security teams to the data exfiltration attempts and possibly stopped them if automation was set up.

It is evident that many notorious attacks were characterized by the failure of multiple security areas. You should consider security as a multi-layered onion, with each layer increasing the difficulty for an attacker to penetrate.

Summary

To sum up this chapter, we have examined the multifaceted security environment of Defender for Office 365, an indispensable weapon in an organization's cybersecurity arsenal. Our discussion began with a detailed overview of the various components that make up this security suite, Safe Attachments, Safe Documents, and Safe Links, which work together to create a strong defense against cyber threats.

By employing a virtual environment, Safe Attachments can proactively detonate and analyze email attachments, effectively detecting any hidden malware and delivering only secure files to end users. Similarly, Safe Documents is a cutting-edge scanning feature that harnesses the power of Microsoft's cloud backend to meticulously analyze Office documents accessed in protected view or Application Guard, bolstering protection against malware intrusion.

Microsoft's ecosystem was brought into focus as the seamless integration of Safe Attachments across SharePoint, OneDrive, and Teams highlighted its robust security measures. Safe Links, the diligent caretaker of hyperlinks, actively scans and reevaluates URLs in emails and Office documents, safeguarding users from dangerous online destinations.

We explored the intricacies of anti-phishing policies that intelligently learn to differentiate between safe and suspicious emails, along with the Report Message add-in that empowers users to directly report suspected phishing attempts, thereby reinforcing collective security intelligence.

The chapter delved into the details of advanced protection for internal mail fortifying the organization's communication channels with an extra level of security. Vigilant surveillance mechanisms, including real-time reports and threat trackers, were introduced to monitor and alert about emerging threats. Advanced threat investigation and AIR were also highlighted as intelligent systems that autonomously monitor and address security incidents, leading to a significant decrease in the time and resources required to resolve threats.

In the next chapter, you will gain insights into pinpointing your organization's security requirements, policies, and procedures. The result of this approach can then be used to evaluate the effective deployment and operation of Microsoft 365 Defender in your unique operational landscape.

References

- Microsoft. (2023). *Safe attachments for SharePoint, OneDrive, and Microsoft Teams*. Retrieved from Microsoft Learn: `https://learn.microsoft.com/en-us/microsoft-365/security/office-365-security/safe-attachments-for-spo-odfb-teams-about?view=o365-worldwide`

- Microsoft. (2023a). *Complete Safe Links overview for Microsoft Defender for Office 365*. Retrieved from Microsoft Learn: `https://learn.microsoft.com/en-us/microsoft-365/security/office-365-security/safe-links-about?view=o365-worldwide`

- Microsoft. (2023b). *Configure anti-phishing policies in Microsoft Defender for Office 365*. `https://learn.microsoft.com/en-us/microsoft-365/security/office-365-security/anti-phishing-policies-mdo-configure?view=o365-worldwide`

- Microsoft. (2023). *Enable the Report Message or the Report Phishing add-ins*. Retrieved from Microsoft Learn: `https://learn.microsoft.com/en-us/microsoft-365/security/office-365-security/submissions-users-report-message-add-in-configure?view=o365-worldwide`

- Microsoft. (2023). *Protection technologies in Microsoft Defender for Office 365*. Retrieved from `https://learn.microsoft.com/en-us/microsoft-365/security/office-365-security/protection-stack-microsoft-defender-for-office365?view=o365-worldwide`

- Microsoft. (2023). *Microsoft Defender for Office 365 overview*. `https://learn.microsoft.com/en-us/microsoft-365/security/office-365-security/mdo-about?view=o365-worldwide`

3

Basic Checks and Balances

In the field of cybersecurity, organizations must use strong security solutions such as Microsoft Defender for Office 365 to protect their digital assets. However, just deploying these tools is not enough. It's important to evaluate their effectiveness and align them with the organization's security strategy. This chapter explores the methods and frameworks for assessing security products within an organization. We will look at well-known frameworks such as ISO 27001, the NIST Cybersecurity Framework, HIPAA, PCI DSS, GDPR, and FISMA. These frameworks help manage cybersecurity risks, ensure compliance with regulations, and maintain customer trust. Implementing these frameworks requires understanding an organization's vision, policies, and procedures. This chapter will guide you in aligning your security strategy with these frameworks and customizing the implementation to meet your organization's unique needs. We will also discuss how to identify **Key Performance Indicators (KPIs)** and metrics relevant to your security goals. These metrics will measure the effectiveness of tools such as Microsoft Defender for Office 365 and highlight areas for improvement.

By the end of this chapter, you will understand how to measure and enhance the performance of your security products, contributing to your organization's overall security posture. Be aware that this chapter is heavily focused on establishing the rationale for implementing defender for office 365 and how to align security tools to the security needs of an organization, not on configuring the tool itself. The contents of this chapter will give you an idea of what metrics to establish to determine if the security tools being deployed align with your organization's security strategy. If you are familiar with security frameworks and how they define the security needs of an organization, you can safely skip this chapter.

This chapter will cover the following topics:

- Ways in which organizations structure their security efforts, such as security frameworks
- Understanding how an organization codifies its security strategy
- How to quantify the needs of an organization into metrics

Let's continue our journey!

Common security frameworks and approaches

Security frameworks equip organizations with structured guidelines, encompassing best practices, standards, and recommendations for efficient management and protection of their information systems. Through the utilization of a methodical approach, they are proficient at managing sensitive company information, implementing imperative security measures to safeguard data and conform to regulatory standards.

Organizations use security frameworks for the following:

- Safeguarding sensitive data against breaches and cyber-attacks
- Ensuring adherence to legal and regulatory obligations
- Implementing a structured approach to risk management
- Improving the trustworthiness and fortifying the protection of systems and data
- Fostering customer trust by safeguarding the confidentiality, integrity, and accessibility of data

Organizations have a plethora of security frameworks to choose from, based on industry, location, security maturity, and executive knowledge of security approaches. Some of the most frequently encountered ones the reader might either peruse or help implement include the ones covered in the following sections.

ISO 27001

ISO 27001, also known as the **International Organization for Standardization 27001**, is a globally recognized standard that provides guidelines for effectively managing information security. It provides a framework for establishing, implementing, operating, monitoring, reviewing, maintaining, and improving an **Information Security Management System (ISMS)**. The primary goal of ISO 27001 is to ensure the protection of information by prioritizing its confidentiality, integrity, and availability. We can effectively implement this system in any industry, particularly in sectors where data security is a major concern (*Infosec Institute, 2023*). To implement ISO 27001 in an organization, we must follow a series of steps:

1. **Understanding the standard**: To begin, you must comprehend the ISO 27001 requirements and their relevance to your organization. This requires understanding the 14 domains this standard covers, which are as follows (*International Organization for Standardization, 2022*):

   - **Information Security Policies**: Ensures alignment of the organization's policies with its information security practices. Policies can be considered as the initial map security teams use to understand what control needs to be accomplished and why. Be aware that policies will typically not go too technically in depth and will instead point at some commonly used security baseline for the technical details (such as the **Center for Internet Security (CIS)** security baselines). This will define what components of Defender for Office 365 will be deployed and how strict.

- **Organization of Information Security**: Provides a structure for the establishment of an information security management system within the organization by focusing on the responsibilities and security roles. This indicates who is responsible for what. Typically, organizations will have a tree-style organization with the CISO at the top, followed by team leads for different groups, but most organizations tend to forget the other aspects, such as those identified in the **Responsible, Accountable, Consulted, Informed (RACI)** model. This will define who makes what decisions and who receives notifications in some scenarios.

- **Human Resource Security**: Ensures that the human resources department can comply with security policies by providing guidance for everyone employed by an organization to understand the process and their responsibilities during all stages of their employment.

- **Asset Management**: Ensures the proper life cycle management of information assets, including their classification and who handles what aspect of the life cycle. This defines all aspects of how long an asset (such as a user account, mailbox, device, etc.) is kept and what actions to take in different situations (such as a breach, police investigation, user leaving the organization, etc.).

- **Access Control**: Ensures that controls are implemented to restrict access to information and its processing facilities to allowed personnel. This control is key for defining who can access and make what changes to different aspects of Defender for Office 365, along with the approval process for changes and how to audit changes.

- **Cryptography**: Addresses the organization's data encryption requirements to ensure that data integrity, confidentiality, and protection are maintained. This control is key in tools that relate to data loss prevention such as Microsoft Purview and is very important to lower the impact of any data leak during a breach.

- **Physical and Environmental Security**: Looks at the physical and environmental controls such as physical access, damage to the organization, and the procedures to protect from damage, theft, and loss. This control relates more to the physical access to different assets, but these assets go beyond just computers. They could also be the controls to prevent unauthorized access to the system's break glass account (a common account with high privileges that is used in emergencies when administrative access is lost).

- **Operations Security**: Provides guidance on controls to use during the processing of data to reduce the risk of data loss. This also aligns with data loss prevention measures such as those implemented by Microsoft Purview and Defender for Office 365; think of alerts on the creation of forwarding rules, mailbox delegation, and so on.

- **Communications Security**: Provides controls for the security of information within the organization's networks. Think of controls related to how the systems are accessed and data transferred, things such as conditional access policies to verify who the user is and encryption of communications.

- **System Acquisition, Development, and Maintenance**: Ensures that information security practices are applied through all parts of the information system components' life cycle. A project that designs security at the early stages tends to be more secure. These controls might define things such as a new user requiring Defender for Office 365 protections in place upon account creation and before initial use.

- **Supplier Relationships**: Provides guidance on the relationship and agreements between the organization and third parties providing services to the organization. Think of third parties that manage your systems that need access to work on proprietary data, and even the platform in which your security tools run; all these organizations should follow a common set of security standards and be audited on a regular basis.

- **Information Security Incident Management**: Identifies responsibilities and processes for the management of information security incidents. Think of the processes in place once an incident occurs: who needs to do what during an incident, who needs to be notified, and even communication with internal and external parties. The security aspects of the Microsoft Defender portal will be a key aspect of these controls.

- **Information Security Aspects of Business Continuity Management**: Addresses the procedures and responsibilities to ensure business continuity, including disaster recovery. Think of what happens if your business suffers downtime, think of the impact on revenue, legal obligations, and even brand impact. Think of the controls to implement to minimize these issues.

- **Compliance**: Ensuring adherence to information security policies, standards, and applicable legal and regulatory requirements. Think of any legal frameworks your organization needs to follow due to their industry; think of what needs to be audited and when, and how long data must be kept.

2. **Management Commitment**: Secure commitment and support from senior management. This is essential for the implementation's success.

3. **Scope Definition**: Specify the boundaries of the ISMS. This involves the identification of locations, assets, technology, and processes to be included.

4. **Risk Assessment**: Conduct a risk assessment to identify the threats and vulnerabilities that could affect the assets in the previously identified scope.

5. **Risk Treatment**: Create a plan to manage or mitigate risks by developing a risk treatment plan.

6. **ISMS Development**: Create the ISMS policy, objectives, processes, and procedures to manage risk and enhance information security aligned with the organization's overall goals.

7. **Training and Awareness**: Ensure someone trained your staff on the ISMS, and promote information security awareness across the organization.

8. **Operate and Monitor the ISMS**: Carry out the ISMS, oversee its performance, and make necessary enhancements.

9. **Internal Audits**: Conduct regular internal audits to assess ISMS compliance with both organizational and ISO 27001 requirements.

10. **Management Review**: The organization's top management should regularly review the ISMS to ensure it remains suitable, adequate, and effective.

11. **Certification Audit**: Obtain ISMS certification from an accredited registrar. The audit process comprises two stages: Stage 1 is an initial, informal ISMS review, while Stage 2 is a formal compliance audit.

Don't forget, ISO 27001 implementation is a continuous process, not a one-time project. All staff members must commit to a cultural shift within the organization.

NIST Cybersecurity Framework

With its origin at the **National Institute of Standards and Technology** (**NIST**), this framework has been specifically designed to enhance cybersecurity measures for critical infrastructure in the United States. This primarily centers on five core functions, which are Identify, Protect, Detect, Respond, and Recover. Although it was originally designed for critical infrastructure industries in the United States, it has gained popularity across different sectors because of its adaptable nature. Implementation requires ensuring that current cybersecurity activities are in line with the desired outcomes outlined in the framework (*NIST, 2023*). The process of implementing the NIST Cybersecurity Framework comprises various steps, including the following:

1. **Establish Organizational Goals**: Comprehend your objectives for your cybersecurity program. This could involve safeguarding customer data, maintaining business operations, or adhering to industry regulations. Think about what needs to be protected and the impact of an attack on the assets.

2. **Understand Your Risks**: Recognize potential risks and weaknesses within your company. It may require a risk assessment or the hiring of a third-party auditor.

3. **Establish Your Baseline**: Gauge your current cybersecurity posture. You need to evaluate your current controls and practices based on the NIST Cybersecurity Framework. This section might require meeting with multiple teams to better understand what controls exist. Be aware that this framework will provide technology-agnostic guidance.

4. **Determine Your Position**: Evaluate how to improve your current cybersecurity posture alignment to your goals and the NIST Cybersecurity Framework. Identify areas that need improvement and gaps.

5. **Implement the Framework**: Create a strategy to close gaps and align cybersecurity practices with the NIST Cybersecurity Framework. This might include implementing new controls, improving current ones, or altering business processes.

6. **Verify Your Success**: Assess and evaluate the efficiency of your cybersecurity program. This might include tracking KPIs or doing regular audits.

7. **Assess Your Current Cybersecurity Measures**: Make sure you regularly review and update your cybersecurity program to maintain its effectiveness and alignment with your organizational goals.

8. **Develop a Cybersecurity Policy**: Record your cybersecurity practices, responsibilities, and procedures. It is important to communicate this policy to all employees and stakeholders.

9. **Implement Security Controls**: Implement the required technical and administrative controls to safeguard your organization. This could include firewalls, intrusion detection systems, or employee training programs.

10. **Train and Educate Your Staff**: Make sure your employees know their responsibilities and receive cybersecurity training. Regular training sessions or awareness campaigns could be part of this.

11. **Monitor and Respond to Cyber Threats**: Implement procedures for identifying, addressing, and bouncing back from cybersecurity breaches. A **Security Operations Center** (**SOC**) or incident response team could be involved in this.

12. **Regularly Assess and Improve Your Cybersecurity Program**: Regularly assess the effectiveness of your cybersecurity program and make necessary improvements. This could include regular audits, penetration testing, or employee feedback.

Keep in mind that adopting the NIST Cybersecurity Framework is an ongoing process of enhancement. Regularly reviewing and updating your cybersecurity program is crucial for effectiveness and alignment with organizational goals.

HIPAA

In the United States, the **Health Insurance Portability and Accountability Act** (**HIPAA**) sets the benchmark for protecting sensitive patient information. For organizations handling protected health information, it's crucial to rigorously adhere to all mandated physical, network, and procedural security measures. The US Department of Health and Human Services oversees the regulation of HIPAA compliance, with enforcement carried out by the Office for Civil Rights (*Proofpoint, 2023*). To effectively integrate HIPAA within an organization, several key steps must be undertaken to ensure adherence and safeguard sensitive health data:

1. **Grasping HIPAA's Essentials**: The primary stage requires a detailed assessment of the mandates delineated in the HIPAA, which are divided into titles. These titles cover the following areas:

   - **Title I, Health Insurance Access, Portability, and Renewability**: This deals with the accessibility, portability, and renewability of health insurance.

   - **Title II, Preventing Health Care Fraud and Abuse**: This deals with the prevention of healthcare fraud and abuse, implementation of medical liability reform, and streamlining administrative processes in healthcare.

   - **Title III, Tax-Related Health Provisions**: This details the various tax-related provisions for medical care.

- **Title IV, Application and Enforcement of Group Health Plan Requirements**: This deals with the application and enforcement of requirements for group health plans.

- **Title V, Revenue Offsets**: This deals with the provisions related to revenue offsets (**National Center for Biotechnology Information (NCBI)**, *2022*).

2. **Risk Analysis Execution**: Conducting a comprehensive risk analysis is vital to pinpoint potential threats to the confidentiality, integrity, and accessibility of **electronic Protected Health Information (ePHI)**. This analysis is key to understanding system vulnerabilities and the impact of various threats (*NIST, 2011*).

3. **Policy and Procedure Formulation and Implementation**: Following the risk analysis, organizations should craft and apply policies and procedures that align with HIPAA's standards. These should encompass privacy, security, and breach notification, conforming to HIPAA's guidelines (**U.S. Department of Health & Human Services (HHS)**, *2022*).

4. **Employee Training**: Training all staff members on HIPAA's policies and procedures is essential. This training should be role-specific and conducted regularly to maintain compliance.

5. **Implementing Physical and Technical Safeguards**: Establishing both physical and technical safeguards is necessary. This includes secure access controls, encryption, and secure data transmission protocols, as outlined by the Department of Health and Human Services (*HHS, 2022*).

6. **Ongoing Auditing and Monitoring**: Regular audits and monitoring of the HIPAA compliance program are critical. This process helps identify compliance gaps and facilitates timely corrective actions, as advised by the Centers for Disease Control and Prevention.

7. **Business Associate Agreement Management**: It's important to ensure that business associates handling ePHI also comply with HIPAA. This involves creating business associate agreements that clearly define ePHI safeguarding responsibilities (*HHS, 2022*).

8. **Incident Response and Reporting Protocol**: Developing an effective plan for responding to potential ePHI breaches is crucial. This should include internal reporting procedures, assessment, mitigation strategies, and notifying affected individuals and relevant authorities about breaches (*HHS, 2022*).

Implementing HIPAA in an organization demands a thorough approach, encompassing understanding regulations, performing risk analyses, developing policies, training staff, ensuring safeguards, conducting audits, managing business associates, and establishing an incident response strategy. These measures are fundamental in protecting sensitive health information and complying with HIPAA standards.

PCI DSS

The **Payment Card Industry Data Security Standard** (**PCI DSS**) is an extensive set of security protocols designed to ensure that any entity dealing with credit card information, whether it's accepting, processing, storing, or transmitting, maintains a highly secure environment. This standard applies to all organizations that handle credit cards from major card brands. Implementing PCI DSS involves a rigorous commitment to a series of security measures and procedures (Japan Card Data Security Consortium).

When it comes to integrating the PCI DSS within an organization, it's a pivotal move for safeguarding cardholder data. This process encompasses various critical steps and compliance with specific criteria. The following is a high-level overview of implementing PCI DSS in an organization:

1. **Understanding the Concept of PCI DSS**: This set of security standards is crafted to ensure that all entities that handle credit card information uphold a secure environment (*Payment Card Industry Data Security Standard, 2024*). You should familiarize yourself with your organization's capabilities in data encryption, network security controls, access controls, monitoring, and security policy establishment.

2. **Assessment**: Begin by evaluating your organization's current handling of cardholder data. Identify where this data is stored and take stock of your IT assets and payment card processing procedures (*Payment Card Industry Data Security Standard, 2024*).

3. **Scope Definition**: Determine the assessment's scope by identifying all systems and networks involved in storing, processing, or transmitting cardholder data (*Payment Card Industry Data Security Standard, 2024*).

4. **Gap Analysis**: Perform a gap analysis to find discrepancies between your existing security measures and the PCI DSS requirements, highlighting areas for improvement (*Payment Card Industry Data Security Standard, 2024*). Now would be an ideal occasion to collaborate with the diverse teams responsible for managing your organization's assets to establish the time and cost required to address any discrepancies.

5. **Implement Missing Controls and Remediate Gaps**: Tackle the gaps found in the gap analysis, possibly by updating software, changing business processes, or improving physical security. Put in place any missing security controls as outlined by PCI DSS, such as setting up firewalls, encrypting data transmissions, and regularly updating antivirus programs.

6. **Documentation**: Keep thorough documentation throughout the implementation process, including details of network architecture, data flows, and all implemented controls.

7. **Training and Awareness**: Create a security awareness program to educate employees about PCI DSS and their role in ensuring compliance.

8. **Regular Monitoring and Testing**: Continually monitor and test security controls to verify their effectiveness and compliance with PCI DSS standards. This monitoring and testing should encompass not only technical aspects but also procedural ones.

9. **Reporting**: Compile and submit as needed any reports, which may include a **Self-Assessment Questionnaire (SAQ)** or a **Report on Compliance (ROC)** for a full audit.

Integrating PCI DSS within an organization is a detailed process that involves assessing the current environment, defining the scope, conducting a gap analysis, addressing gaps, implementing necessary controls, documenting the process, educating staff, ongoing monitoring, testing, and reporting. Strict adherence to PCI DSS is essential for protecting cardholder data and maintaining the confidence of customers and business partners.

GDPR

The **General Data Protection Regulation (GDPR)** is a landmark legislation related to personal data that protects the privacy and rights of individuals in the EU and EEA by standardizing the requirements for the management of this data. One of its key provisions is the need for strong security measures and lawful processing of personal data, with hefty fines of up to 4% of global annual revenue or €20 million for violating the rules (GDPR.eu., n.d.). To comply with the GDPR, organizations need to do as follows:

- **Comprehend the GDPR:** It is crucial to familiarize yourself with the GDPR requirements and how your organization handles data processing. This includes reviewing the processes for collecting and using personal information transparently with explicit consent, giving individuals rights over their data, such as access and the right to be forgotten, and providing timely notifications in case of any data breaches.

- **Data Mapping and Identification**: It's vital to identify the personal data held, its sources, and sharing partners. This involves conducting an exhaustive data audit.

- **Updating Privacy Notices and Policies**: Compliance with GDPR necessitates transparent privacy notices and policies, clearly stating the usage of data and the legal grounds for its processing.

- **Data Protection Impact Assessments (DPIAs)**: These assessments are crucial in situations when data processing may significantly impact individual rights and freedoms.

- **Consent Management**: A critical step under GDPR is the review of consent-seeking, obtaining, and recording processes, ensuring they are clear, specific, and explicit.

- **Upholding Data Subject Rights**: Systems must be in place to verify the identities of data requesters and handle their requests effectively (GDPR.eu).

- **Training and Awareness**: Continuous staff training and awareness are key to maintaining compliance.

- **Data Protection Officers (DPOs)**: Appointing a DPO to oversee compliance and serve as a contact point for data subjects and regulatory bodies is essential (GDPR.eu).

- **Data Breach Response**: Establishing protocols to identify, report, and investigate data breaches, especially those that risk individual rights and freedoms, within 72 hours.

- **Managing International Data Transfers**: Ensuring proper safeguards for data transferred outside the EU.

GDPR is an all-encompassing data protection regulation demanding a considerable commitment from organizations to comply. The implementation process encompasses understanding the regulation, performing data audits, revising policies, managing consent, safeguarding data subject rights, training personnel, appointing DPOs, establishing data breach protocols, and handling international data transfers. These measures not only ensure regulatory adherence, but also foster trust among customers and users through responsible data management.

FISMA

The **Federal Information Security Management Act (FISMA)** of 2002 is a pivotal law in the United States that aims to defend government agency data and infrastructure against threats, ensuring the confidentiality, integrity, and availability of federal information systems and data. This goal is accomplished by requiring federal agencies and those organizations handling federal data to establish, document, and enforce a program for securing and protecting information (NIST Special Publication 800-53). The effective application of FISMA within an organization involves the following steps:

1. **Categorization of Information and Systems**: Initially, it's essential to classify information and the systems managing it according to their sensitivity and the potential impact of a breach on the agency's operations, assets, or individuals. This classification aids in deciding the level of security measures.

2. **Selection of Security Controls**: Organizations then choose suitable security measures based on the prior categorization. They detail the guidelines for these controls in NIST Special Publication 800-53, which advises on the selection and specification of security controls for federal information systems.

3. **Implementation of Security Controls**: The chosen controls are then put into action. This step involves the deployment of required hardware, software, policies, and procedures to safeguard information systems.

4. **Assessment of Security Controls**: After implementation, it's crucial to evaluate these controls to confirm their effectiveness and proper functioning. This evaluation typically involves various testing and review processes.

5. **Authorization of Information System**: A high-ranking official must formally approve the operation of the information system, acknowledging the risks to organizational operations, assets, individuals, other organizations, and the nation, based on the established security controls.

6. **Monitoring of Security Controls**: Ongoing surveillance of the security measures and the overall security status of the information system is essential. Monitoring should extend to changes that might impact the system's security and the periodic reevaluation of controls.

The NIST leads the FISMA Implementation Project by providing standards and guidelines to assist federal agencies in implementation and compliance efforts. Among the guidelines offered, the most well-known one is the NIST Risk Management Framework, which is leveraged by organizations to identify and manage risks as they relate to federal systems.

FISMA is a fundamental framework for securing federal information systems. Its implementation is a systematic process involving the categorization of information, the selection and implementation of suitable security controls, the evaluation of their effectiveness, the authorization of the information system, and the ongoing monitoring of security measures. This methodical approach is key to reducing risks and safeguarding sensitive government data against potential threats.

Acting as a backbone, these frameworks empower organizations to create a secure environment for their data and operations, which is essential in today's digital landscape. Every framework has a particular emphasis and relevance to different industries, and when implementing it, there is usually a comprehensive evaluation of existing procedures, an analysis of gaps, and an incorporation of targeted measures to reduce any identified risks.

What are an organization's vision, policies, and procedures?

Within the realm of cybersecurity, the vision, policies, and procedures of an organization are fundamental factors that collectively shape its approach to protecting digital assets and information. The vision acts as the North Star, delineating cybersecurity's long-term objectives and ambitions. It embodies the organization's dedication to safeguarding its digital infrastructure and the data it possesses. This vision is not merely a testament to determination; it reflects the organization's understanding of the growing cyber threat landscape and its resolve to stay ahead of potential threats.

Policies and procedures are the physical representations of this vision. Policies are formal declarations that prescribe the protocols for an organization and its employees to manage cybersecurity-related affairs. They establish benchmarks, regulations, and guidelines to be adhered to attain the cybersecurity vision. Procedures are the detailed processes that execute these policies. They provide a comprehensive framework for action in a range of scenarios, ensuring the organization's cybersecurity practices are synchronized, successful, and under its overarching vision. Jointly, these elements establish an all-encompassing cybersecurity framework, reconciling strategic visions with workable, implementable principles.

Vision in cybersecurity

The vision for cybersecurity in an organization is its goal or ambition in the field. This vision acts as a guiding light, directing the strategic course and establishing the framework for cybersecurity endeavors. It aligns with the organization's overarching mission and values, showcasing its dedication to safeguarding data, systems, and networks from diverse cyber threats. The vision may include goals such as attaining industry-leading security standards, cultivating a security-conscious culture, or gaining recognition for strong data protection measures.

Companies typically publish their cybersecurity vision on their official websites or in their annual reports. For instance, the **National Cybersecurity Center of Excellence** (**NCCoE**), which is part of the NIST, shares its mission and vision on its official website (www.nccoe.nist.gov). They offer comprehensive guidance, such as a reference design, components list, configuration files, relevant code, diagrams, tutorials, and instructions. This enables system administrators to replicate the example solution and achieve the same outcomes. Some companies may reveal their cybersecurity vision and strategy during industry conferences or in interviews with the media. It is common for companies to discuss their cybersecurity plans and strategies with their board of directors. Proposed cybersecurity regulations will soon mandate companies to report cyber breaches and explain how board members intend to mitigate cyber risk.

Cybersecurity policies

Effectively communicating this vision throughout the organization is imperative, as it significantly impacts the formulation and execution of precise cybersecurity protocols and measures. Cybersecurity policies play a crucial role in safeguarding an organization's IT assets. These policies are official declarations that outline the organization's strategy for cybersecurity. They provide explicit guidelines on various cybersecurity areas, including access control, incident response, and data privacy. They define the duties of diverse stakeholders within the organization and lay down the criteria for acceptable and secure utilization of IT resources. Some common policies found in organizations are as follows:

- **IT Security Policy**: This policy establishes the guidelines and protocols for safeguarding the organization against cyber threats. It covers the proper use of company assets, plans for handling incidents, strategies for business continuity, and the organization's compliance plan.

- **Email Security Policy**: This policy outlines the approved usage of corporate email systems to safeguard the organization from spam, phishing, and malware (such as ransomware) and to prevent misuse of corporate email. It may contain general guidelines for corporate email usage and specific instructions for handling suspicious links and attachments.

- **Bring Your Own Device (BYOD) Policy**: This policy outlines rules regarding the use of personal devices for work. Common security requirements will encompass the use of an endpoint security solution, strong passwords, and a VPN for connecting to corporate networks and IT assets via untrusted networks.

- **Data Protection Policy**: The policy defines confidential data and emphasizes its protection. It additionally specifies protocols for data transfer, guaranteeing security, and deterring unauthorized entry.

- **Incident Response Policy**: The policy provides guidelines on preparing for a cyber incident.

When planning policies, organizations consider their distinct risks, regulatory mandates, and business goals. Policies serve as a structural framework for making well-informed decisions regarding cybersecurity and are essential for ensuring compliance with legal and regulatory standards.

Cybersecurity procedures

Procedures encompass the precise and elaborate instructions or steps required to implement cybersecurity policies. They focus on tactical and operational elements, providing precise instructions on required actions, accountable individuals, and chronological sequence. An effective security program should include detailed procedures in a wide range of tasks, including device configuration such as firewalls and switches, security monitoring such as periodic vulnerability management and anomaly detection, conducting periodic security audits, incident response, and disaster recovery. Some common cybersecurity procedures that organizations often implement are as follows:

- **Control Physical Access**: These procedures guide the organization in restricting physical access to premises and computer networks. Topics covered will include authorized personnel identification and access control methods, such as security guards or badge scanners.

- **Restrict Access to Unauthorized Users**: Application controls limit access to data or services as outlined in these procedures. They could dictate the accessibility of data, such as allowing access to highly sensitive information only from approved locations and devices.

- **Limit Data Copying**: There are restrictions on what can be copied and saved from the system to storage devices.

- **Email Attachments Control**: These procedures identify the limitations on sending and receiving email attachments.

- **Security Awareness Training**: These procedures outline how an organization conducts regular training sessions to educate employees about the latest cybersecurity threats and prevention.

- **Disaster Recovery and Business Continuity**: These procedures outline contact points and steps to be taken before, during, and after a disaster, including strategies for business continuity.

Effective procedures are crucial for the practical implementation of cybersecurity policies, ensuring the consistent and precise execution of the organization's cybersecurity measures. These procedures are often customized to match the requirements and capacities of the organization, and their intricacy can vary depending on the scale and character of the business (*Putrus, 2021*).

Integration and importance

The alignment of the vision, policies, procedures, and cybersecurity framework is essential for a robust security posture. The typical way in which these work together is as follows:

- **Security Vision**: The organization's guiding light regarding its security vision offers a high-level overview of the goals and objectives this vision encompasses. It mainly helps shape the development of security policies and procedures, and determines the appropriate cybersecurity framework that aligns with these goals. An organization's security vision sets the tone for its approach to cybersecurity. It might be something like "*To protect our information assets from all threats, whether internal or external, deliberate or accidental. Our goal is to ensure business continuity, minimize business damage, and maximize return on investments and business opportunities.*"

- **Security Policies**: Security policies are rules that dictate how an organization safeguards its information assets. These policies are based on the organization's security vision and serve as a comprehensive guide to achieving this vision. They encompass multiple areas, including access control, data protection, and incident response. Security policies outline the specific actions that must be taken to ensure the organization's security. For instance, an Access Control Policy could declare that "*Access to information will be determined by business demands.*" Access will be provided, or accommodation will be made for staff, associates, and clients based on their position and the classification of the information they are accessing, with a focus on limiting access to sensitive data.

- **Security Procedures**: Security procedures are essential for implementing security policies by providing detailed instructions. These should ensure a consistent and effective implementation of the security policies throughout the organization and be aligned with specific controls within the cybersecurity framework. This could include things such as defining user roles, guidance on how to assign and review access rights, and providing timelines for incident response efforts.

- **Cybersecurity Framework**: A cybersecurity framework offers a systematic approach to effectively manage cybersecurity risk. It consists of a comprehensive set of standards, guidelines, and best practices specifically designed to address cybersecurity-related risks. When selecting a framework, it is crucial to ensure that it aligns with the organization's security vision. Moreover, the controls provided by the framework should be able to align with the organization's security policies and procedures.

The organization can maintain a strong security posture when these four elements are aligned and functioning collectively. Regular audits and reviews help the organization stay on track with its security vision and effectively manage cybersecurity risk.

Identifying an organization's needs and quantifying these

Familiarizing yourself with the organization's vision, policies, procedures, and security framework is crucial for a successful security product deployment, but further action is required. Consult any project manager and they will confirm that the most expedient route to failure in a project is the absence of a well-defined scope and adherence to it. The team responsible for implementing the solution must ensure that it is striving to achieve a measurable aim. We cannot simply rationalize a deployment based on the need for enhanced security, but we must clearly understand what enhanced security entails. Consequently, we must guarantee that the security project is in line with the organization, and we must track this using metrics.

When crafting metrics, the difficulty lies in formulating metrics that precisely depict measurable objectives. A goal focused on the prevention of malware attacks through email is a good starting point, but it lacks sufficient depth. An outstanding set of metrics would involve impeding the execution of malicious attachments in emails, both prior to and after their receipt, in addition to tracking the effectiveness of harmful file detection before versus after email dispatch.

When analyzing metrics, consider an organization's vision, which establishes an overarching goal and provides direction for its cybersecurity endeavors. Policies set forth the regulations and directives that dictate the execution of security measures. Meanwhile, the security framework provides a structured approach to managing and mitigating risks.

By leveraging these elements, organizations can develop metrics that not only measure the technical aspects of security solutions but also their alignment with business objectives, regulatory compliance, and risk management strategies. This holistic approach ensures that the metrics are not just numbers but valuable insights that drive continuous improvement and strategic decision-making in cybersecurity. Some examples of metrics are as follows:

- **Security Vision Metrics**: If the organization's security vision is to *"guarantee uninterrupted business operations, mitigate business impact, and optimize returns on investments and business prospects,"* then metrics may encompass the frequency of security incidents causing downtime, the financial implications of incidents, and the ROI of security endeavors.

- **Security Policies Metrics**: If the organization has a policy that *"Access to information will be determined by business requirements,"* then metrics could encompass the frequency of access violations, the number of users possessing unnecessary access rights, and the duration of time required to revoke access rights once they are deemed unnecessary.

- **Security Procedures Metrics**: if the organization has an established protocol for consistently assessing and modifying access rights, possible metrics may involve the count of completed access rights reviews, the count of updated access rights, and the count of identified outdated access rights.

- **Cybersecurity Framework Metrics**: If the organization has embraced the NIST Cybersecurity Framework, then metrics could be harmonized with the five core functions of this framework: Identify, Protect, Detect, Respond, and Recover. For example, under the Detect function, metrics could include the number of incidents detected, the time taken to detect incidents, and the percentage of incidents detected by the organization's security systems.

It should *not* be assumed that all security solutions and projects will address all security requirements identified by an organization. The process of creating metrics will always be a tailored approach that seeks to strike a balance between organizational needs and solution capabilities. In addition to maintaining this alignment, it is imperative to regularly review and update these metrics to ensure they consistently offer valuable insights into the organization's security posture and the efficacy of its security product.

Summary

In this chapter, we explored the crucial process of evaluating the effectiveness of implementing a security product in an organization. We focused on integrating and aligning the product with established security frameworks. These frameworks include ISO 27001, the NIST Cybersecurity Framework, HIPAA, PCI DSS, GDPR, and FISMA. Each of these frameworks offers a systematic approach to managing and safeguarding data and information assets. ISO 27001 emphasizes a methodical and risk-oriented approach to managing information security. The NIST Cybersecurity Framework provides recommendations for managing cybersecurity risks. HIPAA establishes regulations for safeguarding confidential patient data, while PCI DSS concentrates on fortifying credit card details. GDPR regulates data protection and privacy within the EU, and FISMA establishes federal data security guidelines.

Implementing these frameworks requires a clear understanding of an organization's vision, policies, and procedures. This understanding is pivotal in formulating efficacious metrics to gauge the success of a security product. Therefore, this chapter guided you through the process of identifying specific objectives that align with the organization's wider goals and compliance mandates. It underscored the significance of establishing quantifiable, pertinent, and attainable metrics. These metrics could include measuring the decrease in security incidents, the improvement in response times to security breaches, and the enhancement of compliance with regulatory standards.

This chapter emphasized the importance of ongoing monitoring and adjusting strategies to maintain the effectiveness and alignment of security measures with the changing cyber threat landscape and regulatory environment. Now that we have a clearer understanding of our organization's goals, we should be able to create a set of metrics to measure against our Defender for Office 365 deployment. Let's apply what we have learned in this chapter as we look at how to perform a basic configuration of Defender for Office 365 in the next chapter.

References

- Infosec Institute. (2023). *ISO 27001 framework: What it is and how to comply.* Available at: `https://resources.infosecinstitute.com/topic/iso-27001-framework-what-it-is-and-how-to-comply/`

- International Organization for Standardization. (2022). *ISO/IEC 27001 (2022nd ed.).* `https://www.iso.org/obp/ui/#iso:std:iso-iec:27001:ed-3:v1:en`

- National Institute of Standards and Technology (NIST). (2023). *Cybersecurity Framework.* Available at: `https://www.nist.gov/cyberframework`

- Proofpoint. (2023). *HIPAA Compliance and Enforcement.* Available at: `https://www.proofpoint.com/us/threat-reference/hipaa-compliance`

- National Center for Biotechnology Information. (2022). *Health Insurance Portability and Accountability Act.* Retrieved from `https://www.ncbi.nlm.nih.gov/books/NBK500019/`

- National Institute of Standards and Technology. (2011). *HIPAA Security Rule.* Retrieved from `https://www.nist.gov/programs-projects/security-health-information-technology/hipaa-security-rule`

- U.S. Department of Health & Human Services. (2022). *Summary of the HIPAA Privacy Rule.* Retrieved from `https://www.hhs.gov/hipaa/for-professionals/privacy/laws-regulations/index.html`

- Payment Card Industry Data Security Standard: Requirements and testing procedures (4.0.1). (2024). *PCI Security Standards Council.* `https://docs-prv.pcisecuritystandards.org/PCI%20DSS/Standard/PCI-DSS-v4_0_1.pdf`

- GDPR.eu. (n.d.). *What is GDPR, the EU's new data protection law?* Retrieved from `https://gdpr.eu/what-is-gdpr/`

- National Institute of Standards and Technology. (n.d.). *NIST Special Publication 800-53.* `https://csrc.nist.gov/pubs/sp/800/53/r5/upd1/final`

- Putrus, R. (2021, August 27). *Best Practices for Setting Up a Cybersecurity Operations Center.* ISACA. `https://www.isaca.org/resources/isaca-journal/issues/2021/volume-5/best-practices-for-setting-up-a-cybersecurity-operations-center`

4

Basics of Configuration

Now that we have a better understanding of what we can achieve with **Defender for Office 365**, it's time to discuss the essential steps needed to set up a basic deployment of this product. In this chapter, we'll cover the necessary preparations and configuration of pre-set security policies. Our approach is based on Microsoft's security recommendations, which have been observed in the field and offer a simplified yet effective strategy.

As part of these preparations, we'll help you determine the appropriate configuration for your organization. We'll guide you through creating a customized plan that matches your organization's specific needs and scale. Additionally, we'll delve into the details of identifying the required licenses, privileged roles, and information needed to make informed decisions during the configuration process.

Moving forward, we'll focus on identifying your organization's risk profile. This step is crucial in understanding the specific threats your organization may face. By doing so, we can develop a more targeted and effective defense strategy. Finally, this chapter concludes with a comprehensive guide to applying these insights to configure Defender for Office 365's pre-set policies. We'll offer practical advice on how to customize these policies to align with your organization's risk profile, ensuring a strong and proactive defense against emerging cyber threats.

This chapter will cover the following topics:

- Preparations and prerequisites
- It is all about the organization's risk profile
- Configuring the preset policies

Let's continue our journey!

Preparation and prerequisites

Licenses and permissions can be perceived as intricate, yet this section is dedicated to simplifying the matter. As we do not want to make for a complex deployment, we want to tailor our efforts to what our organization needs by exploring the available Defender for Office 365 plans that fit your organization's size and security needs. I'll walk you through each option to help you make an informed choice. Next, we'll break down the licenses offered and what you will need. Lastly, we cannot ignore permissions. It's imperative to extend the appropriate permissions to the rightful individuals for smooth operations. We will explore these permissions and any proposed suggestions in detail to enable smooth operation for your team. In the subsequent sections, this information will be crucial in determining our configuration. Let's kick off the process of making your digital workspace as secure as it can be!

Licenses

To comprehend how an organization can obtain Microsoft Defender for Office 365 coverage, it is vital to understand the different licenses offered by Microsoft. In the following subsections, we will examine each license type and the coverage or requirements linked to it at the time of writing. These combinations may undergo alterations before implementation; thus they should only serve as initial references, and you should conduct thorough licensing planning directly with Microsoft.

It's all about the plans

Microsoft Defender for Office 365 coverage is provided under 2 plans called **Microsoft Defender for Office 365 Plan 1 and Plan 2**. The difference between Microsoft Defender for Office 365 Plan 1 and Plan 2 primarily lies in the level of protection and features each plan offers. Both plans safeguard your organization against malicious threats associated with email messages, links (URLs), and collaboration tools (*Microsoft, 2023a*).

Microsoft Defender for Office 365 Plan 1 includes core features such as the following:

- **Threat protection**: It offers protection against malware, viruses, phishing attempts, and other threats via email
- **Safe Attachments and Safe Links**: These features provide real-time protection by checking email attachments and links for malicious content
- **Anti-phishing policies**: Plan 1 includes basic anti-phishing protection
- **Real-time detection**: It provides insights into real-time threats and helps in identifying and blocking malicious emails

Microsoft Defender for Office 365 Plan 2 includes all the features of Plan 1, plus additional advanced capabilities:

- **Enhanced threat protection**: Plan 2 offers more advanced threat protection features, including enhanced phishing protection and impersonation protection

- **Threat trackers and Threat Explorer**: These are advanced tools for threat investigation and response, providing deeper insights into threats and enabling proactive defense strategies

- **Automated Investigation and Response (AIR)**: This feature helps in automating the investigation process and response to threats, reducing the workload on IT staff

- **Attack simulation training**: Plan 2 includes tools for running simulated phishing attacks, which can be a valuable training resource for improving organizational security awareness

While both Plan 1 and Plan 2 provide robust security features for Office 365, Plan 2 offers more advanced capabilities, particularly in terms of threat investigation, automated response, and user training tools. When choosing between plans, organizations must consider their specific security needs, resources, and the level of protection they require. Here's a detailed comparison to help you understand why an organization might choose one plan over the other, along with examples:

- **Basic versus advanced threat protection**:

  - **Plan 1**: This offers essential protection against malware, viruses, phishing attempts, and other threats primarily via email filtering and detection mechanisms. It's suitable for organizations that need a fundamental level of security and are perhaps smaller or have limited security resources. For example, a small startup with limited IT resources might opt for Plan 1 for its straightforward, essential security features.

  - **Plan 2**: This includes all the features of Plan 1 but adds advanced capabilities such as threat investigation and response tools, AIR, and attack simulation training. This plan is ideal for larger organizations or those particularly sensitive to cyber threats, such as financial institutions or healthcare providers, where advanced threat protection is crucial.

- **Threat investigation and response**:

  - **Plan 1**: This does not include advanced threat investigation tools. Organizations with a dedicated IT security team that can manually handle threat investigations might find this level sufficient.

  - **Plan 2**: This offers advanced threat investigation and response capabilities, including tools to analyze, understand, and mitigate threats. A multinational corporation facing diverse and sophisticated cyber threats would benefit from Plan 2's comprehensive tools for identifying and responding to security incidents. Plan 2 includes the capability to proactively search for threats via advanced hunting which can identify a misconfiguration or a bad actor in the environment before any major impact is experienced.

- **Automation capabilities**:

  - **Plan 1**: This lacks automation in threat response, meaning the process needs to be manual and thus can be slower and resource intensive.

  - **Plan 2**: This provides automation capabilities that can help in quickly addressing threats, reducing the time and resources needed for resolution. For instance, a government agency handling sensitive data might prefer Plan 2 for its ability to respond to threats rapidly and efficiently not only thanks to AIR, but also the capability to set automatic responses according to the threat such as isolating devices (if Defender for Endpoint is in use) or even disabling a user.

- **Security education**:

  - **Plan 1**: This does not include security education tools.

  - **Plan 2**: This comes with attack simulation training, helping organizations educate their employees about security threats. A large retail company with many employees could use Plan 2 to train its staff in identifying and avoiding potential security threats, such as phishing.

- **Cost consideration**:

  - **Plan 1**: This is more cost effective than Plan 2. It's a viable option for organizations with a limited budget but still needing a robust security solution.

  - **Plan 2**: This is more expensive, but the additional cost is justified for organizations requiring comprehensive security features, especially those handling sensitive data or operating in high-risk environments.

The choice between Defender for Office 365 Plan 1 and Plan 2 depends on the specific needs and resources of an organization. Plan 1 is suitable for basic protection needs and smaller organizations, while Plan 2 is tailored for larger organizations requiring advanced security features and automation capabilities. The decision should align with the organization's size, budget, nature of data handled, and the complexity of potential security threats.

Selecting the right license

Understanding exactly what license is needed for your organization might look like a tough job because of the number of licenses and combinations available, but it is not impossible. Microsoft provides extensive documentation on these, but the recommended approach is to identify what Plan your organization needs and discuss with an authorized Microsoft reseller for the most up-to-date information. Be aware that depending on the size of your organization and/or relationship with Microsoft, there might be different offers available to your team to lower costs. The following are just some combinations at the time of this writing (*Microsoft, 2023b*):

- **Microsoft 365 E5 or A5**: These licenses include Microsoft Defender for Office 365 Plan 2. Organizations with these licenses don't need additional licenses for Office 365, as Microsoft 365 E5/A5 is a comprehensive offering that includes Office 365.

- **Microsoft 365 E3 with Microsoft 365 E5 Security add-on**: This combination provides Microsoft Defender for Office 365 Plan 2 coverage. The E5 Security add-on enhances the E3 license with advanced security features, including those offered by Defender for Office 365.

- **Microsoft 365 E3 with Enterprise Mobility + Security E5 add-on**: While this combination enhances security, it does not automatically include Microsoft Defender for Office 365. Organizations may need to purchase Defender for Office 365 as a separate add-on.

- **Microsoft 365 A3 with Microsoft 365 A5 Security add-on**: Like the E3/E5 combination, this setup should provide Microsoft Defender for Office 365 Plan 2 coverage, as the A5 Security add-on includes advanced security features.

- **Windows 10 Enterprise E5 or A5 / Windows 11 Enterprise E5 or A5**: These licenses primarily focus on endpoint security and do not include Microsoft Defender for Office 365. Organizations would need to purchase Defender for Office 365 separately.

- **Enterprise Mobility + Security (EMS) E5 or A5**: While EMS E5/A5 includes advanced security features, it does not cover Microsoft Defender for Office 365. This would be a separate purchase.

- **Office 365 E5 or A5**: These licenses include Microsoft Defender for Office 365 Plan 2. No additional purchase is necessary for Defender coverage.

- **Microsoft 365 Business Premium**: This license includes Microsoft Defender for Office 365 Plan 1. It offers core protection but lacks some advanced features found in Plan 2.

- **Microsoft Defender for Business**: This exclusively caters to small to medium businesses, this product stands on its own and focuses on endpoint protection through defender for endpoint. However, it does not include Microsoft Defender for Office 365. A small business may find it more advantageous to acquire the Microsoft 365 Business Premium license.

Microsoft 365 licenses are assigned from the Microsoft admin center, as seen in the following screenshot:

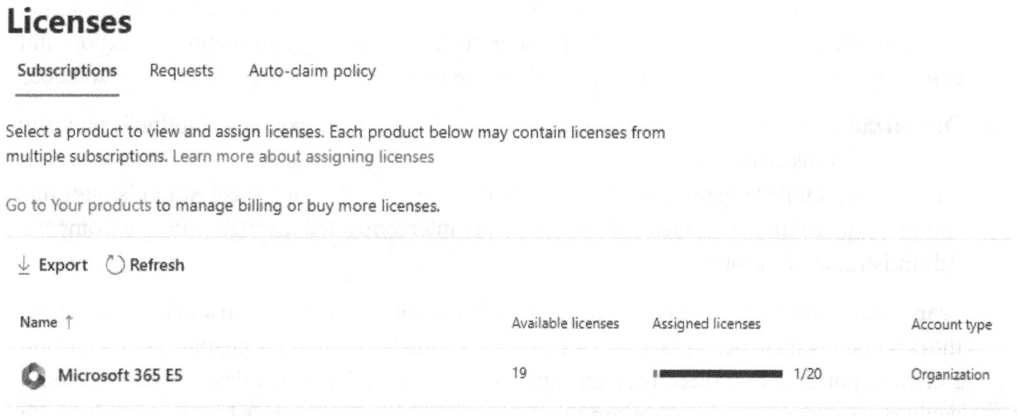

Licenses

Subscriptions   Requests   Auto-claim policy

Select a product to view and assign licenses. Each product below may contain licenses from multiple subscriptions. Learn more about assigning licenses

Go to Your products to manage billing or buy more licenses.

↓ Export   ○ Refresh

Name ↑	Available licenses	Assigned licenses	Account type
Microsoft 365 E5	19	▬▬▬▬ 1/20	Organization

Figure 4.1 – The Licenses section in the Microsoft admin center

Organizations with Microsoft 365 E5/A5, Microsoft 365 E3 plus E5 Security add-on, Microsoft 365 A3 plus A5 Security add-on, and Office 365 E5/A5 licenses already have Microsoft Defender for Office 365 included. For other license types, such as Windows 10/11 Enterprise E5/A5, EMS E5/A5, or Microsoft 365 Business Premium, Microsoft Defender for Office 365 would need to be purchased separately if required.

Permissions required

To effectively configure and employ Microsoft Defender for Office 365, certain permissions need to be obtained. We recommend you prepare the privileges for the personnel to configure the solution before starting any work. It is important to identify the roles that your organization will use and the individuals who will assume these roles once Defender for Office 365 is implemented. In the following subsections, we look in more detail at these permissions.

Roles for personnel performing initial configuration tasks

These roles will have a substantial amount of power to alter your organization and should only be conferred for a limited period. We also strongly advise using the **Privileged Identity Management** (**PIM**) component in Microsoft Entra ID, specifically the PIM for groups feature, which aligns permissions to a group and configures membership eligibility to this group to rely on approval, for a restricted duration per activation, and offers alerts on activation. This approach guarantees streamlined management and minimal consequences in case one of these accounts is breached. The roles required during configuration are as follows:

- **Global Administrator**: This role enjoys nearly limitless access to your organization's settings and most of its data. Global Administrators possess the capability to govern permissions and regulate access to all features in Microsoft 365, including Microsoft Defender for Office 365. They also can alter and reset the passwords of other Global Administrators.

- **Security Administrator**: The Security Administrator role entails configuring and administering Defender security tools, implementing threat protection, managing information protection, and enforcing compliance. This role is essential for the management of security in Microsoft 365.

- **Organization Management in Exchange Online**: This role in Exchange Online is typically given to administrators who are entrusted with the management of a specific division within the Exchange Online organization. It's important to mention that the Global Administrator role and the Organization Management role group are interconnected through a unique Company Administrator role group.

- **Search and Purge**: This role is employed in the Microsoft Defender portal and is necessary for the deletion of email messages. If you aren't assigned the Search and Purge role in the Microsoft Defender portal (either directly or through a role group such as Organization Management), you'll receive an error when you run the `New-ComplianceSearchAction` cmdlet with the **A parameter cannot be found that matches parameter name** message.

Permissions can be assigned via the Microsoft Entra ID portal (`entra.microsoft.com`) as shown in the following figure:

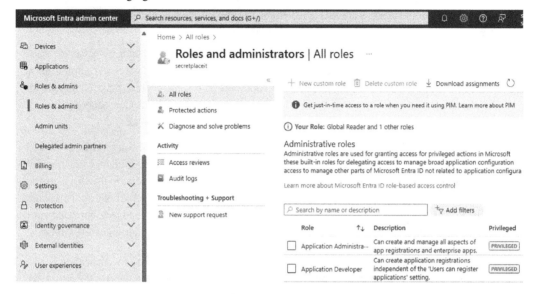

Figure 4.2 – Permissions section in the Microsoft Entra ID portal

As previously stated, after completing the initial configuration of Defender for Office 365, these assignments should be eliminated, and new assignments should be executed based on the roles discussed in the following section.

Roles for personnel performing ongoing maintenance and support tasks

Once Defender for Office 365 has been configured, the following roles will need to be assigned based on your organizational requirements. It is strongly advised to assign these roles using the PIM component of Microsoft Entra, which includes notifications and approval for high-impact roles. The access review feature should also guarantee appropriate life cycle management for these roles. The roles that need to be configured operationally are as follows (*Microsoft, 2023b*):

- The **Global Administrator** and **Global Reader** roles in Microsoft 365 have different levels of access and responsibilities:

  - **Global Administrator**: This role commands the utmost level of permission and can execute all sorts of tasks across the entire suite of Microsoft 365 services, encompassing Defender for Office 365. Global Administrators wield nearly limitless authority over your organization's settings and much of its data. It is advisable to restrict the count of Global Administrators to less than five within an organization and establish time constraints for the duration of their roles. You should allocate this role to a few individuals whom you trust and who require extensive access, such as high-ranking IT staff members.

- **Global Reader**: With this read-only version of the Global Administrator role, you can access and view all Microsoft 365 settings and administrative information. This role is typically assigned to individuals who require monitoring privileges without the ability to make modifications, such as auditors or other stakeholders who need to access but not alter administrative information.

Although both roles are involved in administration, the Global Administrator possesses superior permissions and responsibilities, while the Global Reader is limited to read-only access to administrative information. It is worth noting that, while the Global Reader role is unable to make changes to an organization, it can be exploited by attackers to enumerate your organization. Consequently, this role should be regarded with the same caution and respect as other privileged roles.

- The **Compliance Data Administrator** and **Compliance Administrator** roles in Microsoft 365 are both essential for managing and enforcing your organization's data compliance policies, but each has its own specific focus. These roles should be designated to individuals who bear the responsibility of managing and enforcing your organization's data compliance policies. This could include IT professionals, compliance officers, or other individuals who need to ensure that your organization's data is being handled in accordance with applicable regulations and standards:

  - **Compliance Data Administrator**: Tailored for users with a need to monitor data in the Microsoft Purview compliance portal, Microsoft 365 admin center, and Azure. Additionally, they can monitor compliance data within Exchange. This role is granted the privilege to create and manage compliance data policies and alerts.

  - **Compliance Administrator**: This is equipped with permissions to perform tasks involving content searches across a range of workloads, potentially retrieving emails, chats, OneDrive data, and other relevant materials. They can also manage device settings, implement data loss prevention measures, generate reports, and preserve data.

While both roles are involved in data compliance, the Compliance Data Administrator focuses more on tracking data and managing data policies, while the Compliance Administrator is more involved in content searches and managing settings for device management, data loss prevention, and preservation. When assigning these roles, be aware that the Compliance Administrator role allows a certain level of insight into the data, maybe not directly data stored in a mailbox, but may still show high-level information that could contain **personally identifiable information (PII)**.

- The **Security Administrator**, **Security Operator**, and **Security Reader** roles have different levels of access and responsibilities:

  - **Security Administrator**: This role can view and manage security policies, view, respond to, and manage alerts, take response actions on devices with detected threats, and view security information and reports. This role is typically assigned to IT professionals who are responsible for managing and enforcing your organization's security policies. They need to have a deep understanding of the organization's security needs and technical skills to implement and manage security policies.

- **Security Operator**: This role, while not directly related to Defender for Office 365 management, has an impact on the security incident life cycle management as this role allows for managing security incidents and alerts to include opening and closing these, reviewing security settings, and performing security investigations. If your organization uses a **security operations** (**SecOps**) approach, this role would be assigned to the teams performing **Security Operations Center** (**SOC**) tasks, or those involved in the investigation of resolution of incidents and alerts.

- **Security Reader**: This is a read-only role that allows users to view a list of onboarded devices, view security policies, view alerts and detected threats, and view security information and reports. They can't add or edit security policies, nor can they onboard devices. This is a read-only role that's typically assigned to individuals who need to monitor security but don't need to make changes. This could include managers who want to stay informed about the organization's security status, auditors who need to review security policies and alerts, or other stakeholders who need to view security information.

While all three roles are involved in security, the Security Administrator has the most permissions and responsibilities, the Security Operator is typically involved in ongoing security operations, and the Security Reader has read-only access to security information.

- The **Attack Simulation Administrator** and **Attack Payload Author** roles are designed for users who need to manage and create attack simulations in Defender for Office 365:

  - **Attack Simulation Administrator**: This role allows users to create and manage all aspects of attack simulation campaigns. This includes setting up the simulation, choosing the technique used by attackers, and selecting the users to target.

  - **Attack Payload Author**: This role allows users to create attack payloads that an admin can initiate later. These payloads are the links or attachments in the simulated phishing email messages that are presented to users.

The main difference between these roles is that the Attack Simulation Administrator is responsible for managing the overall simulation campaign, while the Attack Payload Author is responsible for creating the specific payloads used in the simulations. When assigning these roles, focus on the personnel within your organization who are responsible for training and educating staff about potential security threats. For example, a Security Administrator might also be assigned the Attack Simulation Administrator role to manage the simulations, while a member of the security team might be assigned the Attack Payload Author role to create realistic payloads based on the latest threat intelligence.

Other roles related to Microsoft 365

Identifying and assigning (using Microsoft Entra PIM) the security roles directly related to Defender for Office 365 is a significant step toward a solid security foundation, but security is a multiple-layer approach. As attackers might look for other ways to abuse privileges, it is recommended that your team discuss the following roles and assignments. While not all roles need someone assigned to them, at a minimum, assignment approvals and notifications should be configured via Entra PIM to ensure no unauthorized memberships (*Microsoft, 2023c*):

- **Application Administrator**: This role grants permission to manage all aspects of app registrations and enterprise apps in Entra ID. Responsibilities include creating, changing, deleting, and configuring app settings. Examples of app settings include single sign-on, user and group assignments, licensing, application proxy, and consent. Caution must be exercised when assigning this role, as a malicious application could be registered and used in an attack.

- **Application Developer**: The application developer role allows users to create their own app registrations in Entra ID without needing approval from an administrator. However, this role does not grant the ability to manage other aspects of app registrations and enterprise apps. For tasks such as single sign-on, user and group assignments, licensing, application proxy, and consent, the Application Administrator role is required. Assigning the Application Developer role should be done carefully to minimize the risk of registering a malicious application.

- **Authentication Administrator**: The authentication administrator role grants permissions to view, set, and reset authentication method information for non-admin users. These administrators can require users to re-register with non-password credentials, such as **multi-factor authentication (MFA)** or **Fast IDentity Online (FIDO)** keys such as passkeys or yubikeys. They can also revoke the **remember MFA on the device** option, which prompts for MFA during the next sign-in. However, this role does not allow management of other aspects of Microsoft Entra ID, such as users, groups, roles, domains, or policies. Typically, the Authentication Administrator role is assigned to users who need to manage authentication methods and settings for other users in the organization, such as helpdesk staff, IT support staff, or security staff.

- **Authentication Extensibility Administrator**: The role of an Authentication Extensibility Administrator in Entra ID is to grant permissions for managing the authentication methods extensibility feature. This feature enables the integration of third-party authentication methods, such as FIDO2 security keys, with Entra ID. As an Authentication Extensibility Administrator, you can enable or disable the feature, configure settings and policies, and manage partner applications. However, it's essential to remember that this role does not encompass the management of other Microsoft Entra ID elements, such as users, groups, roles, domains, or policies. Typically, the Authentication Extensibility Administrator role is assigned to users in charge of integrating third-party authentication methods with Entra ID.

- **Billing Administrator**: This role is used to manage the billing process for the organization. Some important points to note about this role are its ability to make purchases, handle subscriptions and service requests, and monitor service health. Individuals in this role have authorization to access billing in the Azure portal and perform various billing tasks, such as creating subscriptions, reviewing and paying invoices, and updating payment methods. Billing Administrator roles are typically assigned to members of the finance team to ensure financial compliance and the timely generation, process, and payment of bills. Additionally, individuals in this role are responsible for managing account statuses and balances, as well as identifying any inconsistencies that may arise.

- **B2C IEF Keyset Administrator**: This role allows users to manage secrets in the **Identity Experience Framework (IEF)** in Microsoft Entra ID for customers. The IEF enables users to create and customize user journeys, also known as *custom policies*, in Entra ID. These secrets include keys for token signing and validation, client secrets, certificates, and passwords used in custom policies.

- **Cloud Application Administrator**: The Cloud Application Administrator role confers permissions to oversee and control cloud applications within the Microsoft 365 admin center. This role can execute tasks such as adding, removing, and updating cloud applications, configuring single sign-on settings, assigning licenses, and managing user access.

- **Cloud Device Administrator**: The role of the Cloud Device Administrator includes the ability to delete, disable, or enable devices within Entra ID. Typically, this role is assigned to service desk personnel and other individuals responsible for device-oriented support within an organization.

- **Conditional Access Administrator**: This role can create, change, and delete conditional access policies, as well as view reports on their usage. This role is typically assigned to personnel responsible for managing conditional policies, including their configuration and fine-tuning.

- **Directory Writers**: Users assigned this role can create, read, update, and delete user and group objects in Entra ID. However, they cannot manage other types of objects in Entra ID, such as applications or service principals.

- **Exchange Administrator**: This role allows users to manage email and mailboxes for your organization using the Exchange admin center. Some key tasks include recovering deleted items, setting up archive and deletion policies, configuring mailbox sharing policies, setting up delegates for mailbox access, creating shared mailboxes, and managing email anti-spam protection and malware filters. Exchange Administrators are responsible for maintaining the integrity of the organization's email system. They may also implement new features or services within the Exchange environment.

- **Exchange Recipient Administrator**: This role grants permissions to manage Exchange recipients, which include mailboxes, contacts, and distribution groups, on Exchange Online or Exchange Server. Users in this role can perform tasks such as creating, modifying, deleting, and moving recipients. They also can manage recipient properties and policies.

- **Fabric Administrator**: Responsible for administering Microsoft Fabric, which includes Power BI, Power Apps, Power Automate, and Power Virtual Agents. This role has the power to configure all tenant settings and global policies for Microsoft Fabric. The Fabric Administrator has complete control over organization-wide settings and admin features, except for licensing. They can use the admin portal to configure, monitor, and provision organizational resources. However, it's important to note that this role cannot manage licenses.

- **Groups Administrator**: This role allows users to manage Entra ID security groups without needing Global Administrator permission. Important aspects of this role include using the Microsoft 365 admin center, Azure portal, and other methods to create, edit, delete, and restore groups. Users with this role can also oversee Office 365 group policies, such as creation, naming, and expiration policies. However, this role is specific to Office 365 Groups and cannot manage other group types, such as distribution groups, mail-enabled security groups, or shared mailboxes. It's also not possible to use this role to update a group's email address or change external mail or mail delivery options in the Microsoft 365 admin center.

- **Hybrid Identity Administrator**: The Hybrid Identity Administrator role is designated for users who oversee the synchronization of on-premises identities with Entra ID. Users with this role can handle the installation and setup of Microsoft Entra Connect, which facilitates the synchronization of on-premises identities with Entra ID.

- **Intune Administrator**: This role is usually assigned to individuals who handle corporate and **bring your own device (BYOD)** device policies on Microsoft Endpoint Manager (previously Intune). This role can handle device enrollment, compliance, and app management, as well as create, modify, and delete device management policies, and manage enrollment and compliance policies.

- **License Administrator**: This role allows users to effectively manage licenses within the organization. Important aspects of this role include the ability to read, add, remove, and update license assignments for users. Additionally, users can conveniently manage licenses for groups through group-based licensing. This role grants the capability to edit the usage location of users. It is important to note that this role does not provide the authority to manage other administrative tasks.

- **Message Center Privacy Reader**: This admin role allows users to read data privacy messages. However, it does not grant permission to view, create, or manage service requests. Only Global Administrators and Message Center Privacy Readers can access data privacy messages. This role is specifically for users who need to monitor these messages needing no other admin privileges.

- **Message Center Reader**: This admin role allows users to read and share posts in the Message Center without having additional admin privileges. It is useful for users who only need to monitor messages. Most users with any admin role in Microsoft 365 can view Message Center posts. The Message Center Reader role can be assigned to users who should only be able to read and share posts, with no other admin privileges. This role does not grant access to the Message Center.

- **Office Apps Administrator**: This role allows users to manage Microsoft 365 Apps in the enterprise. They can use the Cloud Policy service to create and manage cloud-based policies and service requests. They also have access to the Microsoft 365 Apps admin center. This center offers modern cloud-based management for administrators who deploy and manage Microsoft 365 Apps in the enterprise. The admin center includes features such as the Cloud Policy service, Office Customization Tool, and Microsoft 365 Apps health.

- **Organizational Messages Writer**: This role enables assigned users to both view and configure organizational messages. It is important to note that this role is specifically designed for individuals who write, publish, manage, and review these messages for end users across Microsoft product surfaces. Organizational messages serve as a means of communication in remote and hybrid work scenarios. They aim to assist employees in adapting to new roles, gaining a better understanding of their workplace, and staying informed about essential updates and required training.

- **Password Administrator**: This role grants users the ability to reset passwords for non-administrative users. This role is useful for users who need to assist other users with password issues but do not require other admin privileges.

- **Power Platform Administrator**: Users with this role can oversee Microsoft Power Apps, Power Automate, Power Pages, and Power Virtual Agents in the Power Platform admin center. It is important to note that these users can sign in to and manage multiple environments. Their access and abilities are independent of their membership in security groups, meaning they can manage environments even if they have not been added to the corresponding security group. Additionally, they hold the System Administrator role in the Microsoft Power Platform, granting them the authority to perform administrative tasks. This role is assigned to users responsible for developing and maintaining an organization's strategy, implementing **data loss prevention (DLP)** policies, managing security groups and user assignments, handling capacity and licensing, facilitating data availability for makers, managing integration and migration processes, and overseeing security measures.

- **Privileged Role Administrator**: The Privileged Role Administrator role in Microsoft Entra ID manages privileged roles and permissions. This role can assign privileged roles to users and groups, create and manage custom roles, and assign permissions. They can also view reports on the usage of privileged roles and permissions.

- **Reports Reader**: This role is for users who need to view reports, schedule reports, view usage data, view activity reports, and view the Microsoft 365 admin center. It's suitable for IT administrators who need to monitor the organization's usage and activity.

- **Search Administrator**: This role allows users to create and manage search result content. It also enables them to define query settings to improve search results within the organization. This role is suitable for IT administrators who manage the organization's search settings and content.

- **Service Support Administrator**: This role is assigned to users who have the responsibility of initiating and handling service requests, as well as accessing and sharing posts from the Message Center. Their responsibility is to ensure efficient resolution of service requests and to distribute significant updates within the organization.

- **Teams Administrator**: This role is for users who need to manage Microsoft Teams and its related services. It has key privileges, including creating and managing Microsoft 365 groups, managing meetings, calling, and messaging policies. This role can also handle configurations, conference bridges, phone number inventory and assignment, and other organizational-wide settings. Mishandling these settings could cause disruptions and, as such, this role should only be assigned to users who need to manage the entire team's workload or have delegated permissions for troubleshooting call quality problems or managing the organization's telephony needs.

- **User Administrator**: This role is for users who need to reset passwords, monitor service health, and manage user accounts, user groups, and service requests. It's suitable for IT administrators who manage the organization's user accounts.

The purpose of these roles is to prevent over-privileged accounts, which are commonly targeted by attackers, from having privileges on the entire platform and decreasing the impact of a breach. Microsoft offers various controls for configuring these roles, based on real-world scenarios. You can find many of these controls suggested in many security publications such as the annual Microsoft Digital Defense Report, and most organizations can apply these with no major issues. These controls vary in their level of difficulty and impact. They can be classified as easy (less than 1 month to implement, minimal impact), medium (1 to 3 months to implement, some planning and testing required), or complex (6 to 12 months to implement, extensive planning and testing required).

The easy-to-implement controls include the following:

- **MFA**: MFA adds an extra layer of security to the authentication process by requiring users to provide at least two forms of identification before access is granted. This typically includes something the user knows (such as a password), something the user has (such as a phone), and something the user is (such as a fingerprint). MFA can be easily implemented with free software such as Windows Hello for Business and Microsoft Authenticator, and hardware FIDO2 keys such as those offered by Yubico (*Microsoft, 2023a*). Microsoft Authenticator can be easily obtained from the Google Play Store or Apple App Store as shown in the following figure:

App Store Preview

This app is available only on the App Store for iPhone and iPad.

Microsoft Authenticator 4+

Protects your online identity
Microsoft Corporation

#3 in Productivity
★★★★★ 4.8 • 312.6K Ratings

Free

Screenshots   iPhone   iPad

Figure 4.3 – Download screen from the Apple App Store for Microsoft Authenticator

- **Just-in-time (JIT) access**: The main component of Microsoft Entra PIM. This principle involves granting access rights only when they are needed and for just long enough to complete a task. This reduces the risk of unauthorized access or the misuse of privileges. For example, let's say you have an IT administrator who occasionally needs to perform high-level tasks, such as configuring the settings of a Microsoft 365 tenant or managing user roles. As the user is not performing these tasks all the time, these high-level privileges should only be available when needed. To manage this, you can use JIT access and have the user start with only the basic permissions necessary for his day-to-day tasks. When a high-level task is to be performed, the user requests elevated access. This request is then either approved automatically based on pre-set policies, or manually by another administrator. After a pre-set amount of time, these permissions are automatically revoked, and the user's account returns to its usual, lower level of access.

- **PIM for Groups**: PIM can simplify the management, assignment, and monitoring of these privileges. Instead of assigning to specific users, users can be made members of a group temporarily that has these privileges. This feature leverages JIT access but at the group level versus the privilege level.

- **PIM notifications**: These are alerts that are triggered when there are changes in the privileged roles, either a change in assignment to a role, a privileged group under PIM for groups, or even activating a role. They help in monitoring and managing privileged access.

- **PIM approvals**: This feature allows for the approval or denial of requests for elevated access. It adds an extra layer of scrutiny to the process of granting privileged access.

- **Access reviews**: As many organizations forget to revoke privileges when users change roles or leave a group, over time it becomes a problem to verify what the actual permissions required for users are. This feature allows specified members in an organization to receive periodic reports on group memberships, access to enterprise applications, and role assignments and determine whether the users should still have these privileges. Policies can be changed as required to handle situations such as no one taking action on an access review.

- **Setting group owners**: Assigning a group owner provides an extra layer of protection to prevent the modification of a privileged group membership.

The controls associated with medium difficulty level of implementation are as follows:

- **Passwordless authentication**: This method provides a more secure and convenient way to sign in. Instead of having to use a password and MFA to log in, a user would just need to enter their credentials and either select a number on their Microsoft Authenticator application (along with fingerprint), use their hardware FIDO2 key, or even use their Windows Hello face unlock or pin unlock features. This method is typically suitable for all users in an organization, as it enhances security while improving user experience. It's especially beneficial for roles that require frequent authentication or have high-security requirements (*Microsoft, 2023a*).

- **Entra Conditional Access**: This is an approach in which access to the cloud platform, tools, and other aspects are controlled via policies that define the acceptable conditions for access. It allows organizations to control how and when users can access resources using conditions such as user location, user risk level, sign-in action risk level, and device compliance. When implementing conditional access, it is recommended that a couple of break-glass accounts be created and excluded from the conditional policies. Their password access should be highly controlled to include an auditing process, alerts, and notifications upon usage, dependent on FIDO2 hardware keys and very limited access to only limited personnel. Microsoft Entra includes premade conditional access policies for the most common scenarios that can be deployed with a few clicks, as shown in the following figure:

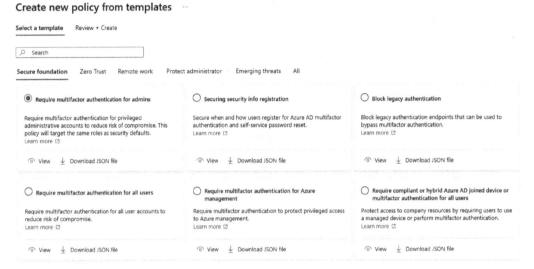

Figure 4.4 – The Microsoft Entra ID Conditional Access | Policies template page

The complex implementation controls include the following:

- **Group Managed Service Account (gMSA) and Entra Workload Identities**: A gMSA is a control used for on-premises Active Directory systems to provide a higher level of security for tasks that require a service account. It addresses many of the risk factors seen in service accounts, such as no password rotation and no granular control over who can use the account. For cloud platforms, a similar control can be obtained via workload identities in Entra ID, which is typically also used for service and resource access to other services and resources. While implementing these identities is a simple process, testing will be needed if any legacy or custom application depends on these.

- **Enterprise access model or tier model**: Known as the enterprise access model for cloud resources and the tier model for on-premises resources, this is a comprehensive access model that has been extensively threat-modeled and tested to create isolation between resources according to their impact on the organization, the identity management system, and the control plane. It addresses many access types used by internal and external users, services, applications, and privileged accounts. While it is not as easy to implement as the other recommendations and it depends on implementing the previous recommendations, this change will provide the most impact toward minimizing attacks and the impact of an attack on an environment. Because of the level of work required to implement this control, it is highly recommended that experienced help such as Active Directory and Entra ID experts are employed to either perform or support this effort.

- **Privileged Access Workstation (PAW)**: These are systems that administrators use to perform privileged tasks. They are designed to provide a highly secure environment for sensitive tasks and are not intended for day-to-day tasks. Things such as web access are strictly restricted on these devices; no productivity applications, such as emails or Microsoft Teams, are installed, the operating system is hardened, and other attack vectors are addressed to minimize the chance of the workstation and any privileged identities used on it being compromised. PAW devices are most effective when used alongside a properly deployed enterprise access model.

These recommendations provide a robust framework for securing privileged roles and access within an organization. They help to protect sensitive data and resources from unauthorized access and potential security threats and will help better support the security posture required. In the next section, we will discuss how to identify an organization's risk profile.

It is all about the organization's risk profile

A **cybersecurity risk profile** is an extensive assessment of the cybersecurity threats faced by an organization. It involves evaluating the organization's operating environment and its ability to manage significant risks that may impact its objectives at various levels and is crucial in guiding the approach to handling cybersecurity risks. Each organization has its own unique cybersecurity risk profile, which serves as a starting point for identifying the most critical cybersecurity issues. It helps in mitigating potential risks by determining the acceptable level of risk that the organization is willing and capable of accepting. A company's cybersecurity risk profile aims to understand how the willingness or aversion to risk will influence decision-making strategies.

The cybersecurity risk profile supports senior management in analyzing priorities and allocating resources effectively by providing a concise overview of key risks for staff, external partners, and decision-makers. When implemented, it helps identify areas of efficiency and potential opportunities, enabling strategic priority setting, resource allocation, informed decision-making, and improved outcomes. Cybersecurity risk profiles can be created in various ways, often starting with a risk profile questionnaire. They can be developed formally or informally, catering to different levels of the organization, such as at the corporate level, within a sector, work unit, or project.

Creating the proper risk profile

In cybersecurity, a risk profile would center on the potential cybersecurity threats an organization may encounter. Fortunately, organizations do not need to start from square one in their endeavors to define these risk profiles as the **National Institute of Standards and Technology (NIST)** releases the **NIST Internal or Interagency Reports (NISTIRs)**, which outline technical research of interest to a specialized audience. The series encompasses interim or final reports on work conducted by NIST for external sponsors, encompassing both government and non-government entities. In cybersecurity, a

NISTIR may offer comprehensive instructions on effectively mitigating a particular type of cybersecurity risk. Some of the commonly used NISTR connected to risk profiles include the following (NIST, n.d.):

- **Ransomware Risk Management Profile (NIST IR 8374):** This profile focuses on the risks associated with ransomware attacks. It includes strategies for preventing such attacks, steps to take when an attack occurs, and methods for recovering afterward.

- **Cybersecurity Framework Profile for Liquefied Natural Gas (NIST IR 8406):** This profile is specific to organizations in the liquefied natural gas industry. It considers the unique cybersecurity risks associated with this sector, such as threats to the physical infrastructure and data security.

- **Cybersecurity Framework Profile for Hybrid Satellite Networks (HSN) (NIST IR 8441 - Draft):** This profile focuses on the risks associated with hybrid satellite networks. It includes threats to both the physical satellite infrastructure and the data transmitted over these networks.

- **Cybersecurity Framework Profile for Genomic Data (NIST IR 8467 - Draft):** This profile considers the risks associated with handling and storing genomic data. It includes both privacy concerns and the potential for data loss or corruption.

- **Cybersecurity Framework Profile for Electric Vehicle Extreme Fast Charging Infrastructure (NIST IR 8473):** This profile focuses on the risks associated with the **Electric Vehicle Extreme Fast Charging (EV/XFC)** ecosystem. It supports each of the four domains: EVs, XFC, XFC cloud or third-party operations, and utility and building networks. The profile provides a foundational profile that may develop profiles specific to the organization and should supplement, not replace, an existing risk management program or the current cybersecurity standards, regulations, and industry guidelines that are in current use by the EV/XFC industry.

Each of these profiles identifies the specific risks associated with their focus area, assesses the potential impact and likelihood of these risks, and proposes mitigation strategies to manage these risks. The specific content and format of a cybersecurity risk profile can vary widely depending on the organization's specific needs and circumstances. Besides the examples provided, here are some more specific elements that might be included in a cybersecurity risk profile:

- **Threat assessment:** This involves identifying the various cybersecurity threats that the organization might face. This could include everything from phishing attacks to advanced persistent threats. Usually, teams create a threat model for this effort.

- **Vulnerability assessment:** This involves identifying any weaknesses in the organization's systems or procedures that a threat might exploit. We might use vulnerability scanners for this effort.

- **Risk assessment**: This involves analyzing the potential impact of each identified threat and vulnerability, and the likelihood of it occurring.

- **Risk mitigation strategies**: This involves identifying strategies to manage each risk, such as implementing specific security measures, purchasing insurance, or accepting the risk.

- **Incident response plan**: This involves outlining the steps the organization will take in the event of a cybersecurity incident. It covers internal and external procedures, including notification, responsibilities, and communication with the media (if necessary).

- **Training and awareness programs**: This involves educating employees about cybersecurity risks and how to avoid them. Some organizations may choose to include realistic exercises based on their maturity level.

- **Regular reviews and updates process**: Cybersecurity threats are constantly evolving, so the risk profile must be regularly reviewed and updated to reflect the current threat landscape.

We should tailor these elements to the specific needs and circumstances of the organization, and the risk profile would inform decision-making and resource allocation in relation to cybersecurity.

Risk profiles in Defender for Office 365

Risk profiles in Microsoft Defender for Office 365 refer to predefined sets of security settings and configurations based on Microsoft's recommended settings that protect against various types of cyber threats. These profiles are tailored to different levels of risk tolerance and security needs while providing a balance between keeping harmful content away from users and avoiding unnecessary disruptions. These profiles are not infinitely configurable such as custom policies and are based on Microsoft's observations in data centers. Using risk profiles is the recommended approach for an initial deployment of Defender for Office 365 and allows for the security team to be up and running while minimizing end-user impact and allows the organization to create custom policies later as needed. There are three types of preset security policies available (*Microsoft, 2023*):

- **Standard preset security policy**: This policy provides a balance of security and productivity and is suitable for most users. Some of the general settings include the following:

  - Changing the phishing email level threshold

  - Enabling impersonated user protection

  - Enabling impersonated domain protection

  - Enabling mailbox intelligence

  - Enabling intelligence for impersonation protection

  - Quarantine messages that are detected from impersonated users

- **Strict preset security policy**: This policy builds on the settings provided by the Standard policy and adds more aggressive limits and settings for security controls that result in more aggressive detections. It also requires higher involvement by the administrators and security team in decisions regarding which blocked emails are to be released to end users. This policy is typically used for executive staff, executive support staff, and historically highly targeted users.

- **Built-in protection preset security policy**: This is the default policy for Safe Attachments and Safe Links protection in Defender for Office 365.

The main difference between these policies lies in their level of aggressiveness in detecting threats and the degree of administrator and security team involvement required. The Standard policy is designed for general use, while the Strict policy is intended for users who require a higher level of protection, even if it means more good mail getting flagged as suspicious. Typical deployments will involve placing most users under Standard policy and high-risk users, such as executives and administrators, under Strict. It all depends on your organizational needs identified in *Chapter 3* and its risk profile. We will go more in depth on providing more granular settings for your organization in future chapters.

Looking deeper into preset policies

The preset policies look at recommendations in specific commonly targeted areas with the intent of providing immediate protection while minimizing the chance of any surprises to the security team. When the product is first deployed, the built-in protection preset security policy will be enabled by default for all users, and it allows your team to configure exceptions as needed. The built-in protection security policy only provides default policies for the Safe Links and Safe Attachments features, which is a good starting point but far from leveraging all the capabilities of the product and establishing a modern security posture.

Assigning the Standard or Strict risk profile to even just a set of users kick-starts the creation of these policies, which involves a combination of **Exchange Online Protection** (**EOP**) and Defender for Office 365 policies. The EOP policies are as follows:

- **Anti-spam policies**: Using several URL block lists holding a vast list of domains known for sending spam helps detect malicious links within messages. This protection is offered for both inbound and outbound messages, helping to protect organizations not only from receiving malicious links but also from being abused by bad actors as a proxy for attacks. The settings changed from the default policy are as follows:

Setting	Standard	Strict
Bulk compliant level (BCL) met or exceeded detection action	Move message to Junk Email folder	Quarantine message
Bulk email threshold	6	5
Spam detection action	Move message to Junk Email folder	Quarantine message

Table 4.1 – Anti-spam policy changes per preset profile as it relates to EOP

- **Anti-malware**: Using multiple anti-malware engines, EOP offers multilayered protection that's designed to catch all known malware. Messages are scanned for malware (viruses and spyware) and if malware is detected, the message is deleted. Further configuration can be performed on the policy to send notifications to senders or admins when deletions occur and no delivery occurs, along with replacing infected attachments with either default or custom messages that notify the recipients of the malware detection. No specific changes occur between Standard and Strict profiles regarding the anti-malware policy.

- **Anti-phishing**: Anti-phishing policies provide anti-spoofing protection via spoof intelligence which searches for evidence of forgery in the From header of an email message and leverages the following email authentication methods for further confirmation:

  - **Sender Policy Framework (SPF)**: This method uses a list to identify what mail servers are authorized to send email

  - **DomainKeys Identified Mail (DKIM)**: This method adds a digital signature to emails, which acts like a seal that confirms the email origin

  - **Domain-based Message Authentication Reporting and Conformance (DMARC)**: This method builds on SPF and DKIM and includes instructions on what to do with the email if it fails SPF and DKIM, including quarantining, rejecting, or letting it be delivered

Settings differences between Standard and Strict risk profiles include the following:

Setting	Standard	Strict
If the message is detected as spoof-by-spoof intelligence	Move message to Junk Email folder	Quarantine message
Show first contact safety tip	Selected	Not selected

Table 4.2 – Anti-phishing policy changes per preset profile as it relates to EOP

The **Show first contact safety tip** feature provides a warning when receiving an email from a new and unfamiliar email address.

The Defender for Office 365 policies runs alongside these EOP policies and, as discussed previously, gets kick-started once an organization assigns a Standard or Strict risk profile. The Defender for Office 365 policies and changes are as follows:

- The Safe Attachments policy does not see any settings differences between built-in protection, Standard, and Strict.

- The anti-phishing policies are improved by adding impersonation protection features and advanced phishing thresholds for both externally and internally sent emails.

- **Impersonation protection** goes beyond just verifying a list of malicious domains or looking for authentication failures or forged email headers. It looks at variations in email addresses and domains for any slight variations that might be used as part of an impersonation attack. Such type of attack will use variations that are difficult to perceive by regular individuals, such as the use of different character sets that look similar (such as with attackers using Cyrillic characters), even if the domain has a valid certificate. This type of attack was seen frequently during the COVID pandemic for sending lots of erroneous information about vaccines and making it appear to originate from legitimate sources. It is important to note that this protection cannot be assigned to the entire organization, so we should assign it to specific recipients such as executives and administrators, as well as specific domains.

The **advanced phishing thresholds** are split into **Standard** (default setting), **aggressive**, **more aggressive**, and **most aggressive**. The more aggressive the setting is, the lower degree of confidence is required to classify an email as a phishing email, with the Standard setting allowing different actions for a low, medium, and high degree, and the most aggressive setting grouping all degrees of confidence and treating them in the same manner as a very high degree of confidence of phishing emails. The setting changes performed in the anti-phishing policies are as follows:

Setting	Standard	Strict
If mailbox intelligence detects an impersonated user	Move message to Junk Email folder	Quarantine message
Phishing email threshold	3 – More aggressive	4 – Most aggressive

Table 4.3 – Anti-phishing policy changes per preset profile as per Defender as it relates to Office 365

- The Safe Links policy receives various changes when moving from built-in protection to either Standard or Strict and allows for stronger protection from unintentionally clicking links, even from internally sent emails. There are no changes in the Safe Links policy between Standard and Strict risk profiles. The changes performed when moving from built-in protection to Standard or Strict are as follows:

Setting	Built-in protection	Standard and Strict
Let users click through to the original URL	Selected	Not selected
Do not rewrite URLs, do checks via Safe Links API only	Selected	Not selected
Apply Safe Links to email messages sent within the organization	Not selected	Selected

Table 4.4 – Safe Links policy changes per preset profile as it relates to Defender for Office 365

The preceding settings focus on the changes to policies when moving from the built-in protection to Standard and Strict but do not cover all aspects of these policies. In later chapters, we will explore each policy and look at how to configure custom policies. For now, let's focus on how to perform our initial configuration using preset policies.

Configuring the preset policies

As we can see from the previous discussions, a big bulk of the effort toward deploying the product properly is the preparation work, not only technical such as preparing permissions, licenses, and user groups, but also on procedures and policies by identifying the actual needs of the organization and getting leadership support to ensure no roadblocks. This following section will focus on a walk-through for implementing preset policies and assumes that you have completed the following actions (*Microsoft, 2023*):

- The desired plan for Defender for Office 365 has been identified and the required licenses have been obtained.

- The `Test user - standard`, `test user - strict`, `early user` (small group to verify settings before the regular users), `regular user`, and `executive/VIP user` groups have been identified and created in Entra ID. It is recommended that any changes are implemented to leverage these groups in a controlled manner to minimize any impact on your user base. These groups will also define who gets Standard versus strict settings.

> **Note**
> When creating these groups, they must be email-enabled (Microsoft 365 Groups and not just security groups) for them to show during configuration.

- All the domains used for email by your organization have been identified along with the email domain for important external organizations that communicate with your organization via email, such as financial and regulatory.

- The privileged roles to be used have been identified and user groups assigned to these roles (assigning roles to groups is preferable to individual user assignments due to the ease of managing groups).

- The preferred way in which policies will be called, also called the organization's policy nomenclature. A typical nomenclature for policies is `DATE_TYPE_VERSION`. For example, `20231130_PHISHING_01`.

- The company's preferred risk profile has been identified. If not enough information is available, the Standard risk profile can be adopted and granular custom policies adopted at a later time.

- At a minimum, the easy-to-implement controls for privileged roles have been configured and there are plans to implement the medium controls. This is key, as security controls on identity will have a major impact on preventing an attacker from disabling any security controls and tools implemented.

- A list of requirements has been identified, which will be used to test that the deployment aligns with the organization's needs (refer to *Chapter 3* if this step has not been performed).

For most of the configuration, we will perform the actions via the web portal. Just follow along with this guide. One important point to remember is that Standard preset policies must be applied first (even if they are just one user) before strict policies can be applied. If a user has both the Standard protection and strict protection applied, the strict protection will take priority.

The administrative portals

We will first visit the administrative portals, the location where you will spend most of your time configuring Defender for Office 365, correcting issues, and tracking product performance:

1. Start a web browser and log in to the Office portal (`portal.office.com`). If your user possesses the needed permissions (as specified in the *Permissions required* section of this chapter), you will see an icon labeled **Admin** on the left side. Click this icon to be taken to the Office Admin portal.

Figure 4.5 – The Office portal page and the Admin icon on the left side

2. In the Microsoft 365 Admin portal (`admin.microsoft.com`), you can oversee all settings for the organization's Microsoft 365 installation. Click on the hamburger icon (three parallel lines) at the top-left corner to make a menu appear.

3. On the menu, select **show all** if not all entries are visible, then, under **Admin centers**, select the **Security** entry.

Figure 4.6 – The Microsoft 365 admin center and the Security button on the left side

4. A new window will open in the Microsoft 365 Defender portal (security.microsoft. com). This is the central location for configuring most security tools in Microsoft Defender. On the left menu, observe the **Email & collaboration** section. This is the section where the setup, monitoring, and maintenance of Defender in Office 365 occurs.

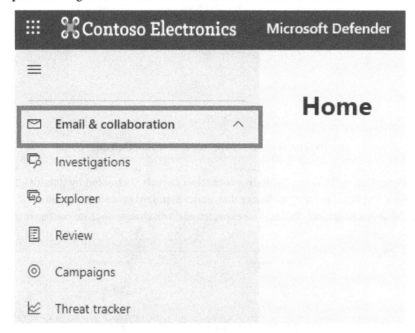

Figure 4.7 – The Email & collaboration menu selection on the Microsoft 365 Defender portal

5. Choose **Policies & Rules** in the **Email & collaboration** section, and on the right side of the page, the **Policies & Rules** section will expand. Click on **Threat Policies** to access the policies. Be aware that if the **Safe Attachments** and **Safe Links** sections show **Premium** next to them, it means that you might not have the correct licenses for Defender for Portal 365.

There are many options on these screens, which we will explore in upcoming chapters, but for now, our primary aim is to establish a fundamental installation of Defender for Office 365.

Enabling Standard protection preset policies

During our configuration, we will apply policies to the test users - standard group. While not mandatory, we highly recommend this approach to test settings and changes before applying it to the rest of the organization:

1. While on the **Policies & Rules** page, click on the **Preset Security Policies** link, and the page to configure these will open:

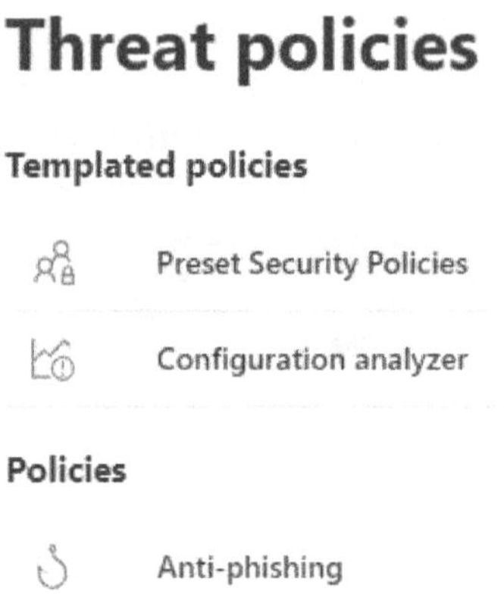

Threat policies

Templated policies

Preset Security Policies

Configuration analyzer

Policies

Anti-phishing

Figure 4.8 – The Threat policies page in the Microsoft Defender portal

2. You will see the options for **Built-in protection** (which is enabled by default), **Standard protection**, and **Strict protection**. Notice that, under **Standard protection**, a switch will show that **Standard protection** is off. This is to be expected and will change once we configure this policy:

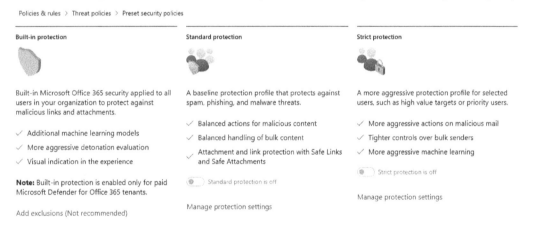

Policies & rules > Threat policies > Preset security policies

Built-in protection

Built-in Microsoft Office 365 security applied to all users in your organization to protect against malicious links and attachments.

✓ Additional machine learning models

✓ More aggressive detonation evaluation

✓ Visual indication in the experience

Note: Built-in protection is enabled only for paid Microsoft Defender for Office 365 tenants.

Add exclusions (Not recommended)

Standard protection

A baseline protection profile that protects against spam, phishing, and malware threats.

✓ Balanced actions for malicious content

✓ Balanced handling of bulk content

✓ Attachment and link protection with Safe Links and Safe Attachments

Standard protection is off

Manage protection settings

Strict protection

A more aggressive protection profile for selected users, such as high value targets or priority users.

✓ More aggressive actions on malicious mail

✓ Tighter controls over bulk senders

✓ More aggressive machine learning

Strict protection is off

Manage protection settings

Figure 4.9 – The Preset security policies page where settings for these are configured

3. Before we enable these, we need to determine who these settings will apply to. Click on **Manage protection settings** and a window will pop up with the **Complete organizational setup** option; click **Yes** for the tenant to be prepared for Defender for Office 365. Be aware that in rare instances you might receive a client error after clicking **Yes**; just wait five minutes, refresh the page, and try again.

4. In the first section, we will configure who the **Exchange Online Protection** settings apply to. Select **Specific recipients** and select our previously configured `test users - standard` group. As you type the name of the group, a search will start for groups in Entra ID; find the correct group, select it, and then click **Next**.

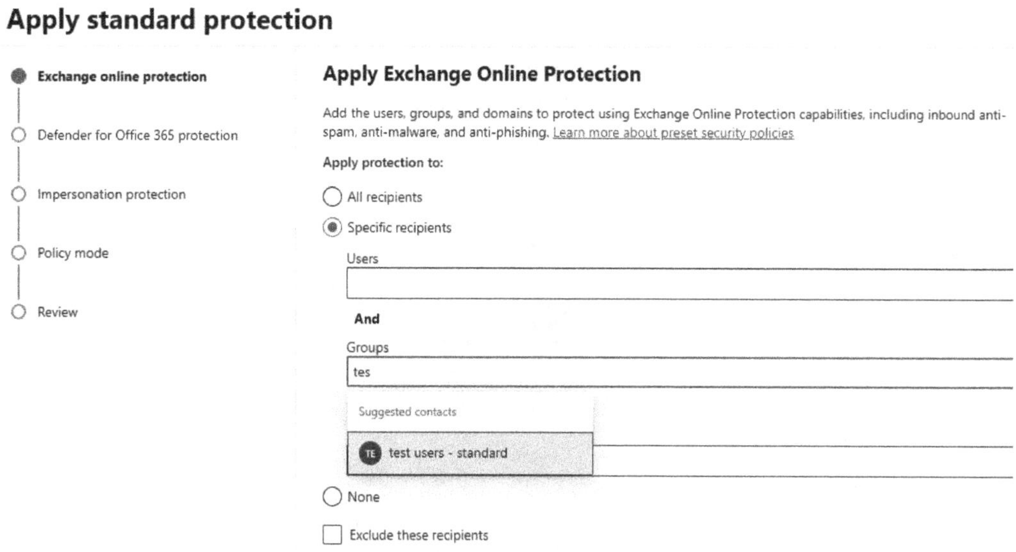

Apply standard protection

Apply Exchange Online Protection

- Exchange online protection
- Defender for Office 365 protection
- Impersonation protection
- Policy mode
- Review

Add the users, groups, and domains to protect using Exchange Online Protection capabilities, including inbound anti-spam, anti-malware, and anti-phishing. Learn more about preset security policies

Apply protection to:

○ All recipients

◉ Specific recipients

Users

And

Groups

tes

Suggested contacts

TE test users - standard

○ None

☐ Exclude these recipients

Figure 4.10 – Configuring the groups to apply the EOP Standard settings

5. In the next section, we will configure who the Defender for Office 365 protection settings apply to. As with the EOP settings, select **Specific recipients** and target the test users - standard group first. Once the test group is selected, click on **Next**:

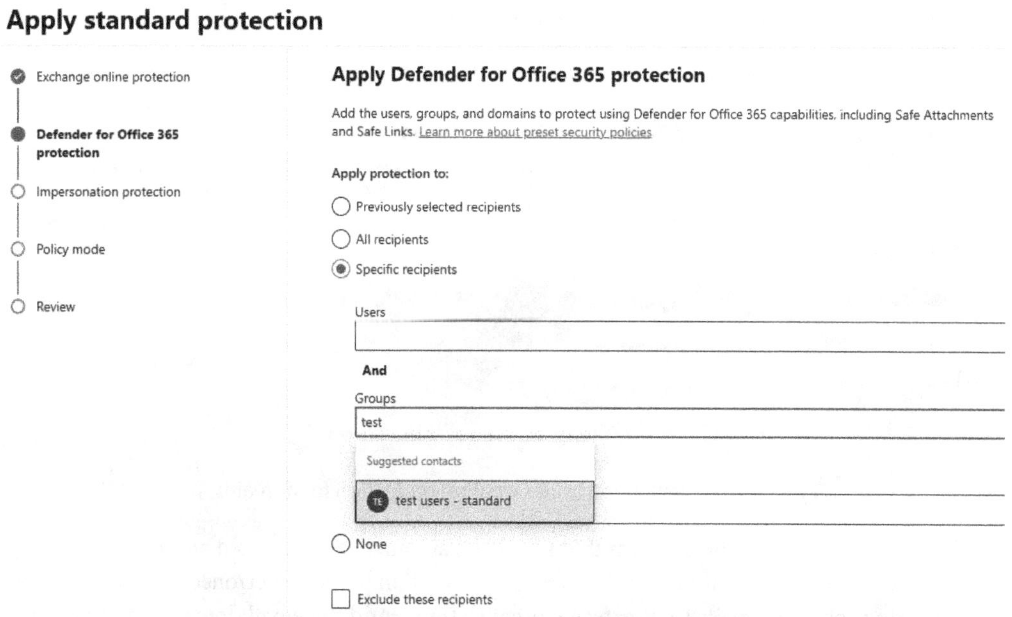

Apply standard protection

Apply Defender for Office 365 protection

- Exchange online protection
- **Defender for Office 365 protection**
- Impersonation protection
- Policy mode
- Review

Add the users, groups, and domains to protect using Defender for Office 365 capabilities, including Safe Attachments and Safe Links. Learn more about preset security policies

Apply protection to:

○ Previously selected recipients

○ All recipients

◉ Specific recipients

Users

And

Groups

test

Suggested contacts

TE test users - standard

○ None

☐ Exclude these recipients

Figure 4.11 – Configuring the groups to apply the Defender for Office 365 Standard protection settings

6. The next section configures the **Impersonation protection** settings. Just click **Next** to configure.

7. We will identify addresses to protect and add the name of a test user here. After we complete testing and are ready to apply these policies to the entire organization, we can add executive names here. After adding users, click **Next**:

Figure 4.12 – Configuring impersonation protection for a user

8. Next, we will configure the domain impersonation settings. Enter the email domains for your organization and the email domains for other important external organizations that communicate with your organization as previously identified. Once the domains have been added, click **Next**. Any messages identified as impersonation emails will be sent to a junk folder or quarantined. During initial testing, these detections can be tracked via impersonation insights and the policy can be adjusted as needed:

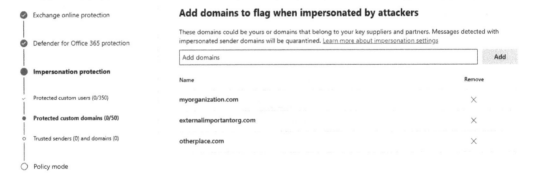

Figure 4.13 – Configuring Impersonation protection for domains

9. Next, we will identify the domains that should be automatically accepted and not flagged for impersonation. You can think of this as an extra protection to prevent erroneous impersonation detections of valid emails from external organizations. Add the email domains of important external organizations previously identified and click **Next**. Any erroneous detections can be tracked via impersonation insights:

Apply standard protection

Figure 4.14 – Configuring impersonation protection flagging excepted domains

10. The next screen will ask whether you want to turn the Standard policy on or leave it off. As we have isolated the protection to a group of test users, it is safe to apply it now. Select **Turn on the policy when finished** and click **Next**.

11. A confirmation screen will appear showing the changes selected; verify that these align with what you entered and then click **Confirm**.

12. A new page will display that the Standard protection has been applied and you can click on **Done**:

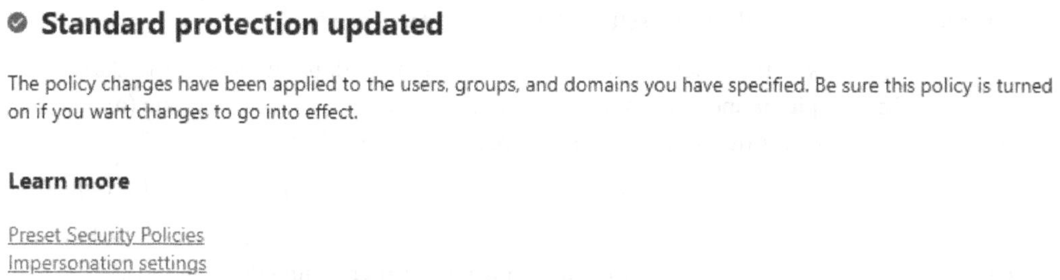

⊘ Standard protection updated

The policy changes have been applied to the users, groups, and domains you have specified. Be sure this policy is turned on if you want changes to go into effect.

Learn more

Preset Security Policies
Impersonation settings

Figure 4.15 – Screen confirming that Standard protection has been applied

13. The initial page for **Preset Security Policies** will show and **Standard protection is on** will show.

If your organization will have all its users under Standard protection, then you have completed the initial setup and can test for a few days. After testing, we can expand the protection to other groups, including executives. When you are ready to expand the coverage, click on **Manage protection settings** again and change the following areas:

- Add more groups to whom the EOP and Defender for Office 365 protection applies to. Once you are ready to apply this policy to the entire organization, you can select **All recipients**. This will ensure everyone receives at least this level of protection and simplifies management. If any recipient requires Strict settings, they can safely receive Standard settings, too, as Strict settings will take priority once applied.

- Add the names of executives, VIPs, administrators, and other high-impact users who might have a high chance of being impersonated to the impersonation protection page.

If your organization only needs Standard protection once testing is completed and all users have been migrated, you can move into the security maturity phase. In this phase, you track the areas that could be improved. This will help identify how policies are to be customized in the future. A well-planned approach that uses real-world data helps you to focus on what really matters to your organization. If your organization needs to configure the Strict preset policies, continue with the next section. If Standard protection is enough, you can safely skip the Strict protection preset policies section.

Enabling Strict protection preset policies

During our configuration, we will apply policies to the `test users - strict` group. While not mandatory, we highly recommend this approach to test settings and changes before applying it to the rest of the organization:

1. While on the **Preset Security Policies** page, under **Strict protection**, select **Manage protection settings**.

2. We will configure who the **Exchange Online Protection Strict** settings apply to. Select **Specific recipients** and select our previously configured `test users - strict` group (or another test group identified for the Strict settings). Once selected, click **Next**.

3. We will configure who the Defender for Office 365 Strict protection settings apply to. As before, select **Specific recipients** and select our previously configured `test users` group (or another test group identified for the strict settings). Once selected, click **Next**.

4. The next section configures the **Impersonation protection** settings. Just click **Next** to configure.

5. We will identify addresses to protect using the Strict settings. Select a test user similar to how we configured the Standard settings during the testing period after adding the user and click **Next**.

6. Next, we will configure the domain impersonation settings. You can enter the same domains as those used when configuring the Standard settings; the main difference is that any emails from these domains identified as an impersonation attempt will be quarantined versus just sent to the junk folder. Once the domains have been added, click **Next**. Any messages identified as impersonation emails can be tracked via impersonation insights and the policy can be adjusted as needed.

7. Next, we will identify the domains that should be accepted and not flagged for impersonation. Add the email domains of important external organizations previously identified and click **Next**.

8. The next screen will ask whether you want to turn the Strict policy on or leave it off. As we have isolated the protection to a group of test users, it is safe to apply it now. Select **Turn on the policy when finished** and click **Next**.

9. A confirmation screen will appear showing the changes selected; verify that these align with what you entered and then click **Confirm**.

    A new page will display that the Strict protection has been applied and you can click on **Done**.

10. The initial page for **Preset Security Policies** will show and **Strict protection is on** will show.

Your organization should now have Strict protection in place for the configured users. We recommend observing for a few days and then moving to expand the protection to other groups. When you are ready to expand the coverage, click on **Manage protection settings** again and change the following areas:

- Add more groups to whom the EOP and Defender for Office 365 protection applies to. If your organization's security needs require all users to be under Strict settings, you can select **All recipients**. This will ensure everyone receives Strict levels of protection and simplifies management. We should leave Standard settings on, as Strict settings will take priority once applied.

- In the impersonation protection page, add the names of executives, VIPs, administrators, and other high-impact users who might have a high chance of being impersonated and require strict settings.

As with the configuration of Standard policies, once we complete testing, you can move into the security maturity phase of this effort. Your organization should be in a good place at this moment while you learn more about the product and learn what works for your organization. Eventually, we will want to move to granular policies, which we will discuss more in future chapters.

Summary

In the chapter, we outlined the essential preparatory steps necessary before configuring Microsoft Defender for Office 365, emphasizing the importance of thorough planning and understanding of the system's requirements. The primary emphasis was on requirements, which encompasses choosing the correct plan for your organization and incorporating the procurement and administration of suitable licenses. This crucial step guarantees access to the essential features and functionalities of Defender for Office 365. We explored the nuances of permissions, including those that are essential during and after configuration, which play a vital role in ensuring security and operational effectiveness. This section of the chapter emphasized providing practical guidance on aligning organizational needs with permission and preparations to ensure a secure foundation in our deployment.

The latter part of the chapter shifts focus to the strategic aspect of security management, specifically on identifying the organization's risk profile. This section plays a pivotal role in guiding readers on assessing their organization's specific security risks and how these assessments directly shape the configuration of Defender for Office 365. The chapter ends by examining the arrangement of pre-established policies in Defender for Office 365, illustrating how we can tailor these policies by using the identified risk profiles. This ensures that the security measures are not only strong but also tailored to the specific requirements and weaknesses of the organization, offering a comprehensive and customized defense strategy while reducing complexity. In the upcoming chapter, we will investigate the methods of tracking issues and configuration drafts to ensure the sustained effectiveness of our deployment.

References

- Microsoft. (2023a, October 24). *Why do I need Microsoft Defender for Office 365?* Microsoft Learn. https://learn.microsoft.com/en-us/microsoft-365/security/office-365-security/mdo-about?view=o365-worldwide

- Microsoft. (2023b, October 23). *Microsoft Defender for Office 365 service description - Service Descriptions.* Microsoft Learn. https://learn.microsoft.com/en-us/office365/servicedescriptions/office-365-advanced-threat-protection-service-description

- Microsoft. (2023a, October 23). *Authentication methods and features - Microsoft Entra ID.* Microsoft Learn. https://learn.microsoft.com/en-us/entra/identity/authentication/concept-authentication-methods

- Microsoft. (2023b, October 24). *Why do I need Microsoft Defender for Office 365?* Microsoft Learn. https://learn.microsoft.com/en-us/defender-office-365/mdo-about?view=o365-worldwide#permissions-needed-to-use-defender-for-office-365-features

- Microsoft. (2023c, November 28). *About admin roles in the Microsoft 365 admin center - Microsoft 365 admin.* Microsoft Learn. https://learn.microsoft.com/en-us/microsoft-365/admin/add-users/about-admin-roles?view=o365-worldwide

- Microsoft. (2023, November 2). *Preset security policies.* Microsoft Learn. https://learn.microsoft.com/en-us/microsoft-365/security/office-365-security/preset-security-policies?view=o365-worldwide

- NIST. (n.d.). *CSRC Topic: cybersecurity framework | CSRC.* https://csrc.nist.gov/topics/applications/cybersecurity-framework

- Microsoft. (2023, November 2). *Microsoft recommendations for EOP and Defender for Office 365 security settings.* Microsoft Learn. https://learn.microsoft.com/en-us/microsoft-365/security/office-365-security/recommended-settings-for-eop-and-office365?view=o365-worldwide

Part 2 - Day-to-Day Operations

In this part, you will learn about the activities performed by a security team once Defender for Office 365 is in place. We will begin by exploring troubleshooting approaches and options for complex issues. Next, we will customize our message quarantine process to minimize user impact and enable users to assist with security efforts. We will then focus on enhancing protection against phishing and spam messages through robust filters and email authentication. To understand how our controls affect users, we will examine the flow of messages through our system and learn how to proactively detect misconfiguration. We will discuss customizing protections against malicious files and links to enhance the user experience. Finally, we will touch on threat hunting, security operations, and alert management, including implementing automation to reduce alert fatigue, detect advanced adversaries, and enhance our security maturity.

This part contains the following chapters:

- *Chapter 5, Common Troubleshooting*
- *Chapter 6, Message Quarantine Procedures*
- *Chapter 7, Strengthening Email Security*
- *Chapter 8, Catching What Passed the Initial Controls*
- *Chapter 9, Incidents and Security Operations*

5
Common Troubleshooting

We have looked at understanding the security threats that pose a risk to our organization and recognizing the role of Defender for Office 365 in protecting us. We have also defined our security goals with the help of our organizational policies, regulations that impact our environment, and well-known security frameworks such as ISO 27001, HIPAA, and PCI DSS, among others. This process shaped and guided our initial Microsoft Defender for Office 365 configuration to include providing a measurable approach to evaluating it against our security requirements. However, setting up Defender for Office 365 is an ongoing process that requires finding the right balance between security and usability. We may encounter challenges and conflicts with other elements in our environment, which may necessitate adjustments and refinements to our policies and settings. Legacy systems and varying priorities in other business units can also impact these adjustments. In this chapter, we will examine our approach to handling these challenges. This includes a comprehensive analysis of the resources and tools at our disposal for troubleshooting and resolving these issues.

This chapter will cover the following topics:

- How to leverage user feedback to fine-tune the configuration
- How to find event logs and error messages to clarify any issues
- Information sources for common errors and seeking Microsoft help on complex issues

Let's continue our journey!

Is this working properly?

Even though we have successfully deployed our product and ensured that it aligns with our security goals, it is important to note that any issues affecting our users may not be immediately apparent to us. To ensure that our issue resolution efforts are effective, it is crucial to establish processes that create a feedback loop. This feedback loop should consider various factors, such as daily business operations, the impact on users, and the security needs of the organization. By striking the right balance, these processes can guarantee an effective approach to issue resolution.

To effectively manage this process, continuous monitoring and evaluation are necessary, taking into consideration the frequency and impact of the issues that arise. Feedback from our users is crucial in understanding how our changes affect them, and we rely on tools such as our help desk ticket system to gain insights into the issues our customers encounter. Thoroughly analyzing these issues requires considering factors such as the number of impacted users and systems, and the overall extent of the problem. To facilitate this analysis, let's discuss some points to keep in mind when looking at issues and developing processes for addressing these.

Users impacted

When assessing the user experience, it is important to consider the prevalence of specific problems or issues among users. This can be determined by analyzing ticket data, which provides information on the frequency and nature of user-reported issues. By utilizing analytics tools or integrating features with the ticketing system, we can summarize and present this data in graphical or numerical formats. For instance, the ticketing system could identify 50% of users who are experiencing delays when receiving messages with attachments; one can then look at what these users have in common such as their department, group, policies applied, and other factors. By identifying patterns and commonalities, support teams can prioritize their efforts and develop targeted solutions. To allocate resources efficiently and resolve issues promptly, we can categorize tickets based on user impact levels.

Systems impacted

To effectively troubleshoot a problem, the first step is to understand how it impacts different systems within the environment. This requires tracing the connections and dependencies among the components that work together. One technique that can be helpful is searching for common patterns or terms in the ticket data. These patterns provide insights into the systems or software often associated with reported issues. By adopting this method, it becomes easier to identify the underlying causes and find the best solutions.

In support of creating a proactive approach to system maintenance and troubleshooting, a taxonomy can be created that categorizes issues based on the IT infrastructure components involved, such as hardware, software, networks, or specific applications. This taxonomy can be a key part of how maintenance is performed and allow for the early recognition of recurring issues, improvement areas, and where changes in resource allocation would prove beneficial. When implemented properly, we can expect reduced downtime and enhancements in the overall system reliability.

Blast radius

The blast radius is a measure of the potential damage that a problem can cause to the surrounding environment. To accurately estimate the blast radius, it is crucial to investigate and understand the chain reactions that occur because of the problem. This involves considering the potential consequences of failures in one system or process on other interconnected systems or processes. We can get valuable information on the scale and impact of an issue by analyzing past ticket data and incident reports. For instance, your organization can establish protocols to identify critical components and assess the potential ripple effects of an issue. By comprehending the blast radius, support teams can effectively prioritize critical incidents, minimize collateral damage, and maintain a resilient IT environment.

Applying the Pareto principle

The **Pareto principle**, also known as the 80/20 rule, says that 80% of effects stem from just 20% of causes, a view that can save your team from being flooded with incidents. An effective approach to implementing it involves first identifying the primary causes of user complaints or system failures. These causes are responsible for a significant portion of the overall impact. By using the Pareto principle to help desk operations, we can direct attention to our most critical causes, which will help prioritize our troubleshooting efforts.

Implementing this approach also aids in prioritizing resources, enhancing customer satisfaction, and reducing future incidents. For example, analyzing ticket data can help pinpoint the top 20% of issues that account for 80% of user problems. Once we have identified these issues, we can allocate resources and prioritize solutions accordingly. By focusing on resolving high-impact problems, support teams can boost their efficiency and, ultimately, enhance user satisfaction.

Analyzing incident impact

The help desk ticketing system is a valuable tool for analyzing incidents, as it provides important incident details that we can quantify and analyze methodically. By examining reported incidents, we can categorize these based on severity and frequency, allowing us to understand their impact on system performance and availability. Ticketing systems can help identify recurring patterns and root causes, which can further help in prioritizing issues based on their impact and urgency and help better allocate resources. This process aids in identifying areas for improvement and implementing preventive measures. For instance, we can establish a framework for categorizing incidents and regularly review and analyze them to identify trends. Support teams can also develop targeted strategies for resolution and prevention that align with system owners.

Continuous monitoring and feedback

To ensure efficient issue resolution, it is crucial to have a continuous monitoring system. This system tracks and monitors the status of each issue, documents the actions taken to address them, and records the outcomes achieved. Users can report issues, request help, and provide feedback through the ticketing system, which allows for a feedback loop that can improve the analysis process. This includes identifying underlying causes, finding best practices for resolving issues, and highlighting areas that need improvement. For example, monitoring tools can track system performance and user satisfaction metrics. We can set up automated alerts for potential issues and gather feedback through surveys or direct communication channels. By staying vigilant and seeking user input, support teams can proactively address emerging problems and make informed improvements to the help desk system.

Where do I find what is wrong?

Let's say our users have reported some issues; how do we know these are related to Defender for Office 365? Rather than guessing, we should examine how the issue affects the user experience, which can lead to some measurable data that can help us find meaningful connections without wasting time or resources. By analyzing user-reported issue data and gathering comprehensive symptom information, we can effectively use various troubleshooting components within Defender for Office 365 to do a comparison between what our tools are doing and what they are impacting. This will ultimately help us gain valuable knowledge and identify the root cause of the issue.

Audit log search

Your first stop when troubleshooting any system is to review the logs to identify patterns in our tools' behavior. An audit log search is the right tool for this job, as it provides a detailed record of everything that happens in your Office 365 environment, including key data point information such as how users work, what changes have been made, and what risks have been detected (*Microsoft, 2024a*).

Administrators can retrieve an extensive array of information by using the audit log search feature, including but not limited to the following:

- **User activities**: Track user logins, file access, and email activities
- **Configuration changes**: Monitor modifications to security settings and configurations
- **Security incidents**: Undertake the detection and thorough investigation of security incidents and possible breaches to strengthen defense mechanisms
- **Data exfiltration**: Pinpoint and track any unusual activities that could signify unauthorized data exfiltration attempts
- **Policy violations**: Track violations of compliance policies and regulations

Audit log search is an invaluable tool for troubleshooting different issues, such as the following:

- **Unauthorized access**: Quickly identify and address instances of unauthorized access or compromised accounts

- **Data breaches**: Investigate and mitigate potential data breaches by analyzing user activities and file access

- **Compliance violations**: Enforce regulatory compliance by identifying and addressing policy infractions

- **Anomaly detection**: Uncover irregular behavior that could indicate potential security threats

To employ the audit log search feature in Microsoft Defender for Office 365, here's a straightforward guide:

1. **Visit the Microsoft Defender portal**: Sign in at `https://security.microsoft.com`.

2. **Go to Audit**: Locate and open the **Audit** section to begin your search. For new environments, you might get a warning that tracking must be enabled; click on the warning and proceed with enablement:

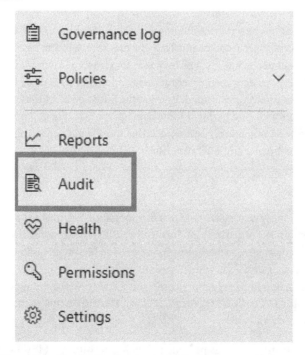

Figure 5.1 – The Audit selection in the Defender portal left menu

3. **Apply filters**: Tailor the search results with filters, specifying criteria such as users involved, date ranges, or the types of activities.

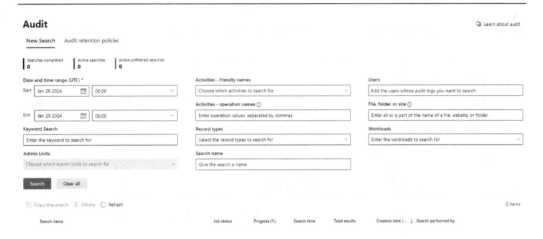

Figure 5.2 – The Audit log search page

4. **Create reports**: Draw up detailed reports that facilitate thorough examination and meet your auditing needs.

This tool has many applications beyond troubleshooting. It can also help create an event timeline with crucial data points, such as the user accounts involved, the files accessed, and any changes made, during breaches and incident response scenarios. For instance, logs may reveal that an employee's account was compromised and sensitive documents were accessed and possibly leaked. An administrator can then proceed to disable the affected account, determine the initial point of the breach, see what other activities the account performed, and perform some damage control. It is important to note that audit logs are massive; therefore, it is highly recommended that loosely defined searches are not used, and instead, limits such as timeboxing and filters are users.

Mailbox auditing

While it stands apart from the functionalities of Defender for Office 365, mailbox auditing is an essential function for any Microsoft 365 organization. This feature diligently records the actions performed by mailbox owners, their delegates, and admins. Its default-enabled status is key for gaining insights into modifications, access, or deletions within mailboxes. Mailbox auditing is pivotal for detecting potential security issues, maintaining regulatory compliance, and troubleshooting within the ecosystem of Microsoft Defender for Office 365 (*Microsoft, 2023b*). Here are a few scenarios where mailbox auditing logs are particularly valuable:

- **Security investigations**: The logs provide a detailed account that is crucial for delving into security-related events, such as breaches of authorization, unusual sign-on activity, or possible account compromises

- **Compliance audits**: They are a tool for organizations to stay in line with data privacy laws, offering a way to track and document access to confidential data

- **Policy management**: Admins find mailbox auditing vital for overseeing and implementing organizational policies, especially when it concerns modifications in mailbox permissions or the setup of delegations

Mailbox auditing logs serve as an indispensable asset, fortifying security measures and enhancing regulatory adherence for users of Microsoft 365.

At the time of this writing, the **Office 365 Security & Compliance** page is being retired and its contents are in the process of migrating to the Defender portal, the Purview portal, and the Exchange admin center. As such, your best way to access these logs is via PowerShell and the Office 365 Management Activity API. Important points about mailbox auditing include the following:

- It does not support all mailbox types. The supported ones are user mailboxes, shared mailboxes, and Microsoft 365 Group mailboxes.

- The actions that can be audited vary between user and shared mailboxes and Microsoft 365 Group mailboxes.

- The `Get-Mailbox` command will provide information on the mailbox configuration, including what actions are logged. The audit configuration varies according to the role as it relates to the mailbox (owner, delegate, or admin).

- The `Set-Mailbox` command allows for changing the mailbox configuration to include what actions are logged.

- In case of issues, the default audit configuration can be set by using the following command: `Set-Mailbox -Identity <MailboxIdentity> -DefaultAuditSet <Admin | Delegate | Owner>`.

- While not recommended, mailbox auditing by default could be turned off for the organization with the following command: `Set-OrganizationConfig -AuditDisabled $true` (change the flag to `$false` to reenable mailbox auditing as a default approach for the organization).

- While not recommended, a mailbox can be configured to bypass mailbox audit logging. It is important to verify this setting as it could indicate possible misconfiguration or a security breach in progress. The command to bypass auditing is `Set-MailboxAuditBypassAssociation -Identity <MailboxIdentity> -AuditByPassEnabled $true`.

Imagine that an organization suspects a security breach in a high-profile mailbox. Using mailbox auditing logs, administrators discover mailbox audit bypass has been configured. Further investigation results in the discovery of a compromised account. The organization takes prompt action to secure the account and avoid potential data breaches.

The Incidents page

While the focus of this chapter is on troubleshooting issues, security incidents could cause some of these. The **Incidents** page helps with troubleshooting and strengthening your organization's security. It merges and presents information, allowing administrators to identify, analyze, and resolve security incidents. Knowing how to use this page is crucial for proactive security management and issue resolution. It provides a centralized view of security events and alerts for Microsoft Defender for Office 365 (*Microsoft, 2024b*). Administrators can gather critical information from this page, including the following:

- **Alert details**: Gain insights into alerts, including the type of threat detected and affected users or systems

- **Timeline of events**: Gain insight into the sequence of incidents to discern potential patterns or pinpoint the underlying cause of security events

- **Affected entities**: Determine which specific users, devices, or mailboxes have been influenced or compromised by security breaches

- **Mitigation recommendations**: Receive guidance on recommended measures and appropriate steps to mitigate and respond to the threats detected

Administrators are equipped with comprehensive information, which is critical for reinforcing security measures and making informed decisions. The **Incidents** page is integral to addressing the array of security woes encountered when utilizing Microsoft Defender for Office 365, such as the following:

- **Phishing attacks**: Swiftly detect and react to schemes designed to bait users into divulging sensitive information or compromising their accounts

- **Malware infections**: Pin down and lessen the repercussions of malicious software that could compromise email exchanges or shared documents

- **Unauthorized access**: Handle events arising from illegal entry or credential compromise

- **Data leakage**: Tackle incidents involving inadvertent exposure or the intentional leak of classified information

By proactively confronting these frequently seen troubles, organizations can shield their digital environments and maintain robust security postures. To leverage the **Incidents** page effectively, take the following strides:

1. **Incidents page navigation**: Enter the Microsoft Defender portal (`https://security.microsoft.com`) and proceed to the **Incidents** page.

2. **Analyze incident specifics**: Delve into the particularities of each incident. Scrutinize affected identities, assess the urgency of the alert, and contemplate the guidance suggested.

3. **Initiate action**: Execute the appropriate countermeasures as guided by the recommendations tailored to each situation.

4. **Unceasing surveillance**: Keep a vigilant eye on the **Incidents** page for any freshly arising alerts and evolving security hazards.

We can see what the **Incidents** page looks like in the following screenshot:

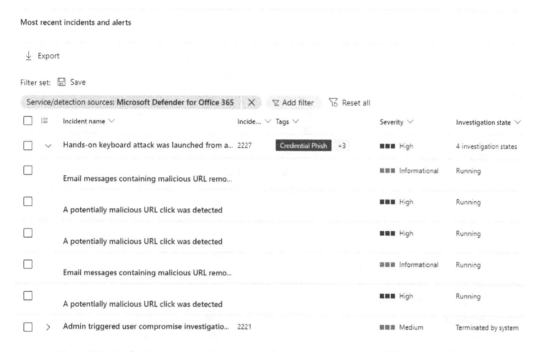

Figure 5.3 – Defender portal Incidents page filtered by Defender for Office 365 incidents

This vigilant methodology guarantees that security personnel can respond promptly to threats, thus curtailing their potential detriment to the organization.

Imagine a case where the **Incidents** page spotlights a phishing campaign aimed at a set of your employees. The analysis reveals that the phishing message includes a perilous link that threatens to capture user credentials. Through the **Incidents** page, administrators can rapidly pinpoint the targeted individuals, grasp the full extent of the threat, and act decisively by sequestering the compromised accounts and amplifying user training on security protocols.

Has this been fixed before?

For reliable insights and effective problem-solving with Microsoft Defender for Office 365, it's essential to turn to credible resources. Many times, the challenges you're wrestling with might have been tackled before, and resolutions are out there. By tapping into the wealth of knowledge from authentic documentation, community discussions, and advice provided by Microsoft, you stand to cut down on the troubleshooting time and dodge a good deal of hassle. During your troubleshooting adventures, the go-to references should include the following.

Step-by-step guides

For hands-on help with setting up or ironing out kinks in Microsoft Defender for Office 365, don't overlook the practical step-by-step guides from Microsoft. These instructions are laid out in an easy-to-follow format, providing clear steps and practical tips to assist users with tasks or to resolve common problems. The aim of these guides is to declutter the setup experience, enabling administrators to put security features into action efficiently (*Microsoft, 2023a*).

For example, administrators encountering issues with email delivery can follow the *Troubleshooting Email Delivery Issues* guide. By systematically following the steps outlined, they can identify misconfiguration or external factors affecting mail flow, ensuring a smooth and secure communication environment.

These guides are available through the Microsoft Learn page (`https://learn.microsoft.com/`) in the **Find technical documentation** section.

Microsoft Service Health Status page

Challenges faced by your users might be traced back to factors outside your immediate control. To keep on top of such situations, the Microsoft **Service Health Status** page is key for both users and administrators. It offers real-time status updates on all of Microsoft's services, including Defender for Office 365, highlighting any service disruptions, active concerns, or approaching system maintenance. Regular monitoring of this page is crucial in maintaining uninterrupted operations of Microsoft services and in anticipating potential issues that could impact those relying on them.

For instance, should there be a reported disruption affecting email flow via Defender for Office 365, you're empowered to swiftly notify your users of the specifics and reassure them that Microsoft's resolution efforts are in motion.

You can access the Microsoft **Service Health Status** page by one of the following methods:

- Check out the Microsoft **Service Health Status** page at `https://status.office.com` for an overview of the health of all Microsoft services:

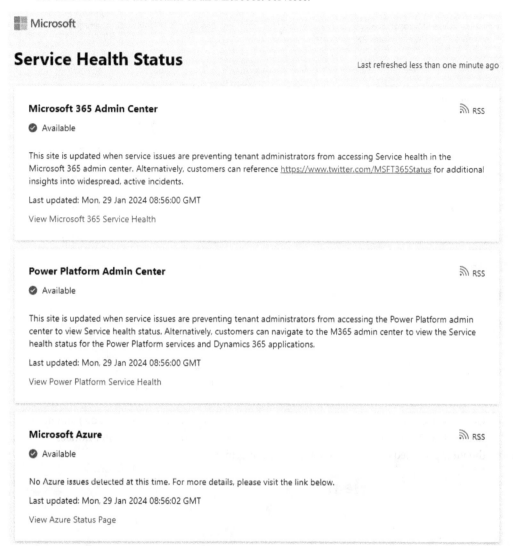

Figure 5.4 – Microsoft Service Health Status page

- Head to the Microsoft 365 admin center (`https://admin.microsoft.com`) as an administrator and navigate to the **Health** section. This will open a comprehensive dashboard of office service statuses, detailed reports on current Microsoft investigations into service issues, and the message center that keeps you updated on platform changes:

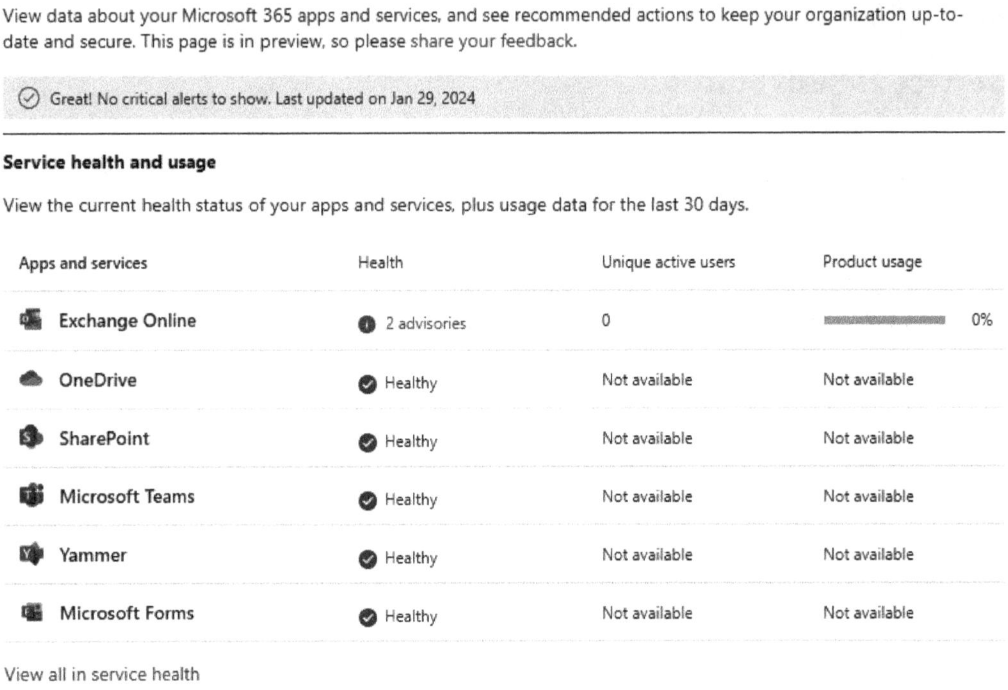

View data about your Microsoft 365 apps and services, and see recommended actions to keep your organization up-to-date and secure. This page is in preview, so please share your feedback.

⊘ Great! No critical alerts to show. Last updated on Jan 29, 2024

Service health and usage

View the current health status of your apps and services, plus usage data for the last 30 days.

Apps and services	Health	Unique active users	Product usage	
◱ Exchange Online	① 2 advisories	0	▬▬▬▬▬▬▬▬	0%
☁ OneDrive	✓ Healthy	Not available	Not available	
◱ SharePoint	✓ Healthy	Not available	Not available	
◱ Microsoft Teams	✓ Healthy	Not available	Not available	
◱ Yammer	✓ Healthy	Not available	Not available	
◱ Microsoft Forms	✓ Healthy	Not available	Not available	

View all in service health

Figure 5.5 – The Microsoft admin center Health dashboard

- Go to the Defender Security portal (`https://security.microsoft.com`) and navigate to the **Health** section to access service condition updates and message center notices, such as the view provided by the Microsoft 365 admin center:

Health

Name

Service health

Message center

Figure 5.6 – The Health page in the Defender portal provides links to the same
sections as in the Microsoft 365 admin center Health section

As a vital tool in your IT toolkit, the Microsoft **Service Health Status** page is your go-to for keeping ahead of any issues with Defender for Office 365. Staying informed through this page is a smart move for any administrator looking to mitigate service hiccups promptly and offer users an uninterrupted experience.

Microsoft 365 troubleshooting library

The Microsoft 365 troubleshooting library stands as an essential asset, especially for those utilizing Microsoft Defender for Office 365. Acting as a detailed troubleshooting manual, it aids in bolstering security across the Microsoft 365 framework. The resource is tailored to help users tackle security predicaments, offering deep dives into the workings of Microsoft Defender for Office 365. Spanning various subjects from current service conditions, device integrity, and security analytics to concise briefings on incidents, the library is consistently refreshed to provide the most current data, ensuring that users are well equipped to sustain a well-fortified computing atmosphere. We can see how the page looked like at the time of this writing in the following screenshot:

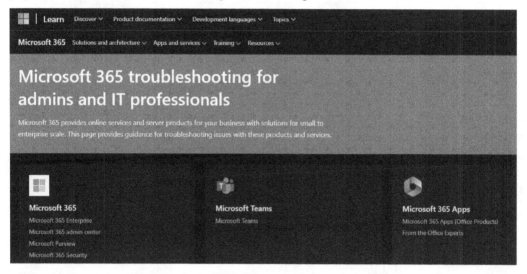

Figure 5.7 – The Microsoft 365 troubleshooting library

To access the Microsoft 365 troubleshooting library, users can navigate to the official Microsoft 365 troubleshooting for admins and IT professionals website (`https://learn.microsoft.com/en-us/microsoft-365/troubleshoot/`). The library offers sections related to each of the Microsoft products offered to the public library, including productivity and security.

Microsoft Community Hub

The Microsoft Community Hub serves as a digital gathering space that promotes dialogue, wisdom exchange, and collective problem-solving for users of Microsoft products. It cultivates a cooperative online community where individuals can find support, narrate their journeys, and add to the shared repository of insight. In the following screenshot, we can see how the Microsoft Community Hub looked at the time of this writing:

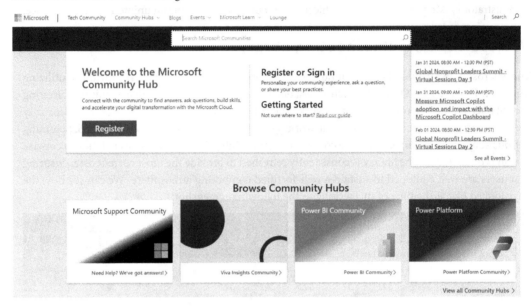

Figure 5.8 – The Microsoft Community Hub

Users can visit the hub by navigating to the Microsoft Community Hub page (`https://techcommunity.microsoft.com`). Once on the platform, users can sign in with their Microsoft account to take part in discussions and access a wide range of resources.

Ways to leverage the Community Hub include the following:

- **Search for relevant topics**: Users can start by searching for specific topics related to Defender for Office 365 issues within the Community Hub. For example, searching for *emails wrongly flagged as spam* can yield relevant discussions and solutions to issues related to blocked emails.

- **Engage in discussions**: Users are encouraged to jump into lively interactions, posing questions or recounting their encounters with specific challenges. The community, alongside Microsoft specialists, frequently steps in, offering walk-throughs, optimal methods, and valuable perspectives.

- **Explore knowledge base articles**: The Community Hub typically hosts a collection of knowledge base articles. Users can explore these articles to find in-depth guides, tutorials, and troubleshooting steps for common Defender for Office 365 issues.

Depending on your issue and the impact on your environment, free resources might not be enough, in which case there is paid support to the rescue.

Microsoft paid support

Microsoft offers comprehensive support options for users of Microsoft Defender for Office 365 through its official paid support services, primarily the Microsoft 365 commercial support team and Microsoft professional support (pay-per-incident). These services assist organizations in effectively using and troubleshooting Microsoft Defender for Office 365, addressing any issues or challenges they may encounter. Let's delve into the details of each support option.

The Microsoft 365 commercial support team represents Microsoft's commitment to assisting commercial clients, spanning from small businesses to large corporations, schools, and government entities. With a specialized focus on the distinct security and compliance demands of each customer segment, this team delivers customized resolution strategies to meet varying needs.

Some key aspects of this support option include the following:

- **Scope of support**: Microsoft's 365 commercial support team is prepared to address a comprehensive suite of challenges associated with Microsoft Defender for Office 365. This includes assistance with setup, diagnostic troubleshooting, and guidance for fine-tuning performance.

- **Access to experts**: Customers gain direct access to a cadre of connoisseurs with deep expertise in Microsoft 365 solutions. They stand ready to deliver precise and efficient support for any Microsoft Defender for Office 365 queries.

- **Service access**: Support inquiries can be funneled through the online Microsoft 365 Support portal, which offers a unified location for lodging and monitoring requests for assistance.

- **Response timeliness**: The promptness of responses can fluctuate, contingent upon the criticality of the reported issue and the specifics of the **service-level agreement** (**SLA**) applicable to your plan. Issues deemed critical are typically escalated and attended to with expedited care.

For instance, if an organization faces persistent missed emails or many emails get wrongly identified as malicious, they can reach out to the Microsoft 365 commercial support team for expert guidance and resolution.

Reaching out to the Microsoft 365 commercial support team is straightforward, offering users a few different contact methods. The most direct way is via the official Microsoft Support portal, where users can log in with their Microsoft credentials to submit a detailed support ticket about any troubles they're facing with Defender for Office 365.

For more specialized assistance, Microsoft caters to subscribers on certain plans such as Microsoft 365 Business and Enterprise with dedicated support phone lines. These contact numbers can be found within the user's **Microsoft 365 Subscription** interface or on the Microsoft Support web page.

While Microsoft ensures baseline support for all its offerings, including Defender for Office 365, scenarios might necessitate special paid support plans. These subscription levels guarantee quicker response times, prioritized issue resolution, and the provision of more sophisticated support tools. Those subscribed to Microsoft 365 Business Premium, various Microsoft 365 Enterprise tiers, or holding additional support plans are likely to benefit from these premium support services. It's advisable for users to check their subscription information to understand the scope of support their plan includes.

Summary

This chapter presented a detailed roadmap for navigating the complexities of maintaining and refining the security posture through Microsoft Defender for Office 365. Acknowledging the dynamic interplay between security and operational functionality, the chapter furnishes a toolkit of methodologies and best practices, enabling you to adeptly balance protection measures with usability.

From leveraging user feedback to dissecting event logs, and tapping into the robust resources provided by Microsoft, this chapter equips administrators with the knowledge to proactively address and rectify issues, ensuring that Defender remains a stalwart guardian of their organization's cyber fortifications. We delved into the practical use of tools such as audit log search, mailbox auditing, and the **Incidents** page, exemplifying their vital roles in the swift identification and management of security events and service health issues.

Furthermore, we explored the utilization of the Pareto principle and various strategies to analyze user and system impact effectively. The chapter underscored the importance of continuous monitoring and setting up feedback loops, thus fostering an environment where issues can be nipped in the bud, maintaining system resilience and user satisfaction.

By methodically employing the recommended approaches and practices, drawing from pre-existing solutions and Microsoft's extensive support networks, IT professionals can rest assured that the advice distilled in this chapter is not just theoretical but also pragmatically actionable. These recommendations and resources culminate in an empowering compendium, designed to demystify the troubleshooting process and reinforce the user's confidence in managing Defender for Office 365. In our next chapter, we will dig deeper into day-to-day operations and how to fine-tune our configuration to manage email message quarantines and ensure it aligns with our user base needs.

References

- Microsoft. (2023a, August 24). *Microsoft Defender for Office 365 step-by-step guides and how to use them*. Microsoft Learn. https://learn.microsoft.com/en-us/microsoft-365/security/office-365-security/step-by-step-guides/step-by-step-guide-overview?view=o365-worldwide
- Microsoft. (2023b, December 12). *Manage mailbox auditing*. Microsoft Learn. https://learn.microsoft.com/en-us/purview/audit-mailboxes

- Microsoft. (2024a, January 17). *Audit log search in the Microsoft Defender portal.* Microsoft Learn. `https://learn.microsoft.com/en-us/microsoft-365/security/office-365-security/audit-log-search-defender-portal?view=o365-worldwide`

- Microsoft. (2024b, January 19). *Security Operations Guide for Defender for Office 365.* Microsoft Learn. `https://learn.microsoft.com/en-us/microsoft-365/security/office-365-security/mdo-sec-ops-guide?view=o365-worldwide`

6

Message Quarantine Procedures

Securing your organization's inbox and Teams channels requires a multi-pronged approach. By leveraging quarantine and inbound anti-spam policies, you can create a digital shield, filtering out spam, phishing attempts, and malicious content. This chapter will guide you through configuring these policies and provide the best practices for implementation. We will cover email and Teams message protection, ensuring security across your communication channels. Whether you're an experienced IT professional or a security-conscious administrator, this chapter will give you the knowledge and expertise to configure and refine these policies for optimal email hygiene and digital resilience.

This chapter will cover the following topics:

- How do quarantine and spam policies work?
- The meaning of the different settings in these policies
- Introduction to the different policies available
- How to best implement and test these policies?

Let's continue our journey!

I stopped the message. Now what?

Quarantined email messages in **Exchange Online Protection** (**EOP**) and Defender for Office 365 as well as quarantined Teams messages represent a proactive measure to shield organizations from common attack vectors originating from both external attackers and malicious insiders. By default, Defender for Office 365 scans all incoming messages for signs of spam, phishing attempts, or policy infractions. Any message flagged by these security measures is diverted into quarantine—a special holding area outside users' primary inboxes or Teams chats, where they can't cause harm or provoke action from unsuspecting recipients.

Identifying quarantined email messages

Administrators play a key role in managing quarantined messages. They receive notifications or use security dashboards within the Microsoft 365 Defender portal to oversee quarantined items. For individual users, depending on the organization's policies, they might receive a digest email that summarizes quarantined items, or they may have permission to check their quarantine area directly within their email settings.

Verifying the validity of quarantined messages

The process of verifying the legitimacy of quarantined emails involves several steps:

1. **Review sender information**: Often, looking at the sender's email address offers the first clue. Organizations train their staff to recognize signs of suspicious email addresses, such as slight misspellings of legitimate company names or the use of public email services for business communications, which are uncommon.

2. **Analyze email content**: Content that urges immediate action, especially involving sensitive information or financial transactions, warrants scrutiny. Legitimate organizations rarely ask for personal information via email.

3. **Check links and attachments**: Hovering over links without clicking can reveal misleading URLs disguised as legitimate ones. Treat with caution any attachments from unknown senders or unexpected attachments from known senders.

4. **Use external verification tools**: Tools such as VirusTotal or URLVoid can scan URLs and attachments for malicious content, providing an added layer of verification.

5. **Consult whitelists and blacklists**: Comparing the sender's domain against known whitelists (safe senders) and blacklists (prohibited senders) helps in assessing credibility. Sites such as `spamhaus.org` and `cleantalk.org` offer domain blacklist databases that can be used to verify whether the sender domain has been previously reported as sending spam or malicious emails.

It is imperative that the teams reviewing these emails define a quantifiable methodology to use, which will lead to improved identification processes and decisions not based on emotion.

How Defender for Office 365 quarantines emails

Defender for Office 365 leverages various policies and settings to customize the frontline defenses according to an organization's requirements, including sensitivity, actions, and even what users may do regarding spam and phishing. Effective protection requires understanding quarantine and anti-spam policies and aligning these to the organization's risk appetite and how hardened systems are expected to be. Remember that the stricter protection is, the more secure an environment might be, but the more frustrated the audience of these tools might become. As such, it is important to strike a balance. Let's break down how Defender for Office 365 protects users through its quarantining functionalities, focusing on the policies' functions and proper implementation.

Quarantine policies

Defender for Office 365 anchors its email protection strategy in its nuanced quarantine policies. When the system's safeguards mark an email as suspicious or potentially harmful, it's not directly sent to the addressee's primary inbox. Instead, it gets rerouted to a designated quarantine zone. This strategy achieves a twofold advantage: keeping the user's inbox both safer and more organized, and offering the chance to manually scrutinize the suspicious email to ascertain whether it was incorrectly flagged or whether it's genuinely dangerous (*Microsoft, 2023e*).

The adaptability of the quarantine process stands out. Administrators have the capability to craft policies that detail which kinds of emails should be quarantined. They could, for example, elect to quarantine emails that fail either **DomainKeys Identified Mail (DKIM)** or **Sender Policy Framework (SPF)** tests. They could also choose a more focused approach, quarantining only those emails that carry certain malware types or have been identified as phishing attempts. This ability to customize offers a sharper and more efficient way to combat email threats.

Once in quarantine, emails can be handled in several ways. Administrators can review them and decide to either release them to the intended recipients, remove them if they're verified as threats, or report them to Microsoft for further analysis. The system allows for setting up notifications, so when emails are quarantined, both administrators and users can be informed, enabling quicker manual review and decision-making.

Anti-spam policies

Anti-spam policies and quarantine measures join forces to form a robust barrier against unwanted and possibly malicious emails in Defender for Office 365's arsenal. Through stringent anti-spam policies, Defender for Office 365 scrutinizes each incoming email for spam-like traits. It evaluates factors such as the sender's reputation, the content of the email, and familiar spam hallmarks to calculate a spam score for every email. Emails tipping over a preset spam score limit then undergo the procedures prescribed in the quarantine policies.

The platform's anti-spam policies boast remarkable flexibility, empowering organizations to calibrate their defense intensity to match their specific requirements and appetite for risk. I have supported multiple organizations on their transition to mature security, and one of the points end users tend to appreciate is a security system customized for the environment and the user group that minimizes impact on daily work. For example, organizations can implement a policy to sift through emails originating from previously uncontacted new domains, a common trait of spam campaigns. Furthermore, these policies offer the ability to be tailored to different segments within an organization, affording more stringent filters for teams more susceptible to attacks, such as those in finance or executive roles (*Microsoft, 2023c*).

The vigilance against spam extends to outgoing messages as well, preventing the organization from unintentionally turning into a spam distributor. This involves tracking for an abnormal surge in sent emails, which could indicate a compromised account. Such comprehensive anti-spam efforts demonstrate Defender for Office 365's commitment to safeguarding both inbound and outbound email communication.

Quarantining Teams messages

Just as email communications can serve as conduits for sharing malicious content or links, Teams messages also present an avenue for attackers to exploit unsuspecting users. These malevolent messages might pack malware, phishing schemes, or various security hazards capable of compromising confidential data, interrupting business activities, or tarnishing an organization's standing. The strategy, akin to handling email messages, revolves around intercepting these potentially harmful messages before they make their way to the end user. This proactive approach substantially slashes the odds of security infringements, while furnishing IT administrators with valuable insights to evaluate and address threats within a managed setting. It's crucial, however, to strike a balance between rigor and risk tolerance to avoid unduly bothering users while still slimming the likelihood of a significant breach.

Although both email and Teams messages can be quarantined, it's important to note the distinctions in how each process operates and its capabilities:

- **Email quarantine**: This typically involves redirecting flagged emails to a separate, secure area for review, allowing IT administrators to scrutinize suspected content without exposing end users. This process benefits from a mature set of protocols and practices, given email's long history as a communication tool.

- **Teams message quarantine**: Handling suspicious content within instant messaging platforms such as Teams poses unique challenges, given the real-time nature of these communications. Quarantine methods might include temporarily restricting the message from being seen by the recipient until it's reviewed by an administrator or offering tools for real-time inspection and decision-making without impeding the flow of legitimate conversations.

- Both scenarios underscore the need for dynamic security measures that can adapt to the different contexts and inherent functionalities of email and Teams messaging. Implementing a balanced, comprehensive security framework that safeguards against threats without hindering communication requires a nuanced understanding of each platform's technicalities, user behavior, and potential risk areas. Next, let's look at how quarantine occurs from a more technical perspective from trigger to release, and see areas to consider:

- **Triggering quarantine**:

  - **Emails**: Primarily triggered by anti-spam policies based on factors such as spam score, phishing indicators, exceeding bulk complaint levels, or malicious attachments identified by Safe Attachments policies. Zero-hour purge is also available for quarantine actions.

  - **Teams messages**: Triggered by the following:

    - **Zero-Hour Auto Purge (ZAP) for malware**: Any messages detected to have malware after delivery are quarantined automatically

    - **ZAP for high-confidence phishing**: Any messages with a high probability of being phishing attempts are quarantined automatically

    - **Manual actions by admins**: Admins can quarantine suspicious teams' messages manually

- **Storage and access**:

  - **Emails**: Stored in a central location accessible by both users and admins. Depending on the policies' configuration, users can browse, release, or delete quarantined emails.

  - **Teams messages**: Stored within the Microsoft Defender portal, only accessible by admins. Users cannot directly access or manage quarantined Teams messages.

- **Notification**:

  - **Emails**: Depending on policy settings, users receive a notification about quarantined emails, allowing them to review and potentially recover legitimate messages.

  - **Teams messages**: Users do not receive notifications about quarantined Teams messages. They might suspect a missing message but lack direct confirmation. Admins review and notify relevant users about quarantined messages.

- **Release mechanisms**:

  - **Emails**: Depending on policy configuration, users can release quarantined emails themselves, choosing the delivery location (`Inbox` or `Junk` folders).

  - **Teams messages**: Only admins can release quarantined Teams messages, potentially notifying the message initiator or adding it back to the channel.

- **Other differences**:

  - **Reporting**: Users can report false positives for quarantined emails directly to Microsoft. Teams message reporting through users is not currently available.

  - **Search**: Users can search for quarantined emails within the quarantine interface. Searching for Teams messages in quarantine requires admin access to the Microsoft Defender portal.

- **Additional points**:

  - **ZAP offers different actions**: Choose between automatic deletion, moving to the `Junk` folder, or quarantine for both emails and Teams messages based on your needs.

  - **Consider user awareness**: Educate users about quarantine for emails and teams' messages to manage their communication effectively.

While both email and Teams message quarantine aim to isolate potentially harmful content, the process, user involvement, and functionalities differ. ZAP plays a crucial role in both but with distinct applications for emails and Teams messages. Understanding these differences helps you choose the quarantine options for your specific needs and ensures effective communication security across your organization.

Everyone gets a seat at the quarantine table

Managing quarantine emails and team messages in Defender for Office 365 is a critical aspect of protecting an organization's digital communication from threats such as spam, malware, and phishing attempts. Because of the volume of messages that an organization sees daily, trying to validate quarantine messages might be a full-time job, which might lead to delays that impact the business. Defender for Office 365 allows users to have some control over quarantine messages, hoping to strike a balance between security and productivity. Depending on the risk appetite of the organization, this approach might include allowing users to do the following:

- Have a say in what they consider to be a threat or false positive, improving the system's accuracy

- Quickly address and resolve cases where legitimate emails are mistakenly quarantined, ensuring that important communications are not missed

- Empower users to learn and recognize potential threats, contributing to a culture of security awareness within the organization

To allow for user participation in these efforts, multiple approaches can be configured according to the organization's risk appetite.

Options available for the user management of quarantined messages

Under Defender for Office 365, the default policies provided in the standard and strict preset policies offer different levels of protection. Options will also vary not only depending on the policy configuration but also depending on whether it is an email message or a Teams message.

Email

By default, quarantine actions available for users only cover the anti-spam (for a good deal of spam except for high-confidence phishing) and anti-phishing policies along with allowing the user to view, release, or delete the quarantine message. No user options are available by default in the anti-malware policy, Safe Attachment polices, and the mail flow rules. For the anti-malware policy, a custom quarantine policy can be configured depending on organizational needs. Options include the following:

- **Review details**: See why the message was quarantined, and the sender, recipient, subject, and date

- **Release**: Deliver the message to your inbox

- **Delete**: Permanently remove the message

- **Report false positive**: Inform Microsoft about a mistakenly flagged message

The level of control a user can exercise on quarantine messages should be controlled and implemented in phases to include periodic user education along the way (*Microsoft, 2023b*).

Teams messages

Options for the user management of quarantined Teams messages are limited compared to email:

- **Review details**: View the sender, recipient, channel, date, and quarantine reason

- **Report false positive**: Inform Microsoft about a mistakenly flagged message

For both emails and Teams messages, it is always a balancing act between organizational needs and user convenience. For a default implementation, we should remember that **standard preset policies** will typically provide a balanced level of security without overly aggressive filtering, which might quarantine legitimate emails. **Strict preset policies**, designed for environments where maximum security is a priority, might limit the options to release or report emails depending on the organization's settings, emphasizing the prevention of potential threats entering the communication ecosystem.

Step-by-step actions to manage quarantined messages

Let's have a look at the different procedures involved in quarantine management and how these are performed.

Reviewing quarantined messages

Reviewing quarantined messages is the first step toward understanding how a system is performing. Users should be provided with ways outside the tools to provide feedback on the system's performance. A common process the security team can use daily to review quarantined messages is as follows:

1. **Access the Quarantine page**: The **Quarantine** page is in the Microsoft Defender portal (`https://security.microsoft.com`). Go to **Email & collaboration | Review | Quarantine** or visit `https://security.microsoft.com/quarantine`, as shown in the following screenshot:

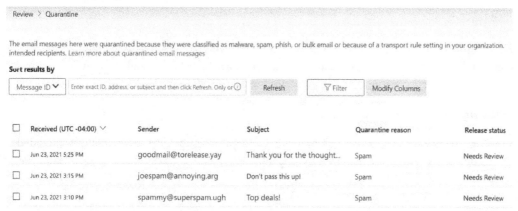

Figure 6.1 – The Quarantine page

2. **Select the target message to inspect**: Select either the **Email**, **Files**, or **Teams messages** tab depending on the target to inspect.

3. **Examine the message**: Users can filter and search the quarantine to locate specific messages. Clicking on a message will reveal details about why it was quarantined and the threat it potentially poses, as shown in the following screenshot:

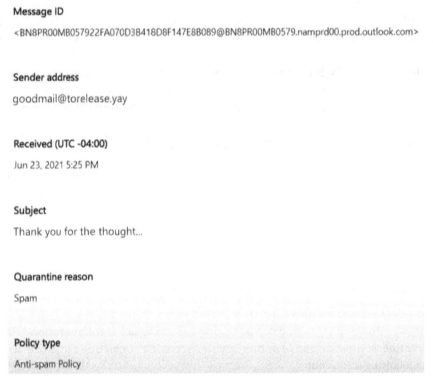

Message ID

<BN8PR00MB057922FA070D3B418D8F147E8B089@BN8PR00MB0579.namprd00.prod.outlook.com>

Sender address

goodmail@torelease.yay

Received (UTC -04:00)

Jun 23, 2021 5:25 PM

Subject

Thank you for the thought...

Quarantine reason

Spam

Policy type

Anti-spam Policy

Figure 6.2 – Message details

Understanding what to release and what to leave as is will take users time to understand. It is recommended that sources of information to make an informed decision should be provided to users periodically to have them develop a habit of detecting unusual messages.

Releasing quarantine messages

For wrongly identified messages, depending on the organization's policies, users might be provided with the option to release the message themselves, or they might need to request for release. The latter option, while safer, will cause delays while an administrator reviews the request and approves the release. Security teams can follow these steps to release quarantine messages:

1. **Select the message**: Within the **Quarantine** view, select the message(s) you have determined to be safe.

2.   For organizations allowing users to release messages, choose **Release Message**. After selecting the message, choose the option to release it. Depending on policy settings, you may need to confirm that you understand the risks of releasing the message.

3.   For organizations requiring approval for release, choose **Request release**. The message will be flagged for review and your IT or security team will be notified to inspect the message, as shown in the following screenshot:

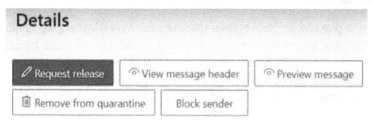

Figure 6.3 – Options in the Details section

4.   **Await decision**: The security team will review the flagged message and decide whether to release it or keep it in quarantine, based on its assessment. A notice will appear to the user, as shown in the following screenshot, letting know the status.

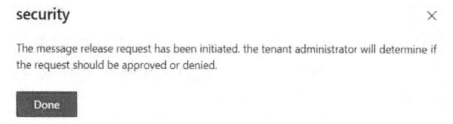

Figure 6.4 – User notification when requesting a release of a quarantined message

These steps will happen a lot during the initial few months as the organization learns to use Defender for Office 365 and policies are fine-tuned.

Deleting quarantined messages

There will also be times when the quarantine message is indeed very damaging to the environment and must be deleted as soon as possible. For these times, we need to understand how to delete messages:

1.   **Select the message**: Similar to releasing a message, start by selecting the message(s) you believe should not be released.

2.   **Choose Remove from Quarantine**: With the message selected, opt to delete it. This action will remove the message from quarantine, and it will not be delivered to any inbox.

This structured approach to managing quarantined messages in Defender for Office 365 enables users and organizations to maintain control over their email and messaging environments, enhancing both security and efficiency. By empowering users to take part in quarantine management, organizations can fine-tune their security measures and reduce the risk of legitimate communications being incorrectly marked as threats.

Administrative actions

The previous section provides multiple options that users can have when managing quarantine, but this will always depend on how policies are configured and the risk appetite of the organization. There are several potential benefits to allowing users to manage some quarantined messages themselves, although complete self-management comes with risks. Here are some points organizations can use to determine the best approach for their environment.

Advantages of effective user management

Don't let your administrative team waste time on repetitive tasks that can be better performed by the email recipient. I have seen many organizations that end up ignoring proper security approaches due to the sheer volume of emails to review, and other organizations with many frustrated users as administrators never resolve false detections. Here are some examples of why we should opt for more user involvement:

- **Swift resolution**: By enabling users to swiftly identify and retrieve legitimate emails wrongly marked as spam, this minimizes disruptions and bolsters productivity. Immediate access ensures that important communications remain uninterrupted, enhancing the overall workflow.

- **Lighter administrative burden**: User management effectively reduces the strain on administrators by delegating the handling of low-risk items to users themselves. This shift allows IT teams to dedicate their focus and resources to tackling more critical security issues and strategic tasks, optimizing operational efficiency.

- **Enhanced individual autonomy**: Providing users with the ability to directly manage their inboxes empowers them with personalized control. This empowerment allows individuals to prioritize their messages according to their personal or professional importance, ensuring a tailored email experience.

- **Elevated security engagement**: By involving users in the email security process, an enhanced sense of ownership and accountability is cultivated. This active participation not only builds a more security-conscious culture within the organization but also strengthens trust in the implemented security measures.

These benefits collectively contribute to a more secure, efficient, and user-centric email management environment. They highlight the importance of integrating user management features within email security strategies to achieve optimal outcomes (*Microsoft, 2023d*).

Navigating the complexities of user-managed email quarantines

It is not all rainbows and butterflies when users are involved in these tasks; for every benefit, there is a problem that occurs. It is important to understand these issues and create ways to mitigate them:

- **Threat of security violations**: Granting users the autonomy to retrieve emails from quarantine introduces the risk that they may mistakenly whitelist harmful communications. Such incidents can sidestep established security frameworks, jeopardizing the organization's cybersecurity fortress.

- **Uneven security practices**: Individual differences in handling quarantined emails can lead to a mosaic of security postures within an organization. This lack of uniformity may inadvertently forge vulnerabilities, inviting exploitation by digital adversaries.

- **Knowledge deficiency**: The reality that not every user has the technical acumen to accurately discern safe from dangerous emails amplifies the chances of menacing emails infiltrating beyond quarantine defenses.

- **Increase in help requests**: Authorizing users to govern their quarantined messages might inflate the volume of support requests, potentially overburdening IT support teams. This scenario is particularly likely if users find navigating quarantine protocols challenging.

To judiciously incorporate user management into your organization's email security strategy while mitigating associated risks, ponder the following mitigatory tactics:

- **Selective user privileges**: Limit user privileges to releasing only low-risk quarantined content, such as conventional newsletters or promotional emails, keeping items with a higher risk index for IT administrator review.

- **Invest in education**: Fortifying user competency through intensive educational programs on the nuances of email security, and the prudent release of emails, can substantially decrease the likelihood of accidental security breaches.

- **Establish and support clear guidelines**: Crafting unambiguous user guidelines for quarantine management, supplemented by accessible support mechanisms, empowers users to make informed decisions within a safety net of guidance and assistance.

Balancing the empowerment of users with the imperatives of maintaining a robust security architecture demands careful consideration of these elements, blending user independence with the safeguarding oversight of IT professionals.

Policies

Policies are the meat of Defender for Office 365 and where you will spend the most time during the initial months. It is important to understand the impact of these policies as they relate to inbound messages.

Quarantine policy

Quarantine policies provide a modular way to control what actions are taken on quarantine messages and files. Depending on the maturity of the organization, different quarantine policies can be applied depending on the environment (production versus test), user base (test users, regular users, and VIP), and even the use of groups of devices. Quarantine policies also work with other policies, including the anti-spam policy (discussed in the next section) and the anti-malware policies (discussed in *Chapter 8*). Let's review the key settings, best practices, and user impact when customizing a policy.

Recipient message access options

This section specifies the access and actions users can perform on quarantine messages. We can see the options in the following screenshot:

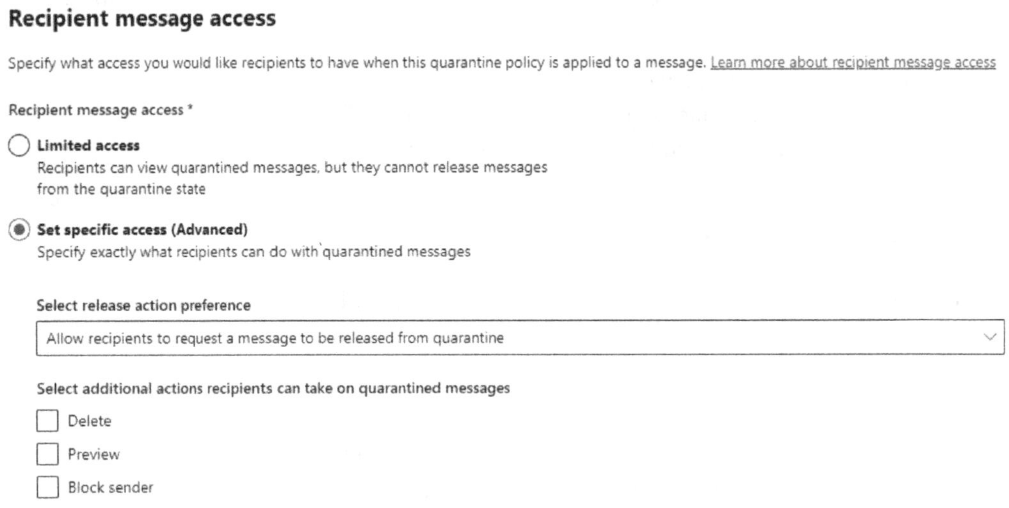

Recipient message access

Specify what access you would like recipients to have when this quarantine policy is applied to a message. Learn more about recipient message access

Recipient message access *

○ **Limited access**
Recipients can view quarantined messages, but they cannot release messages from the quarantine state

◉ **Set specific access (Advanced)**
Specify exactly what recipients can do with quarantined messages

Select release action preference

Allow recipients to request a message to be released from quarantine	⌄

Select additional actions recipients can take on quarantined messages
☐ Delete
☐ Preview
☐ Block sender

Figure 6.5 – Quarantine message access options

The importance of this section cannot be understated as this is where we will define how users interact with quarantined messages once they receive the alert, which, in turn, will impact user engagement in our security efforts:

- **Limited access:**

  - **User impact:** Allows users to only preview quarantined messages, forbidding the release or other actions.

  - **Best practices:** Ideal for scenarios with significant risk, where message release could be detrimental. Be mindful that this could lead to an uptick in support requests due to restricted user capabilities.

- **Set specific access (Advanced):**

  - **User impact:** Grants specific, controllable access for users over quarantined messages.

  - **Best practices:** Provides a balance between flexibility and control, necessitating detailed setup to ensure proper use.

- **Select release action preference** (if **Set specific access** is selected):

  - **Allow recipients to request message release:**

    - **User impact:** Enables users to petition admins for the release of a message awaiting administrative approval.

    - **Best practices:** Creates a middle ground between security and user autonomy. Ensure that the procedure for requests and expected timelines are clearly communicated.

  - **Allow recipients to directly release messages:**

    - **User impact:** Permits users to unilaterally release messages, bypassing admin oversight.

    - **Best practices:** Convenient but potentially risky. Use alongside stringent quarantine measures and educate users on discerning malicious content.

- **Select additional actions recipients can take on quarantined messages** (if **Set specific access** is selected):

  - **Delete:**

    - **User impact:** Authorizes users to permanently remove messages.

    - **Best practices:** Employ with caution to prevent the loss of important emails. An *undelete* feature or restrictions may be prudent.

  - **Preview:**

    - **User impact:** Allows message content viewing without releasing it.

    - **Best practices:** Encourages making informed decisions. Advise using this feature before deletion or blocking the sender.

  - **Block sender:**

    - **User impact:** Prevents receiving future messages from the sender, risking the loss of legitimate communications.

    - **Best practices:** Apply judiciously to avoid blocking important contacts inadvertently. Whitelisting and user education on this feature are advisable.

While access is important, users must also be notified when something is quarantined or risk missing the action.

Quarantine notification

This option enables the sending of notifications to users upon a message they are the recipient of being quarantined. Enabling these is as simple as checking the box, but it is an important part we should not forget:

- **Enable**:

    - **User impact**: Alerts users to quarantined messages, enabling further action.

    - **Best practices**: Keeps users informed without overwhelming them. Customize notification settings based on origins or severity.

Ticking this checkbox is only part of the approach; we also need to configure the global notifications, which will be explained in the next section.

Best practices for configuring quarantine policies

We can't just create quarantine policies at random. Some important points are discussed next:

- **Start conservatively**: Initiate with limited access or request to release settings, adjusting as required based on security and user feedback

- **Balance security with autonomy**: Integrate controlled options such as **Preview** and **Request release**, coupled with user education on safe practices

- **Monitor and adapt**: Continuously evaluate policy performance, tweaking as necessary to align with evolving security landscapes and user needs

This framework ensures an effective quarantine policy that judiciously balances security concerns with user independence, adapting dynamically to changing threats and organizational feedback.

Quarantine notifications, global settings

These global notification settings are not configured in the quarantine policy itself; we instead need to open the quarantine policy page global settings to configure these. This allows you to configure how and when information about quarantined items is communicated to both administrators and users. Some of the options available can be seen in the following screenshot:

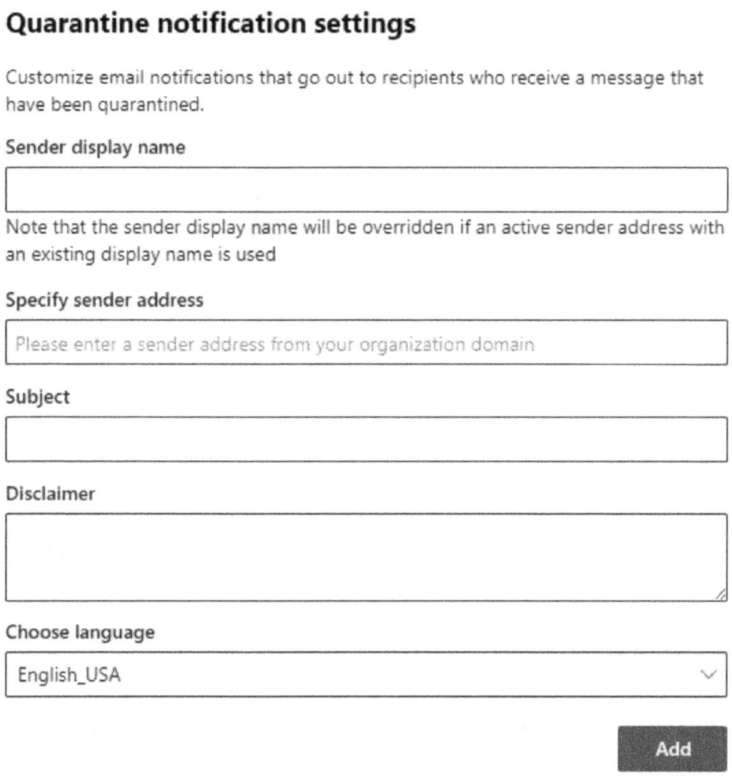

Quarantine notification settings

Customize email notifications that go out to recipients who receive a message that have been quarantined.

Sender display name

Note that the sender display name will be overridden if an active sender address with an existing display name is used

Specify sender address

Please enter a sender address from your organization domain

Subject

Disclaimer

Choose language

English_USA

Add

Figure 6.6 – Quarantine notification settings panel

These settings are global in nature and help us to define how the notification message will appear to the user according to how we complete the field, as we will discuss next.

The following fields offer a way to customize the message the user first sees upon a message being quarantined; whatever format is selected, it is important that users are informed:

- **Sender display name**: Configures a custom name to be displayed for the sender in the notification (for example, *Security Team* instead of *John Smith*):

  - **User impact**: Provides a more recognizable sender name, improving understanding and potential action by users

  - **Best practice**: Use a clear and informative name that reflects the sender's identity

- **Specify sender address**: Configures a custom email address displayed for the sender in the notification (for example, `securityteam@mycompany.com`):

  - **User impact**: Offers technical details for reference or potential whitelisting by trusted senders

  - **Best practice**: Configure it for transparency and allow users to identify potential spam senders

- **Subject**: Allows for configuring a custom subject line for the notification:

  - **User impact**: Helps users recognize and potentially recall the content of the quarantined message

  - **Best practice**: Configure it for context and faster identification of relevant messages

- **Disclaimer**: Configures a disclaimer message to be added to all spam quarantine notifications:

  - **Best practices**:

    - Include a clear and concise disclaimer stating that the information in the notification is intended for informational purposes only and should not be used for making security decisions. If possible, consult your legal and human resource departments to see whether they have any pre-made disclaimers.

    - Specify that users should contact the IT department or a designated administrator for any concerns or questions about quarantined items.

- **Choose language**: For international organizations, it offers the template used for notifications in other languages:

  - If your quarantine policy offers multilingual notifications, provide a drop-down menu or other selection method for users to choose their preferred language

  - Ensure that the chosen language accurately reflects the content of the notification and is easily understandable by the recipient

- **Use my company logo**: Allows the use of any previously uploaded company logo to replace the Microsoft logo used in the spam quarantine notifications:

  - **Best practices**:

    - Including your company logo in the notification can enhance brand recognition and build trust with users

    - Ensure the logo is appropriately sized and displayed clearly

- **Send end-user spam notifications**: Configures the frequency of notifications about spam to users. Options include **Within 4 hours**, **Daily**, and **Weekly**.

Effective quarantine notification settings are crucial for balancing security, user experience, and communication regarding quarantined items. By carefully configuring these settings and ensuring clarity, you can empower users, improve transparency, and ultimately enhance your organization's email security posture.

Inbound anti-spam policy

The inbound anti-spam policy within Defender for Office 365 is crucial for sifting through incoming messages and directing quarantined emails. Serving as an essential barrier, it scrutinizes incoming emails by checking the sender's reputation, analyzing the content, and searching for specific keywords. When an email is marked as spam based on these evaluations, and quarantine is the selected course of action within the anti-spam policy, quarantine protocols are then engaged. Delving into the inbound anti-spam policy, we uncover pivotal settings, best practices, and their implications for users (*Microsoft, 2024*).

Bulk mail threshold

This setting operates as a sieve for high-volume emails, often promotional or marketing communications distributed en masse. Analogous to the spam identification process, it measures incoming emails against a **Bulk Complaint Level** (**BCL**) scoreboard. This grading considers the sender's standing, recipient complaint rates, and detailed content scrutiny. Emails that surpass the set BCL score threshold are tagged as bulk mail and managed according to the predetermined policy actions. We can see some of the options available in the following screenshot:

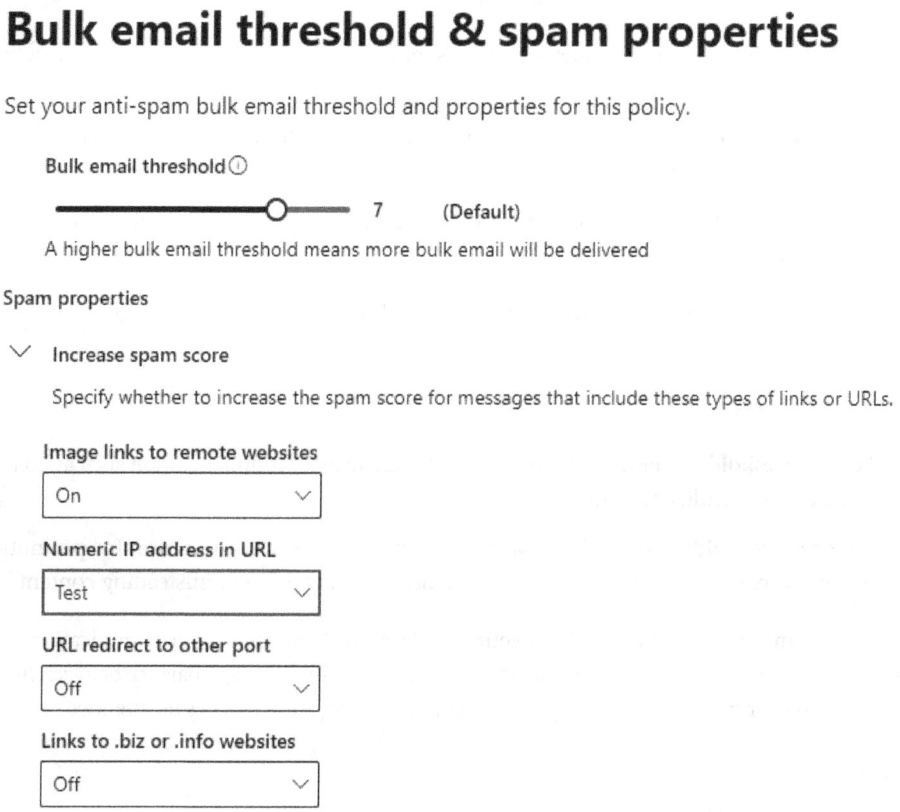

Bulk email threshold & spam properties

Set your anti-spam bulk email threshold and properties for this policy.

Bulk email threshold ⓘ

⬤━━━━━━━━━━○━━ 7   (Default)

A higher bulk email threshold means more bulk email will be delivered

Spam properties

⌄ Increase spam score

Specify whether to increase the spam score for messages that include these types of links or URLs.

Image links to remote websites

| On | ⌄ |

Numeric IP address in URL

| Test | ⌄ |

URL redirect to other port

| Off | ⌄ |

Links to .biz or .info websites

| Off | ⌄ |

Figure 6.7 – The Bulk email threshold & spam properties section

These options, while simple, can have a major impact as follows:

- **Available options**:

  - Ranging from **1** (most strict) to **9** (least strict).
  - Lower numbers signify stricter filtering, catching more potential bulk mail but also increasing the risk of false positives (legitimate mass emails mistakenly quarantined).
  - Higher numbers allow more bulk mail through, reducing false positives but potentially overwhelming inboxes with unwanted promotional messages.

- **Best practices**:

  - **Start with a moderate setting (4 to 6)**: This balances filtering effectiveness with minimizing disruption to legitimate mass communication. Many security professionals tend to approach filtering like a firewall, using an allowed list and banned list approach, or setting to high or low and adjusting, but this tends to be counterproductive.
  - **Adjust based on your organization's needs**:

    - **Strict environments**: Consider lower thresholds (**1** to **3**) for tighter control over unsolicited bulk mail
    - **Flexible environments**: Explore higher thresholds (**7** to **9**) if accepting some promotional emails is acceptable
    - **Consider industry norms**: Tailor the threshold based on typical bulk mail volume in your sector
  - **Monitor and adapt**: Regularly review bulk mail reports and user feedback to assess false positives and adjust the threshold accordingly.

- **Impact on users**:

  - **Lower thresholds**: Users might miss out on legitimate mass communication such as company newsletters or industry updates.
  - **Higher thresholds**: Users might experience cluttered inboxes with unwanted promotional emails, impacting productivity and potentially exposing them to misleading content.

The optimal bulk mail threshold depends on your specific security needs, user communication habits, and organizational culture. Carefully evaluate your priorities and find the right balance between filtering unwanted bulk mail and ensuring that important mass communication reaches its intended recipients.

Spam properties

Two individual options work together to influence how potential spam emails are identified and handled.

Increase spam score

Adjust the sensitivity of the spam filter by assigning additional points to emails that exhibit certain characteristics commonly associated with spam:

- **Available fields**:

  - **Image links to remote websites**: Increase the spam score for emails with excessive or unusual numbers of image links, especially those linked directly from the sender's domain. This discourages spammers who embed images from external sources to bypass content inspection.

  - **Numeric IP address in URL**: Increase the spam score for emails containing URLs with unconventional formats or predominantly numerical components. Many spammers use IP addresses instead of domain names to avoid reputation-based filtering.

  - **URL redirect to other port**: Increase the spam score for emails containing URLs that redirect to non-standard ports (for example, anything other than HTTP port 80 or HTTPS port 443). Redirects to unusual ports might indicate phishing attempts or malicious content delivery.

  - **Links to .biz or .info websites**: Increasing the spam score for all links might be too broad, but you can target unsolicited emails with links to free email providers' domains often used by spammers.

Instead of specific options for the mentioned indicators, you can achieve similar results by adjusting the spam score based on combinations of characteristics commonly found in spammy emails. Analyze email samples and observe spam trends to identify effective configurations for your organization.

Mark as spam

This configures the spam filter to mark a message as spam even without a high spam score if any of these common aspects are seen in the message:

- **Marking specific content and attributes**:

  - **Empty messages**: While there's no direct option, consider increasing the spam score for emails with very low content size or exceeding a specific character limit, potentially showing spam messages with minimal content

  - **Embedded tags in HTML, JavaScript/VBScript in HTML, form tags, frame/iframe tags, and object tags**: Microsoft's anti-spam filtering already analyzes email content for malicious scripts, forms, or hidden elements

  - **Web bugs**: While not a specific option, increasing the spam score for emails with excessive tracking pixels or invisible images might help flag suspicious messages

Sender reputation and authentication

The following settings are related to components configured by the sender and Defender for Office 365 and are used to further classify inbound messages:

- **SPF record – hard fail, Sender ID filtering – hard fail**: Defender for Office 365 leverages these sender authentication protocols. Mark emails with failing authentication checks as spam by enabling the **SPF hard fail** action and configuring the Sender ID policy to reject or quarantine messages failing authentication.

- **Backscatter**: While not directly configurable, Microsoft anti-spam filters analyze sender reputation and IP addresses to identify and reject backscatter emails.

Many organizations forget to properly configure these, so don't be surprised if you see logs related to these quarantine settings.

Advanced filtering

Your policies can even go the extra mile and be fine-tuned for very specific things seen in spam campaigns. These settings should only be used after extensive experience with inbound message filtering:

- **Sensitive words**: This leverages a dynamic but non-editable list of words used in spam and phishing attempts maintained by Microsoft to block malicious emails. However, use this cautiously to avoid blocking legitimate communication.

- **Contains specific languages**: This allows for configuring marking as spam messages in certain languages. Remember, blocking solely based on language might be discriminatory and should be considered carefully.

- **From these countries**: Defender for Office 365 utilizes location information for sender IP addresses. However, directly blocking based on the origin country is not recommended due to potential discrimination and limitations in accurately pinpointing the sender's location. Security teams should instead focus on sender reputation, email content analysis, and what the organization's human resources regulations might say.

Carefully consider the potential impact on legitimate communication when using these settings. Regularly monitor user feedback, false positives, and spam trends to refine your configurations and ensure optimal email security without unnecessary disruptions.

Test mode

This offers a valuable option for evaluating new policy settings before applying them to your entire organization. It acts as a simulation environment to assess the potential impact of your configuration changes without directly affecting users' inboxes. When enabled, test mode logs the results of anti-spam filtering based on your configured settings. It does not actually move, delete, or quarantine emails. You can access these logs to see which emails would have been flagged as spam, high-confidence spam, phishing, or bulk mail under the chosen policy. By default, this mode is off, and in previous versions of EOP, it required understanding **Advanced Spam Filter** (**ASF**) settings before enabling it. At the time of writing this, Microsoft migrated many of the Exchange admin center functions to the Microsoft 365 Defender portal (`security.microsoft.com`) and simplified the approach to configuring these. When enabling test mode, the administrator should know that test mode can be applied to all the settings in the anti-spam policy, except for **SPF record: hard fail**, **Sender ID filtering: hard fail**, **Backscatter**, **Contains specific languages**, and **From these countries** (*Microsoft, 2023a*).

For environments that want to automate actions, test mode can be enabled via PowerShell by following these steps:

1. Open a PowerShell window and connect to your EOP tenant by using `Connect-ExchangeOnline`.

2. Run the following command, replacing "MyASFPolicy" with the actual name of your policy and [ASFSetting] with the specific ASF setting you want to test:

    ```
    Set-HostedContentFilterPolicy -Identity "MyASFPolicy"
    -[ASFSetting] Test
    ```

    For example, to test the anti-spam setting for image links to remote websites in a policy named MyTestPolicy, use the following command:

    ```
    Set-HostedContentFilterPolicy -Identity "MyTestPolicy"
    -IncreaseScoreWithImageLinks Test
    ```

The ASF settings are as follows:

Category	Setting	ASF setting to use in PowerShell
Spam Score	Image links to remote websites	`IncreaseScoreWithImageLinks`
Spam Score	Numeric IP address in URL	`IncreaseScoreWithNumericIps`
Spam Score	URL redirect to other port	`IncreaseScoreWithRedirectToOtherPort`
Spam Score	Links to .biz or .info websites	`IncreaseScoreWithBizOrInfoUrls`
Mark as Spam	Empty messages	`MarkAsSpamEmptyMessages`
Mark as Spam	Embedded tags in HTML	`MarkAsSpamEmbedTagsInHtml`
Mark as Spam	JavaScript or VBScript in HTML	`MarkAsSpamJavaScriptInHtml`
Mark as Spam	Form tags in HTML	`MarkAsSpamFormTagsInHtml_`
Mark as Spam	Frame or iframe tags in HTML	`MarkAsSpamFramesInHtml_`

Category	Setting	ASF setting to use in PowerShell
Mark as Spam	Web bugs in HTML	`MarkAsSpamWebBugsInHtml_`
Mark as Spam	Object tags in HTML	`MarkAsSpamObjectTagsInHtml_`
Mark as Spam	Sensitive words	`MarkAsSpamSensitiveWordList_`

Table 6.1 – ASF settings to use with PowerShell

Remember, manipulating spam filtering settings can significantly impact legitimate email delivery. It's essential to proceed cautiously, test thoroughly, and understand the potential consequences before making any changes.

Available options for test mode

Only one option can be selected per anti-spam policy for what occurs when test mode is triggered, as seen in the following screenshot:

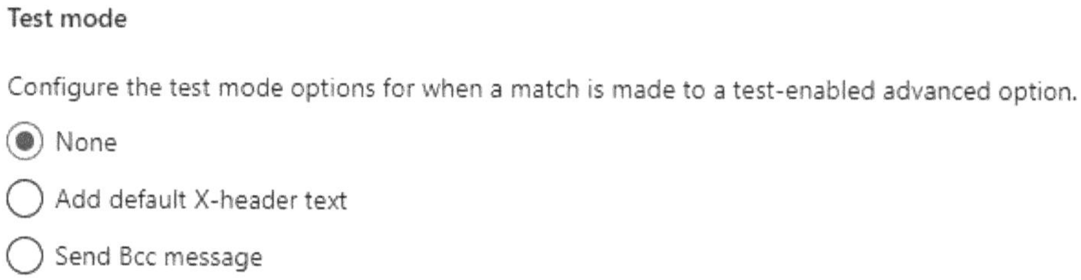

Test mode

Configure the test mode options for when a match is made to a test-enabled advanced option.

◉ None

◯ Add default X-header text

◯ Send Bcc message

Figure 6.8 – Test mode options

These settings will be your best friend when fine-tuning policies, but be aware that enabling test mode prevents the setting from taking part in quarantine efforts and just provides logs:

- **None**: In this mode, test mode simply logs the results of anti-spam filtering in the background. You won't receive any notifications or visual indicators about which emails would have been flagged:

  - **When to use**: This option is suitable for simple tests where you only need to analyze the logs later without immediate feedback.

  - **Best practices**: Combine **None** with thorough log analysis afterward to assess the impact of your settings.

- **Add default X-header text**: When enabled, test mode inserts a specific header, `X-MS-Exchange-Organization-Filtering-Level`, into the email headers of messages processed during the test. This header shows the filtering level that would have been applied:

  - **When to use**: This option is helpful for quickly identifying messages affected by specific policy settings within your email client. You can search for the header text to view potentially flagged emails.

  - **Best practices**: Use this option when you want a quicker, in-client overview of filtering impact without detailed logging.

- **Send Bcc message**: In this mode, test mode sends a **blind carbon copy (bcc)** message to a designated email address whenever an email would have been flagged as spam, high-confidence spam, phishing, or bulk mail. This allows you to review specific examples of emails impacted by your settings:

  - **When to use**: This option is ideal for in-depth analysis of potential false positives or to understand how specific senders or email types are handled by the policy.

  - **Best practices**: Use this option cautiously, as it generates additional messages and might require a dedicated test mailbox to avoid cluttering regular inboxes. Choose the recipient's address carefully and ensure email retention policies are compatible with the test messages generated.

Test mode can save you a lot of time when things just seem off and can help point you in the right direction, but always remember to limit its target audience.

Things to remember about test mode

Here are some further points to remember when using test mode:

- Choose the test mode option that best aligns with your testing goals and desired level of feedback.

- Consider combining different options for comprehensive evaluation.

- Start with simpler options such as **None** or **Add default X-header text** for preliminary tests, and use **Send Bcc message** for detailed analysis when needed.

- Always disable test mode before deploying the policy for daily use to avoid affecting users' inbox behavior.

- Test mode is a crucial tool for ensuring your anti-spam policy works effectively and minimizes inconvenience for users. Use it diligently to achieve optimal email security and a smooth user experience. Many organizations tend to ignore testing and end up with unexpected user impact and negative sentiment toward security efforts.

One can think of test mode as an audition of the quarantine setting. It is helpful to discuss with your security team and define policies and procedures on what will happen when a setting is triggered. This will help define the actions to configure in the policy.

Actions

These dictate what happens to emails categorized under specific spam classifications by the filter. This allows you to tailor your response to different types of unwanted messages, balancing security with user experience. Most fields in this section will use the following options:

- **Move message to Junk email folder**: This is the default action, placing flagged emails in the user's Junk folder for review. It helps identify potential spam while minimizing disruption to legitimate communication.

- **Add X-header**: This allows for inserting a custom X-header text in messages classified as spam.

- **Prepend subject line with text**: This adds a prefix to the subject line of flagged emails, making them easily identifiable in the recipient's inbox. Useful for customizing how spam messages are visually distinguished.

- **Redirect message to email address**: This forwards flagged emails to a designated email address for centralized monitoring or analysis. Suitable for collecting and reviewing spam samples for further action.

- **Delete message**: This permanently deletes flagged emails without user intervention. Use this cautiously, as it might lead to data loss for legitimate emails accidentally marked as spam.

- **Quarantine message**: This allows for the selection of a quarantine policy to apply. Different message actions allow for different quarantine policies.

It might be beneficial to create multiple policies with different actions in different scenarios; this will help you discover what works best.

Best practices when selecting any of these options

If you have no idea how to approach selecting actions to take, here are some points to get you started:

- Start with moderate settings such as **Move to Junk folder** or **Quarantine message** to balance security and user experience.

- Use **Delete message** cautiously and consider alternative actions when possible.

- Use **Add X-header** or **Prepend subject** for clearer identification of spam messages.

- The **Redirect** option might be useful for specific monitoring or analysis needs.

- Regularly review and adjust settings based on user feedback, spam trends, and security needs.

Remember, effective customization of the **Spam** option in your anti-spam policy requires careful consideration of your organization's specific requirements and security posture.

Fields in the Actions section

We can see many of the options available in the **Actions** section in the following screenshot and bullet points:

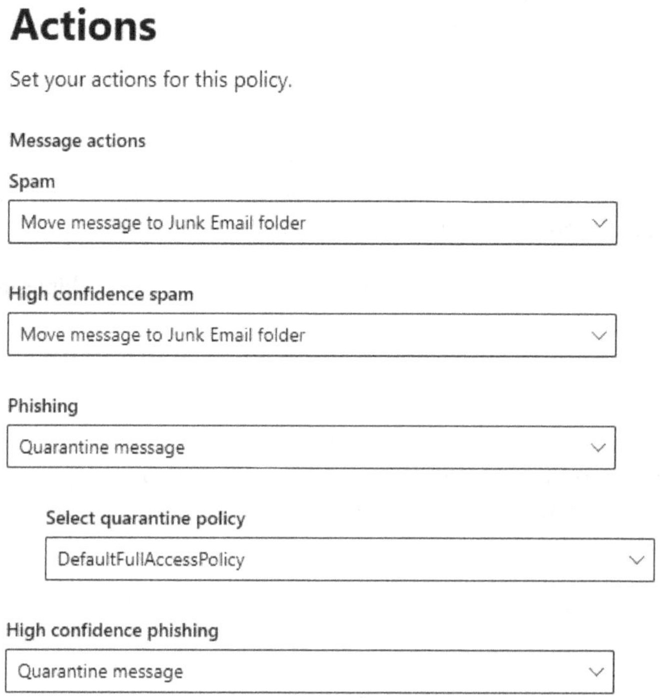

Actions

Set your actions for this policy.

Message actions

Spam

Move message to Junk Email folder

High confidence spam

Move message to Junk Email folder

Phishing

Quarantine message

Select quarantine policy

DefaultFullAccessPolicy

High confidence phishing

Quarantine message

Figure 6.9 – The Actions section

- **Spam**: Shows what actions to take when messages that exceed the spam threshold or are marked as spam are encountered. It uses the options identified previously.

- **High confidence spam**: Offers different actions compared to the regular **Spam** classification. It allows you to handle emails deemed highly likely to be spam with even stricter measures. It uses the options identified previously.

- **Phishing**: Shows what actions to take when messages are identified as phishing attempts. It uses the options identified previously.

- **High confidence phishing**: As with **Phishing**, it offers actions to take on messages deemed highly likely to be a phishing attempt with even stricter measures. Options are limited to **Quarantine message** and **Redirect message to email address**.

- **BCL met or exceeded**: Shows the actions to take on messages that met or exceeded the BLC threshold configured. It uses the options identified previously and adds an option called **None**, which takes no action.

- **Intra-organizational messages**: This option focuses on targeting inbound messages from within the organization and allows the administrator to select which phishing and spam categories to take action.

- **Retain spam in quarantine for this many days**: Determines how long quarantined emails are stored (between 1 and 30 days) before automatic deletion occurs.

- **Safe Tips**: Enabling this option configures a color-coded warning on messages warning about potential spam and phishing messages.

- **Zero-Hour Purge (ZAP)**: Enabling this option automatically takes action on messages classified as spam or phishing after they have already been delivered to user mailboxes. It acts retroactively, providing an additional layer of protection against malicious emails that evade initial detection.

I hope you noticed a trend here of the same settings being configured in multiple areas; this is why defined procedures are important before configuring these policies.

Allow & block list

This offers granular control over email filtering by allowing or blocking messages based on specific senders, domains, URLs, or files. It provides an alternative to relying solely on automated spam filtering and can be useful in various scenarios. The options available in this section can be seen in the following screenshot:

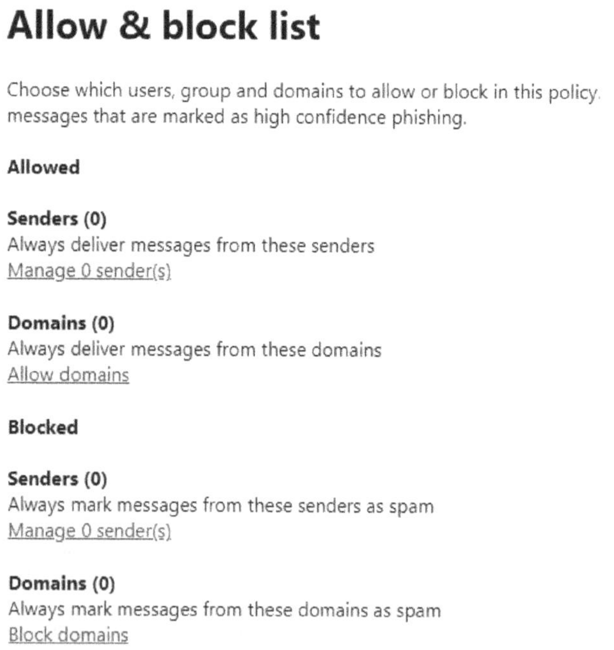

Allow & block list

Choose which users, group and domains to allow or block in this policy.
messages that are marked as high confidence phishing.

Allowed

Senders (0)
Always deliver messages from these senders
Manage 0 sender(s)

Domains (0)
Always deliver messages from these domains
Allow domains

Blocked

Senders (0)
Always mark messages from these senders as spam
Manage 0 sender(s)

Domains (0)
Always mark messages from these domains as spam
Block domains

Figure 6.10 – The Allow & block list section

The optimal configuration depends on your organization's security needs, user tolerance for false positives, and communication culture. Regularly review and adjust settings based on user feedback, spam trends, and evolving security threats.

Implementing policies and simplifying your approach

Understanding what each policy does is just part of the equation. Any administrator must understand how to apply these policies, build custom policies that are easy to understand and implement, and know how to implement these in a way that minimizes user impact.

Order of implementation

While you can approach configuration in any order desired, the following order proves to be the easiest to do in most organizations:

1. **Quarantine policy**: Sets the foundation for managing quarantined items, so configure this first
2. **Quarantine notifications policy**: Adjusts communication preferences related to quarantined items
3. **Anti-spam policy**: Fine-tunes spam filtering and handling to balance security and false positives

Always observe this order when planning and deploying policies, as it will allow for faster implementation and minimize misconfiguration.

Configuring the quarantine policy

The quarantine policy modular approach allows for configuring a set of policies and reusing these in multiple configurations, thus simplifying management. When working on other policies, such as the anti-spam or anti-malware policies, you can select which quarantine policy to use. Modification of existing policies or creation of new ones is performed as follows:

1. Navigate to the Defender portal (`security.microsoft.com`) | **Email & collaboration** | **Policies & rules** | **Threat Policies** | **Rules** | **Quarantine policy**. The **Quarantine policy** page can be seen in the following screenshot:

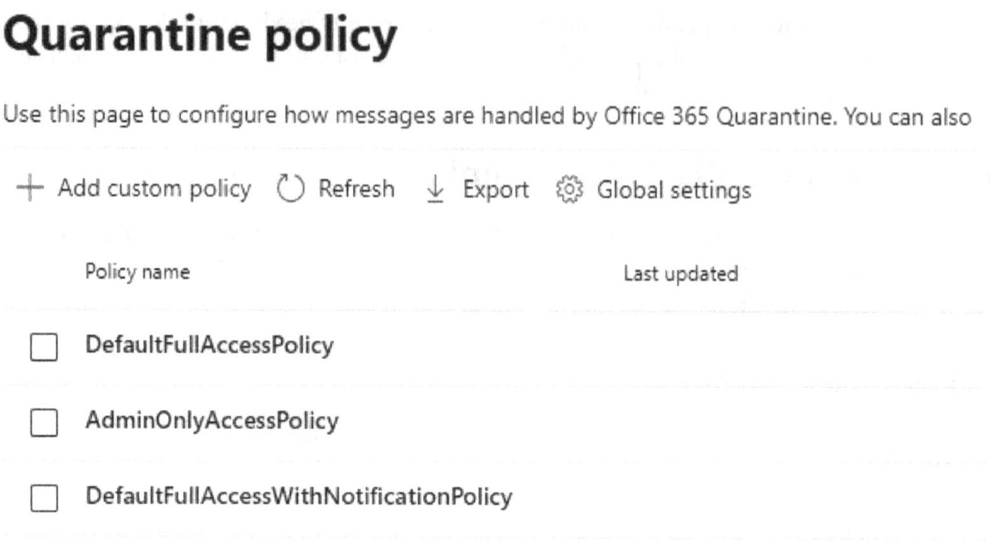

Quarantine policy

Use this page to configure how messages are handled by Office 365 Quarantine. You can also

$+$ Add custom policy   ↻ Refresh   ↓ Export   ⚙ Global settings

Policy name	Last updated
☐ DefaultFullAccessPolicy	
☐ AdminOnlyAccessPolicy	
☐ DefaultFullAccessWithNotificationPolicy	

Figure 6.11 – The Quarantine policy page

2. Select from any of the existing policies and select **Edit policy** to change it.

> **Note**
> Default policies such as **DefaultFullAccessPolicy**, **AdminOnlyAccessPolicy**, and **DefaultFullAccessWithNotificationPolicy** cannot be changed.

3. If desired, you can create a custom policy by selecting **Add custom policy** on the top menu, and the quarantine policy creation wizard will start. Be aware that you will not need to specify a target audience for the quarantine policy as these policies are attached to other policies instead.

> **Note**
> Once a custom policy has been created, the name cannot be changed.

The modular approach to quarantine policies also means that one must standardize the approach of naming these policies to ease management. A common approach to naming these involves the following approach: `<target group - impact>`, for example, `VIP-FullAccess&notification`.

Quarantine notification settings

Ensuring a standard approach to user notifications not only makes users smarter and improves security but also helps you detect when your configuration might need fine-tuning. Configuration of these settings is at a global level and not per policy:

1. Navigate to the Defender portal (`security.microsoft.com`) | **Email & collaboration** | **Policies & rules** | **Threat Policies** | **Rules** | **Quarantine policy**.

2. Select **Global settings** on the top menu.

3. Modify these settings as needed and click **Save** to apply the changes.

International organizations always aim at using localized language templates, as it will make for less confused users and decrease help desk calls.

Anti-spam policy

Anti-spam policies allow for the configuration of spam detection on both inbound and outbound traffic. For this chapter, we will focus on inbound traffic:

1. Navigate to the Defender portal (`security.microsoft.com`) | **Email & collaboration** | **Policies & rules** | **Threat Policies** | **Policies** | **Anti-spam policies**.

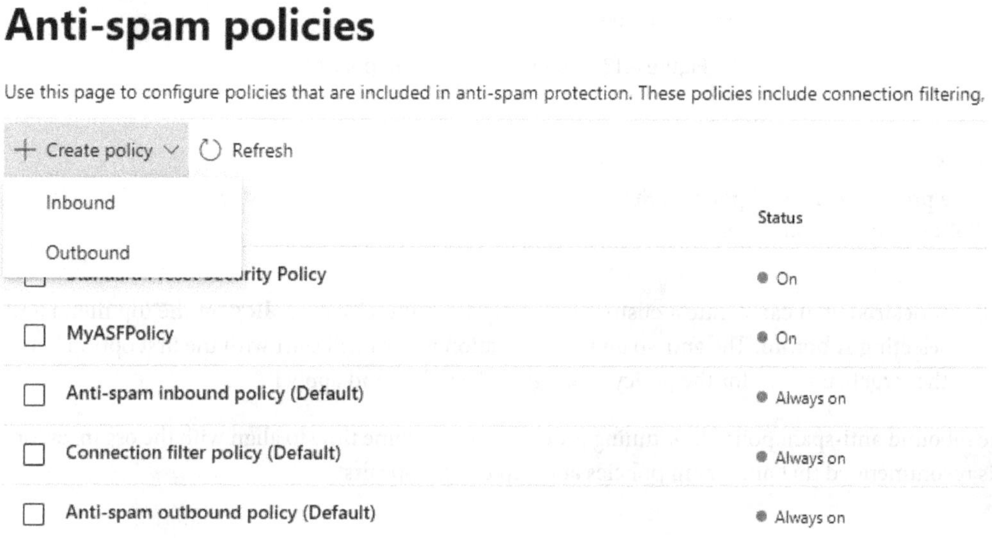

Anti-spam policies

Use this page to configure policies that are included in anti-spam protection. These policies include connection filtering,

	Status
+ Create policy ∨ ○ Refresh	
Inbound	
Outbound	
.... rity Policy	● On
☐ MyASFPolicy	● On
☐ Anti-spam inbound policy (Default)	● Always on
☐ Connection filter policy (Default)	● Always on
☐ Anti-spam outbound policy (Default)	● Always on

Figure 6.12 – The Anti-spam policies page

2. Select any of the existing policies and a window will appear to allow for editing the policy settings:

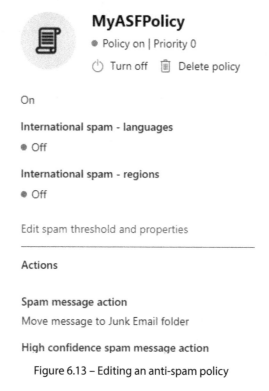

MyASFPolicy

● Policy on | Priority 0

⏻ Turn off   🗑 Delete policy

On

International spam - languages

● Off

International spam - regions

● Off

Edit spam threshold and properties

Actions

Spam message action
Move message to Junk Email folder

High confidence spam message action

Figure 6.13 – Editing an anti-spam policy

> **Note**
> The present generated policies, **Strict Preset Security Policy** and **Standard Preset Security Policy**, cannot be changed.

3. If desired, you can create a custom policy by selecting **Create policy** on the top menu and selecting **Inbound**. The anti-spam policy creation wizard will start with the first option being the target audience for the policy. Configure the policy and apply it.

The inbound anti-spam policy fine-tuning process will take some time to align with the organization. It is recommended that anti-spam policies apply to test groups first.

Summary

In this chapter, we focused on quarantine and anti-spam policies, which lay a solid foundation for enhancing email and messaging security. Starting with the identification and verification of quarantined emails, we've learned the importance of distinguishing legitimate communications from potential threats, a vital step in maintaining smooth operations. We've walked through the automated processes behind email and Teams message quarantining, highlighting the empowerment of users through options for managing quarantined messages. These options, including reviewing, releasing, and deleting messages, when combined with detailed insights into quarantine and anti-spam policy configurations, underscore the balance between user autonomy and robust security measures. The discussion also illuminated the strategic implementation of policies and the benefits of enabling a test mode for anti-spam efforts to ensure minimal disruption and optimal security adjustments.

As we conclude, it's clear that managing quarantine and anti-spam policies within Microsoft Defender for Office 365 is a fine balance between leveraging advanced technological safeguards and applying meticulous policy management. This balance is critical for protecting against sophisticated digital threats while ensuring uninterrupted business operations. The chapter emphasized the need for ongoing vigilance and adaptation to security practices—encouraging organizations to engage in regular policy reviews and to foster a culture of security awareness. The evolving nature of digital threats requires a proactive and informed approach to security, making the insights from this discussion not just a conclusion but also a starting point for ongoing security engagement and enhancement. In the next chapter, we will look at controls to further control spam and phishing attempts.

References

- Microsoft. (2023a, June 12). *ASF settings in EOP*. Microsoft Learn. `https://learn.microsoft.com/en-us/microsoft-365/security/office-365-security/anti-spam-policies-asf-settings-about?view=o365-worldwide`

- Microsoft. (2023b, August 2). *Find and release quarantined messages as a user*. Microsoft Learn. `https://learn.microsoft.com/en-us/microsoft-365/security/office-365-security/quarantine-end-user?view=o365-worldwide`

- Microsoft. (2023c, October 24). *Anti-spam protection*. Microsoft Learn. `https://learn.microsoft.com/en-us/microsoft-365/security/office-365-security/anti-spam-protection-about?view=o365-worldwide`

- Microsoft. (2023d, November 2). *Manage quarantined messages and files as an admin*. Microsoft Learn. `https://learn.microsoft.com/en-us/microsoft-365/security/office-365-security/quarantine-admin-manage-messages-files?view=o365-worldwide`

- Microsoft. (2023e, November 16). *Quarantine policies*. Microsoft Learn. `https://learn.microsoft.com/en-us/microsoft-365/security/office-365-security/quarantine-policies?view=o365-worldwide`

- Microsoft. (2024, January 2). *Configure spam filter policies*. Microsoft Learn. `https://learn.microsoft.com/en-us/microsoft-365/security/office-365-security/anti-spam-policies-configure?view=o365-worldwide`

Strengthening Email Security

Previously, we looked at how to protect our organization from inbound spam messages using inbound anti-spam policies. While these policies decrease the attack vectors to our users dramatically, it doesn't mean these are the only vectors that attackers can exploit in our environment. In many cases, a lack of proper controls for outbound traffic might leave the door open for a free attack venue that can be leveraged by bad actors to remain hidden, with no cost for the attacks leveraged. Closing these gaps via anti-phishing, email authentication, and outbound anti-spam controls is the key to ensuring a more holistic security approach and minimizing the risk of an impact on our organization's reputation.

This chapter will cover the following topics:

- How to configure and leverage anti-phishing policies
- How to configure email authentication
- How to configure outbound spam controls.

Let's continue our journey!

Phishing – a danger both inbound and outbound

Phishing, a social engineering technique we are seeing more and more in the news, targets individuals and organizations to gain unauthorized access to sensitive information. It is proving to be a formidable threat in the digital age. This method of attack is perilous not only due to the inbound messages that unsuspecting recipients might interact with—tricking them into divulging passwords, credit card numbers, or other private data—but also due to the peril of outbound messages. Compromised systems can unknowingly disseminate malicious content, potentially harming an organization's reputation and security posture. For example, you might receive a message from someone acting as your company's help desk, asking you to verify your identity due to suspicious activity. This email would include a link to click and would urge prompt action in less than 24 hours if you do not want to risk having your account disabled. Once you clicked the email, you would be taken to a fake login page where you would enter your credentials, receive a thank-you message, and go on with your day. In this example, the attackers would have obtained your credentials, and you would not be aware until these were used in an attack. In this section, we will delve into the mechanisms of phishing attacks from both inbound and outbound perspectives and explore the strategic implementation of anti-phishing policies. These proactive measures are essential to bolster an organization's security infrastructure, effectively mitigating the risk of phishing campaigns and safeguarding against the multifaceted dangers they present.

Anti-phishing policies

Organizations equipped with Defender for Office 365 and Exchange Online mailboxes, as well as those exclusively using **Exchange Online Protection** (**EOP**), have access to advanced options for configuring anti-phishing defenses. EOP establishes a solid baseline with critical anti-phishing components—a standard policy for basic protection, mechanisms to prevent spoofing, and safety tips on first-time interactions. Further enhancing these measures, Microsoft Defender for Office 365 introduces stringent policies against impersonation and more refined criteria for identifying phishing efforts, ensuring a more comprehensive security stance (*Microsoft, 2023a*).

As with safe link policies, depending on whether you have configured the preset security configuration, you will find up to three policies already configured: the Office365 **AntiPhish Default policy**, which covers the most common phishing attacks, as well as the **Standard** and **Strict preset policies**, which align with common Microsoft best practice recommendations. We can see these policies in the following figure:

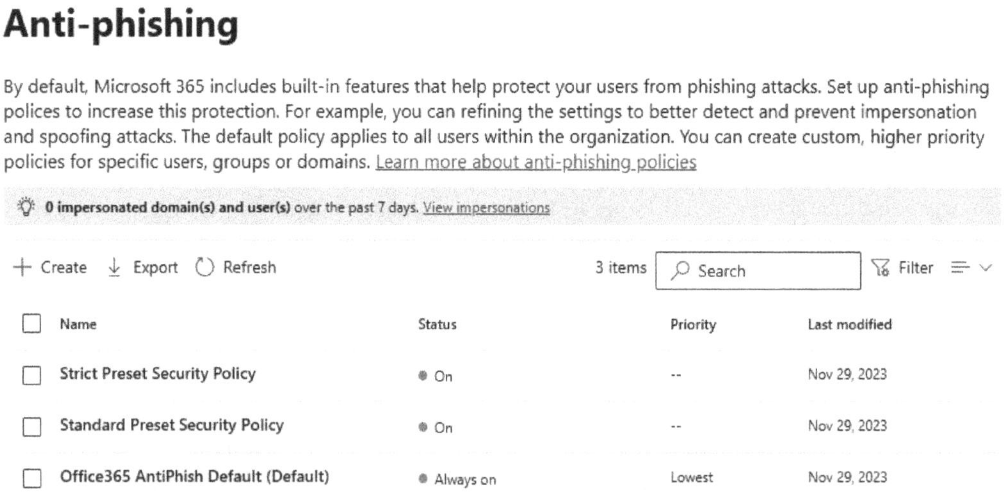

Anti-phishing

By default, Microsoft 365 includes built-in features that help protect your users from phishing attacks. Set up anti-phishing polices to increase this protection. For example, you can refining the settings to better detect and prevent impersonation and spoofing attacks. The default policy applies to all users within the organization. You can create custom, higher priority policies for specific users, groups or domains. Learn more about anti-phishing policies

💡 **0 impersonated domain(s) and user(s)** over the past 7 days. View impersonations				

Name	Status	Priority	Last modified
Strict Preset Security Policy	● On	--	Nov 29, 2023
Standard Preset Security Policy	● On	--	Nov 29, 2023
Office365 AntiPhish Default (Default)	● Always on	Lowest	Nov 29, 2023

+ Create   ↓ Export   ○ Refresh   3 items   🔍 Search   ⅄ Filter ≡ ∨

Figure 7.1 – The anti-phishing policies page

A major difference between safe links and anti-phishing policy configuration is that you can perform changes on the default policy (Office365 AntiPhish) alongside any custom policies. Standard and strict preset policies cannot be changed. As for policy priority, for both default and preset policies the changing of the included policies priority is not allowed, with preset policies always having higher priority and the default policy having the lowest. For custom policies, you also have the option of enabling, disabling, deleting, and changing its priority (not to be higher than preset policies or lower than the built-in policy) (*Microsoft, 2023b*).

Policy sections and fields

As with any other policies in Defender for Office 365, it is important to understand the fields and sections in these policies before configuring them. Understanding them will allow you to determine which capabilities can be changed and which areas you can align with organizational requirements.

Section 1 – policy name

As with previous policies, this section should help with policy management and identification. The **Name** field should have a name that aligns with the organization's naming standards and helps identify the impact, target audience, and policy version. The **Description** field should provide a brief explanation of what the policy is intended for. It should include the target and impact.

Section 2 – users and domains

As with other policies, in this section, you can specify the target of the policy. This can be a combination of a set of users, groups, and domains, as well as the exclusion of some users, groups, and domains, or a combination of all of these fields. The idea is to allow you to configure granular policies. However, always remember that the more granular a policy is, the more policies will be required to cover the entire organization, which will increase the administrative workload.

Section 3 – phishing threshold and protection

This section describes how email messages are classified as phishing emails. The focus of this section is on the phishing score threshold, and other features such as impersonation protection and spoof protection to include specifying allow and block lists.

The phishing email threshold relates to the phishing confidence score that the email message receives and the threshold to identify the email as a phishing email. **1 - Standard** is the default value and covers the most common user cases. Higher levels include **2 - Aggressive**, **3 - More Aggressive**, and **4 - Most Aggressive**. These will lead to more emails being classified as phishing emails. It is recommended to start with the default threshold to prevent user frustration and only adjust this if too many phishing emails are landing in user mailboxes. We can see the phishing threshold setting in the following figure:

Phishing threshold & protection

Set your phishing thresholds and desired impersonation and spoof protections for this policy. Learn more

Phishing email threshold ⓘ

◯━━━━━━━━━━━━━━━━━━━━━━━━━━━ **1 - Standard**

This is the default value. The severity of the action that's taken on the message depends on the degree of confidence that the message is phishing (low, medium, high, or very high confidence).

Figure 7.2 – The phishing threshold setting

The impersonation area focuses on a common approach used by attackers to impersonate **Very Important Persons (VIPs)** in an organization as part of their social engineering efforts. Techniques such as changing one character in a domain or email address along with the name of the sender to match the impersonated person are part of an attack vector that is easily missed by even trained security professionals. When adding a checkmark in the checkbox next to **Enable users to protect**, and adding up to 350 internal and external users, allows for these users to enjoy impersonation protection. Domains owned and custom domains can also be protected by adding a checkmark next to **Enable domains to protect** along with checkmarks next **to Include domains I own** and another checkmark next to **Include custom domains**. If protecting domains, verify the listed domains to ensure none are missed. It is important to note that when protecting domains (only the domain is checked and not the entire email address).

This section also includes the **Add trusted senders and domains** option, which allows for trusted senders and domains to be excluded from being flagged by the phishing policy. It is recommended to leave this section blank and only use it in rare instances such as testing new settings or during troubleshooting because of a domain receiving a high number of false positives. If the automatic impersonation protection checkbox was previously left blank, you will also be allowed to configure it if you wish to enable mailbox intelligence (enabled by default on the automatic protection option), which performs extra impersonation verifications during the first time an address contacts a mailbox. Another checkbox, **Enable intelligence for impersonation protection**, is also enabled by default when using automatic impersonation protection that focuses on frequent contacts and uses artificial intelligence to identify patterns used in impersonation attacks. As with other options available when not using automatic impersonation protection, you will need to configure what occurs in the **Actions** section of the policy.

The spoof area of this section focuses on another common technique used by attackers in which a message will appear to originate from someone or somewhere other than the actual source. This differs from impersonation attacks, as the email's From header is changed in the message body versus just trying to trick the user by changing a character or two in the email address. As email clients use this field to present recipients with information on the sender, it is often an easy way to trick users. There are times when spoofing is valid, such as using a third-party company to generate advertising, administrative assistants' duties, company surveys sent in bulk to internal employees, and even communications between companies. In valid scenarios, there is a list of allowed and blocked spoofed senders that can be configured at the tenant level by modifying the **Spoofed senders** tab in the **Tenant Allow/Block Lists** in the **Threat policies** page under the **Rules** section, or by opening `https://security.microsoft.com/tenantAllowBlockList?viewid=SpoofItem`.

Section 4 – actions

This section focuses on the actions that occur when a message is detected to be a phishing email. If impersonation protection was enabled in *Section 3*, the options related to the actions to be taken when encountering user impersonation, domain impersonation, or mailbox intelligence will be set to **Don't apply any action**, but will allow for modification. If any of the impersonation protection options was not enable, the field for that section will be greyed out and **Don't apply any action** will be shown. In the fields enabled, you will be allowed to set the action to take from a drop down menu with options available as follows:

- **Redirect the message to other email addresses**
- **Move the message to the recipients' Junk Email folders**
- **Quarantine the message**
- **Deliver the message and add other address to the Bcc line**
- **Delete the message before it's delivered**
- **Don't apply any action**

We can see this section in the following figure:

Actions

Set what actions you'd like this policy to take on messages. You may access all available policy actions.

Message actions

If a message is detected as user impersonation

Quarantine the message	∨

We'll quarantine the message for you to review and decide whether manage quarantined messages

Apply quarantine policy

AdminOnlyAccessPolicy	∨

If a message is detected as domain impersonation

Move the message to the recipients' Junk Email folders	∨

Move the message to the recipients' Junk Email folders

Figure 7.3 – The Actions section in the anti-phishing policy

The **Actions** section also includes actions regarding spoof emails. An option is given to honor the **Domain-based Message Authentication, Reporting, and Conformance** (**DMARC**) record policy configured by the sender's domain. This even allows for modifying the action taken to differ from the sender's domain's DMARC record instruction to either quarantine, reject, or move the email to a user's folder. If spoof intelligence was enabled in the previous section, you will also be presented with similar options for detections performed by spoof intelligence.

Finally, the safety tips and indicators area of the **Actions** section provides indicators to show users in emails. The intent is to help users identify possible spoof emails if the spoof intelligence does not have enough confidence in the email being spoof. Options include presenting a question mark when the sender fails authentication methods such as DMARC and showing a tag on emails when these are sent from a domain that is different from the one the email was originally sent from. A security team that periodically educates users can leverage these options to identify spoof emails that might have passed the protections and look at fine-tuning these types of emails to prevent future attempts.

Configuring anti-phishing policies

As with other policies, anti-phishing policy configuration is a simple process that mimics how other policies are configured for Defender for Office 365:

1. Access the Microsoft Defender portal (`security.microsoft.com`).

2. In the left-hand menu, navigate to the **Email & collaboration** section and click on **Policies & rules**. On the **Policies & rules** page, select **Threat policies** as shown in the following figure:

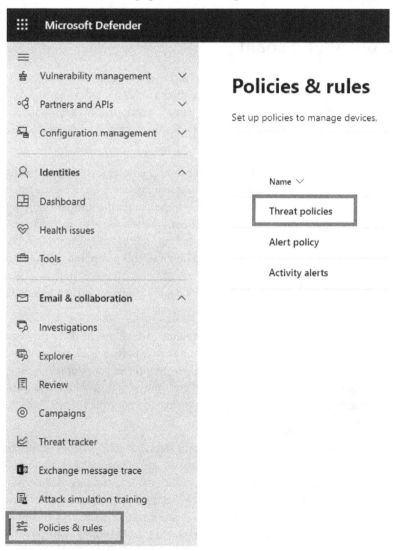

Figure 7. 4 – The Policies & rules page with the Threat policies option

3. On the **Threat policies** page, navigate to the **Policies** section and click on **Anti-phishing**.

4. On the **Anti-phishing policies** page, click on **+ Create** on the top menu to start the custom policy creation.

5. Once the policy creation is completed, verify that it has the correct priority level.

As with other policies, remember policy priority. Also, remember that preset policies are always applied first (if the target is included in the policy targeting settings) and that processing stops after the first policy.

Stopping outbound spam

In a previous chapter, we discussed the impact and configuration of inbound spam protection. While this is a significant step in protecting your organization from attacks, it fails to address what happens if a bad actor uses your infrastructure to attack third parties. Neglecting this area could result in your security team overlooking a bad actor hiding in your environment, and your organization could be held liable for any attacks involving your infrastructure.

Implementing outbound spam controls is a simple multi-step process involving the configuration of email authentication and the configuration of outbound spam policies. Implementing these controls brings multiple positive effects to your environment, which justifies the time and effort spent. These include improved email deliverability, as legitimate emails are less likely to end up in spam folders. This ensures timely communication and avoids frustration. Additionally, there is a reduced risk of phishing, as unauthorized senders cannot spoof your domain. This protection safeguards users from phishing attacks that aim to steal sensitive information. Lastly, implementing outbound spam controls increases trust and confidence among users. They can be more confident in the authenticity of emails received from your domain, which fosters trust and engagement.

Administrative impact includes **reduced spam complaints**, which means that fewer emails will be misidentified as spam. This, in turn, minimizes the administrative burden of addressing user complaints. Moreover, there is an **enhanced brand reputation** when you protect your domain from spoofing, as it builds brand reputation and prevents damage from phishing attacks. Additionally, **better resource management** is achieved as the reduced spam filtering load on your mail server frees up resources for other critical tasks.

In the realm of legal consequences, abiding by regulations is fundamental. Various sectors are governed by standards such as HIPAA and GDPR, which mandate the authentication of emails to safeguard user information. Ignoring these rules can result in significant fines and legal troubles. Additionally, curtailing the dangers of phishing related to one's domain decreases legal hazards, mitigating the chance of facing lawsuits from individuals hurt by spoofing attempts. What's more, DMARC reports can significantly bolster litigation efforts by serving as crucial proof in cases of email forgery or impersonation.

The process of setting up measures to prevent outbound spam also requires careful thought. Getting everything up and running demands both time and dedication to set up and maintain surveillance. Nevertheless, the effort pays off in the long run. Educating users about these protocols enhances their comprehension and trust, which is equally vital. Plus, consistent testing and surveillance are imperative to ensure that everything works as intended, preventing legitimate emails from being mistakenly blocked.

Let's start with authentication

At the core of securing sent emails lies the critical practice of authenticating emails properly. This authentication is achieved through the implementation of SPF records, DKIM signatures, and DMARC enforcement. These measures are pivotal in confirming that emails sent from your domain are legitimate, and in making it easier to spot any unauthorized ones.

SPF records play a vital role in this defense system by being incorporated into DNS records to guard emails on their way out. They operate by listing the mail server IP addresses authorized to send emails on your domain's behalf. When an email arrives, the recipient's server consults the sender's DNS for the SPF record to check for any red flags regarding spoofing.

Alongside this, DKIM signatures further fortify the security of outbound emails. These signatures act as a safeguard against anyone attempting to tamper with the **From** address by attaching a digital signature to each legitimate outgoing email. This allows the recipient to use a publicly available key from the sender's DNS to verify the email's authenticity. Should this signature be altered or absent, the DKIM protocol deems the email as *Fail*.

To round off this security trifecta, DMARC enforcement stipulates how outbound emails should be treated if they don't pass SPF or DKIM validations. It gives the receiving side guidance on whether to reject, quarantine, or ignore the email, thereby ensuring a higher level of protection against unauthorized or malicious emails.

Configuring SPF, DMARC, and DKIM

The following steps will focus on configuring email authentication on a custom Microsoft 365 domain. We will implement these controls in order of ease, with SPF being first, DKIM second, and lastly, DMARC. We will also provide guidance on controls that can be implemented when using `*.onmicrosoft.com` domains.

SPF

SPF acts as a safeguard for emails originating from your Microsoft 365 environment, essentially blocking impersonators who might attempt to conduct cyber threats such as **Business Email Compromise** (**BEC**), ransomware, or various phishing scams. SPF's core role is to authenticate the origins of emails for a domain. This is achieved through a TXT record in the domain's DNS, delineating the approved email sources for that domain. Upon receiving an email, the receiving email system references this

SPF TXT record to ensure that the sender's address in the **Simple Mail Transfer Protocol (SMTP)** protocol is recognized and sanctioned for the domain. For example, if you use the `exampledomain.com` domain within Microsoft 365, you must establish an SPF TXT record for `exampledomain.com` in your DNS, listing Microsoft 365 as a legitimate email source for this domain. Consequently, email recipients will consult the SPF TXT record for `exampledomain.com` to ascertain that the incoming emails are indeed from a legitimate source corresponding to `exampledomain.com`.

When diving into the nuances of setting up SPF for your email domains, there are a few critical points to bear in mind, especially concerning how your Microsoft 365 operates with domains. If your organization uses the default **Microsoft Online Email Routing Address (MOERA)**, which would look like `exampledomain.onmicrosoft.com`, you're in luck—there's nothing extra you need to do. The SPF TXT record comes pre-configured because Microsoft manages the `onmicrosoft.com` domain, along with its DNS records.

However, the scenario shifts when you bring custom domains into the mix, such as `exampledomain.com`. During the Microsoft 365 setup process, you're guided to either create or update an SPF TXT record in your DNS, tagging Microsoft 365 as a trusted sender. This step is crucial but becomes slightly more complicated if your organization uses external email services for purposes such as sending bulk emails or conducting surveys. These third-party services might not be directly manageable by your team, and they can potentially affect your main domain's email reputation negatively.

To mitigate this risk, it's advisable to deploy a subdomain exclusively for these types of email activities. For every subdomain put into use for emailing, ensure that you establish an SPF TXT record. Similarly, for any subdomain that you're not utilizing for emails, set up an SPF TXT record too, but this time, indicate within the record that no emails should be sent from this domain. This strategic approach helps fortify the integrity of your email domain's reputation while efficiently managing the sender validation process (*Microsoft, 2024c*).

Creating the SPF TXT record

The SPF TXT record will be created in your DNS server. In the case of environments with separate DNS servers for external and internal traffic, these should be created on the external DNS servers. The format of the record is as follows:

```
v=spf1 [<ip4>|<ip6>:<PublicIPAddress1>
<ip4>|<ip6>:<PublicIPAddress2>... <ip4>|<ip6>:<PublicIPAddressN>]
[include:<DomainName1> include:<DomainName1>... include:<DomainNameN>]
<-all | ~all>
```

The initial entry, `v=spf1`, identifies the DNS record as an SPF TXT record. `PublicIPAddress` is the public IP address of your organization's email server and you can add as many as needed, separated by a space. These addresses will also require a prefix to indicate whether they are IPv4 or IPv6 addresses. `DomainName` is the domain name of the service sending emails. At minimum, a Microsoft 365 organization will require `include:spf.protection.outlook.com` along with any other third-party official email sending domain. Finally, you can select what happens to the emails that originate from a server IP address that is not in the SPF TXT record by using the `all` switch. The `-all` switch will cause a hard failure, meaning the recipient would reject these. The `~all` switch will cause a soft fail, so the recipient will either send the email to a junk folder or quarantine it. We can see an example of an SPF record for an organization that only uses Microsoft 365 for emails in the following figure:

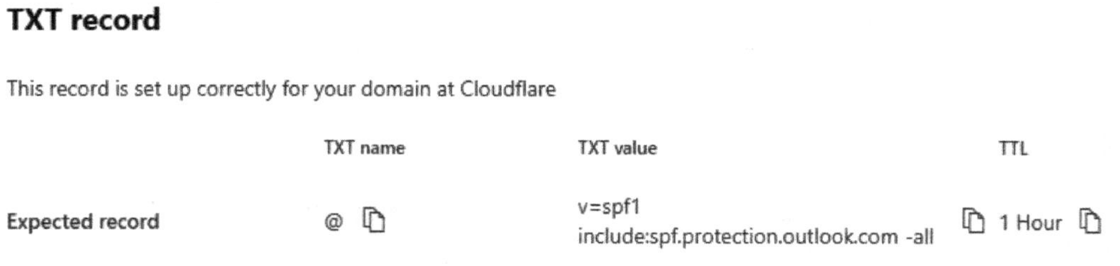

TXT record

This record is set up correctly for your domain at Cloudflare

	TXT name	TXT value	TTL
Expected record	@	v=spf1 include:spf.protection.outlook.com -all	1 Hour

Figure 7.5 – An SPF TXT record

Some example records would look like this (we will use internal addresses for our examples, but in your real-world deployments, use external addresses):

```
v=spf1 ip4:192.168.50.50 ip4:192.168.100.30 include:spf.protection.
outlook.com -all
v=spf1 ip4:192.168.200.10 ip4:192.168.210.40 include:spf.protection.
outlook.com include:thirdpartyemailexample.com -all
```

Be aware that when using the Microsoft 365 **Government Community Cloud High** (**GCC High**) and the Microsoft 365 Department of Defense environments, the included domain will be `spf.protection.office365.us`. When using a partner such as 21Vianet in China, the domain will be `spf.protection.partner.outlook.cn`. This is because of these environments residing in separate cloud infrastructures.

For subdomains that are not to be used for emails, the difference in the SPF record is just that no email sender IP address is added. An SPF TXT record for a subdomain not used for email would look something like this:

```
v=spf1 -all
```

When configuring SPF, some things to always remember include the following:

- Never configure more than one SPF TXT record per domain or subdomain, as this could cause a **DNS lookup loop**, which will lead to SPF failures.

- Aim to use nine or fewer addresses in the SPF TXT record, as 10 or more can cause DNS lookup issues. When using nested resources, such as an IP from a load balancer that points to other IP addresses, confirm that the total number of IP addresses, including nested addresses, does not exceed nine.

- As addresses are evaluated from left to right and the evaluation stops upon the first validated IP address, always aim to place the most-used IP addresses first in the record.

- If your environment also configures DMARC, the distinction between hard fail and soft fail is gone and all failures are treated as SPF failures with the recipient deciding what to do.

Be aware that DNS records might take some hours to propagate, so ideally, these changes should be performed during low outbound email traffic times. Always validate your SPF records after modification using many of the free online tools available such as `dnschecker.org` or `spf-checker.org`.

DKIM

DKIM is a great addition to any domain using SPF. While SPF provides references that recipients can use to verify valid sender email senders, DKIM provides a way to verify that a message hasn't been altered in transit. To accomplish this, for each domain, one or more private keys are generated, which are used by the sender's email system to digitally sign some of the header fields and the message body in the outbound message. A field is added to the outbound message header called **DKIM-Signature**, which identifies the signing domain with a `d=` value. The sender DNS records will keep either a **Canonical Name (CNAME)** record for Microsoft 365 systems or, sometimes, a TXT record for some third-party systems, in which the corresponding public keys can be found to validate the DKIM signature in the message header. It is important to note that the signing DKIM domain does not have to match the sending domain, as DKIM is only focused on ensuring that the message is not changed during transit. In fact, some messages could have multiple DKIM signatures, as with some hosted email services that sign the message, and then sign it again using the customer domain.

Before diving into the setup of DKIM, it's crucial to grasp a few key details, similar to how we approach SPF. For those using the default MOERA, such as `domainexample.onmicrosoft.com`, for their email needs, there's no extra step required on your part. Microsoft has got you covered by auto-generating a 2048-bit public-private key pair for your initial `*.onmicrosoft.com` domain. This automation extends to outbound messages, which are seamlessly DKIM signed with the private key, while the corresponding public key takes its place in a DNS record. This setup allows the recipient email systems to authenticate the DKIM signature of the messages without fuss. However, the scenario changes a bit if you've moved beyond Microsoft's MOERA domain, venturing into the territory of custom domains or subdomains. For these, you're tasked with setting up DKIM signatures individually. Echoing the sentiment with SPF, if your organization uses third-party email services for things such as sending out surveys, opting for a subdomain for these activities is a wise move. This approach helps keep your main domain's reputation untarnished by any potential mishaps originating from these services. Lastly, there's a notable difference from SPF when it comes to subdomains that are not active in email dispatch. In such cases, bypassing DKIM configuration on these dormant subdomains is the recommended route (*Microsoft, 2024b*).

Creating the DKIM CNAME record

As with SPF, a DNS record needs to be created in the DNS servers that are used for external traffic. The format of the record involves two parts: a selector, which points at the order of the record, such as `selector1` or `selector2`, and a CNAME record for the public key. Two DKIM selector records are published on the DNS server. The format of these records is as follows:

```
Host name or alias: selector1._domainkey
Value: selector1-<CustomDomain>._domainkey.<InitialDomain>
Hostname or alias: selector2._domainkey
Value: selector2-<CustomDomain>._domainkey.<InitialDomain>
```

For these records, the only values to change are `CustomDomain`, which will point to your organization's domain with the dots replaced by dashes, and `InitialDomain`, which will be the original `onmicrosoft.com` domain that was configured when initially configuring your Microsoft 365 tenant, but it will keep the periods (.) in the address. If the same tenant hosts multiple custom domains, the `InitialDomain` value for these will all be the same. For example, for an original tenant called `original.onmicrosoft.com` that hosts the custom `exampledomain.com` domain, the DKIM selector records would look like this:

```
Host name or alias: selector1._domainkey
Value: selector1-exampledomain-com._domainkey.original.onmicrosoft.com
Hostname or alias: selector2._domainkey
Value: selector2-exampledomain-com._domainkey.original.onmicrosoft.com
```

If we use a third-party DNS service provider, such as Cloudflare, these records should not be proxied, and the configuration of both SPF and DKIM can be performed right from the Microsoft 365 admin center. In the following figure, we can see a DKIM configuration in Cloudflare:

Type ▲	Name	Content	Proxy status
CNAME	selector1._domainkey	selector1-████████-com._domainkey.sec...	DNS only
CNAME	selector2._domainkey	selector2-████████-com._domainkey.sec...	DNS only

Figure 7.6 – A DKIM record configuration in Cloudflare

To allow Microsoft 365 to perform the configuration on the third-party DNS service provider, follow these steps:

1. Visit the Microsoft 365 admin center (`admin.microsoft.com`).

2. On the left-hand menu, select **Settings**, and under **Settings**, select **Domains**.

3. Select the domain to create the SPF and DKIM records in and a new page will open related to the domain. On the new page, select the **DNS records** tab.

4.   A line will appear reading: **To manage DNS records for <custom domain name>, go to your DNS hosting provider:** and the third-party DNS service provider name will be displayed along with a pencil icon. Ensure this is the correct DNS provider. If not, you might need to click on the pencil icon to correct it. If you are making this change, be aware that email service might be interrupted, so ensure that you verify and plan this change accordingly.

5.   Below this section, a menu with options related to the DNS records will be shown; select **Manage DNS**.

6.   A new wizard will start. In the section that says **Let Microsoft add your DNS records**, click on **More options**. Ensure that **Let Microsoft add your DNS records (recommended)** is selected and click **Continue**.

7.   The **Add DNS Records** page will appear showing the records to be added related to the exchange and EOP will be initially visible, including the entry for the SPF TXT record. Ensure there is a checkmark next to **Exchange and Exchange Online Protection**.

8.   To add the DKIM record, click on **Advanced options**. Other options will appear; ensure that DKIM is enabled. Click on **Add DNS Records**.

9.   A new window will open to log you in to the third-party DNS service provider. Log in and follow the instructions. If your account has enough permissions, the records will be updated. If you receive notice that it has failed, verify the DNS records on the third-party DNS service provider tools. Be aware that this process might sometimes show as failed yet be implemented due to latency in communications.

It is highly recommended to use the Microsoft 365 admin center for automated configuration of DNS records. This approach will ensure that all the records follow the best practices set by Microsoft and will minimize any changes or issues. Be aware that for custom domains, enabling DKIM signing for outbound messages switches DKIM signing from using the initial `onmicrosoft.com` domain to having the custom domain do the signing. Therefore, along with the DNS record, we must ensure that our systems perform the signing by following these steps:

1.   Visit the Microsoft Defender portal (`security.microsoft.com`).

2.   On the left-hand menu, navigate to the **Email & collaboration** section and click on **Policies & Rules**. On the **Policies & Rules** page, select **Threat policies**.

3.   On the **Threat Policies** page, navigate to the **Rules** section and click on **Email authentication settings**.

4.   On the email authentication page, select the **DKIM** tab to see a list of the domains under your tenant. Click on the domain name for the domain in which you want to enable DKIM.

5.   A new window will appear for the domain. Ensure that **Sign messages for this domain with DKIM signatures** is enabled. You will also have the option to rotate the DKIM keys as needed, as Microsoft maintains and manages these.

As with other DNS records, propagation could take some hours. Once the records have propagated, you can perform a help check with multiple free tools, such as dnschecker.org. You can also test that outbound messages are using DKIM by sending an email to a test address and looking at the header using an email client such as Outlook. Copy this header and paste it into a message header analysis site such as `https://mha.azurewebsites.net`, which will provide different values to include `d=` for the sending domain, `s=` for the selector used, and `DKIM=pass` or `DKIM=OK` to signify that the email has passed DKIM checks.

DMARC

DMARC enhances the security provided by SPF and DKIM by focusing on the verification of sending domains, a critical step that is not covered by the other two protocols. It ensures that the domain mentioned in the email message corresponds with the actual sender's domain. This verification happens at two levels: the domain used for communication between SMTP email servers, which also acts as a fallback for undelivered emails (known as the **RFC5321.MailFrom** address), and the domain visible in the sender's address in the email `From` header (referred to as the **5322.From** address). DMARC utilizes SPF to confirm that the email originates from an accredited source of the domain specified in the **MAIL FROM** address. It also checks for a match between the domains in the MAIL FROM address and the `From` address in the email. Additionally, DMARC employs DKIM to ascertain whether an email was appropriately signed and to verify the integrity of the signature, ensuring that the email's content remains unchanged (*Microsoft, 2024d*).

Through these checks, DMARC conducts a thorough three-step examination of email messages to establish their authenticity. The protocol can be configured to reject or quarantine messages, or to take no action on messages that fail these inspections, offering flexibility in handling potential threats. Another notable advantage of DMARC is its ability to produce detailed reports about these verifications, aiding in the ongoing battle against email fraud and spoofing.

Before diving into the DMARC configuration, it's essential to grasp some critical insights. Unlike SPF and DKIM, employing the MOERA domain for your emails (such as `domainexample.onmicrosoft.com`) necessitates the creation of a DMARC TXT record for the `*.onmicrosoft.com` domain via the Microsoft 365 admin center. With custom domains and subdomains, it's imperative to set up SPF and DKIM signing beforehand, as these elements are foundational for DMARC to function correctly. Another distinction between SPF and DKIM is that a single DMARC record can suffice for a domain and inherently cover all its subdomains.

For any domains or subdomains that are not utilized for email purposes (including the `onmicrosoft.com` domain), it's advisable to establish a DMARC record indicating that no emails are anticipated from these domains or subdomains. Lastly, if your organization avails services that alter emails during transmission, consider implementing a trusted ARC sealer. This adjustment ensures that your organization's emails seamlessly pass DMARC verification, maintaining the security and authenticity of your communications (*Microsoft, 2024a*).

Configuring DMARC

As with SPF and DKIM, there is a specific format for the DSN record when configuring DMARC. This format is as follows:

```
Hostname: _dmarc
TXT value: v=DMARC1; p=<reject | quarantine | none>;
pct=<0-100>; rua=mailto:<DMARCAggregateReportURI>;
ruf=mailto:<DMARCForensicReportURI>
```

The values that we will change in this format will be `p=`, which denotes what happens to the email that fails DMARC verification (rejection, quarantine, or nothing), and `pct=`, which denotes how many of the email messages that fail DMARC verification are subject to the action identified in `p=`. This setting should be `100` unless testing is being performed. The next two fields are optional, but provide a lot of useful functionality. They are `<DMARCAggregateReportURI>`, which is the email address to where the DMARC aggregate report is sent (typically daily and showing which emails in your organization failed delivery), and `<DMARCForensicReportURI>`, which is the email address to where the DMARC failure report is sent (typically after each failed message sent by your organization). We can see a DMARC record for an environment that uses Cloudflare for its aggregate reports in the following figure:

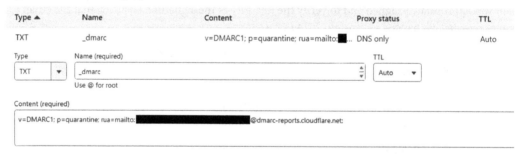

Figure 7.7 – A DMARC record using Cloudflare for reports

In contrast to SPF and DKIM, there is not an option to automatically configure DMARC. On a positive note, configuring a DNS record on the domain and subdomain used for sending emails is all that is required, with one configuration per domain covering all the subdomains under that domain. Best practice dictates that any domains and subdomains that are not used for email should have a DMARC record that rejects all emails sent from that domain with the same hostname as other DMARC records (`_dmarc`) and a value of `v=DMARC1; p=reject;` to include the `onmicrosoft.com` domain if it is not used for emails. When DMARC record configuration is completed, it is recommended to verify that the record was created correctly as DNS propagation has occurred. Many tools are available to verify DMARC records. These include `dnschecker.org`, `mxtoolbox.com`, `easydmarc.com`, and various others. For DMARC reports, there are many free tools available, such as `valimail.com` and `cloudflare.com`, which will require the modification of the DMARC record's `rua=mailto:` field to include an address that these services provide.

Outbound spam filtering

Outbound spam filtering is a thorough procedure that meticulously evaluates and blocks emails originating from your Exchange Online or Defender for Office 365 configuration. Its main goal is to stop the spread of harmful content or violations of organizational regulations. This process is vital for preventing accidental disclosure of sensitive data. It also prevents phishing scams, where emails imitate your domain to deceive people into sharing sensitive information or engaging with harmful links. Additionally, it prevents unsolicited spam and malware, safeguarding the reputation of your brand. By taking measures to prevent your domain from being linked to spam, you can establish trust and ensure that your emails are delivered successfully.

In Microsoft 365 setups that use Exchange Online mailboxes or solely rely on EOP, all outgoing emails are automatically scanned for spam and unusual sending behavior. When outbound spam is detected, it indicates a security breach in a user's account within your organization. To safeguard the service's reputation and prevent Microsoft 365's email servers from being blacklisted, suspicious emails are routed through a dedicated high-risk delivery pool, irrespective of their **Spam Confidence Level (SCL)**. The high-risk pool is specifically designated for emails of less credibility, such as spam. It operates on a separate IP address pool to avoid tarnishing the reputation of the standard outbound email IP pool, which is used for high-quality communications. This helps reduce the risk of these IP addresses being blacklisted. It is worth mentioning that some email systems may not accept messages from this high-risk pool, so being able to deliver them to the intended recipients is not always certain.

Administrators are promptly notified of any suspicious outbound email activities and the users responsible via alert policies. All senders are subject to a default policy for handling outbound spam. Nevertheless, if you want more precise control, you can create personalized outbound spam policies to specifically target certain users, groups, or domains. This allows for a tailored approach to your organization's email security strategy (*Microsoft, 2023c*).

Outbound anti-spam policies

Protection in EOP and Defender for Office 365 is managed through policies. For inbound spam protection, you can use the built-in configuration, standard and strict preset configurations, or customize your own policies. However, for outbound spam protection, you can only choose between the built-in policy or custom policies for all outbound messages. On the **Anti-spam policies** page, you will only find the **Anti-spam outbound policy (Default)** option available by default. We can see the anti-spam policies page, including the built-in anti-spam outbound policy in the following figure:

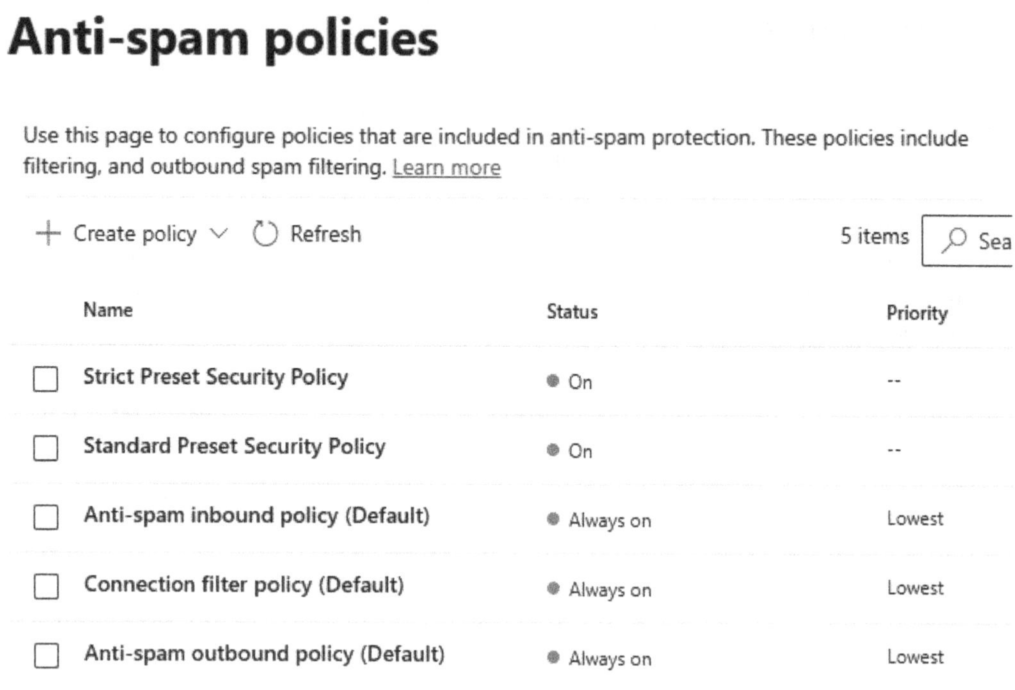

Anti-spam policies

Use this page to configure policies that are included in anti-spam protection. These policies include filtering, and outbound spam filtering. Learn more

+ Create policy ∨   ○ Refresh   5 items   🔍 Sea

Name	Status	Priority
☐ Strict Preset Security Policy	● On	--
☐ Standard Preset Security Policy	● On	--
☐ Anti-spam inbound policy (Default)	● Always on	Lowest
☐ Connection filter policy (Default)	● Always on	Lowest
☐ Anti-spam outbound policy (Default)	● Always on	Lowest

Figure 7.8 – The anti-spam policies page

The built-in policies cannot be turned off and have the lowest priority to ensure that all users have some level of protection. Custom policies allow you to prioritize outbound protection policies, but only the highest priority policy is processed for each user.

Policy fields

Before we begin deploying the outbound anti-spam policies, let's review their composition. Understanding the fields to be completed and the options available helps create a strategy and rationale on how to configure these policies in a manner that aligns with organizational requirements and policies.

Section 1 – naming your policy

This is a common section in all policies that allows you to name the policy in an easy-to-understand manner. Best practices dictate that names should be descriptive, align with any approved naming standards, and include the impact, target audience, and version. An example would be `OutboundAntiSpam_ restrictive_testusers_20240210`. Descriptions provided in the optional **Description** field should provide the policy's purpose and scope. An example would be `Applies the restrictive outbound anti-spam protection configuration to the test users' groups.`

Section 2 – users and domains

In this section, you can define the policy's target audience. The target can be as small as an individual user or as big as a group, a domain, or a combination of all of these. To allow for further flexibility, specific users, groups, and domains can also be excluded.

Section 3 – protection settings

In this section, we zero in on a strategy that attackers commonly deploy in spam campaigns: mass emailing to a broad audience simultaneously. It's possible to set parameters to manage the volume of outgoing messages, including establishing caps for both external (**Set an external message limit**) and internal (**Set an internal message limit**) recipients on an hourly basis, as well as instituting a maximum count of addressees permitted per day (**Set a daily message limit**). These limits offer a wide range and are adjustable from 0 (which denotes adherence to the service's default thresholds) up to a limit of 10000, aligning with the constraints inherent in pre-existing policies. We can see the configurable limits in the following figure:

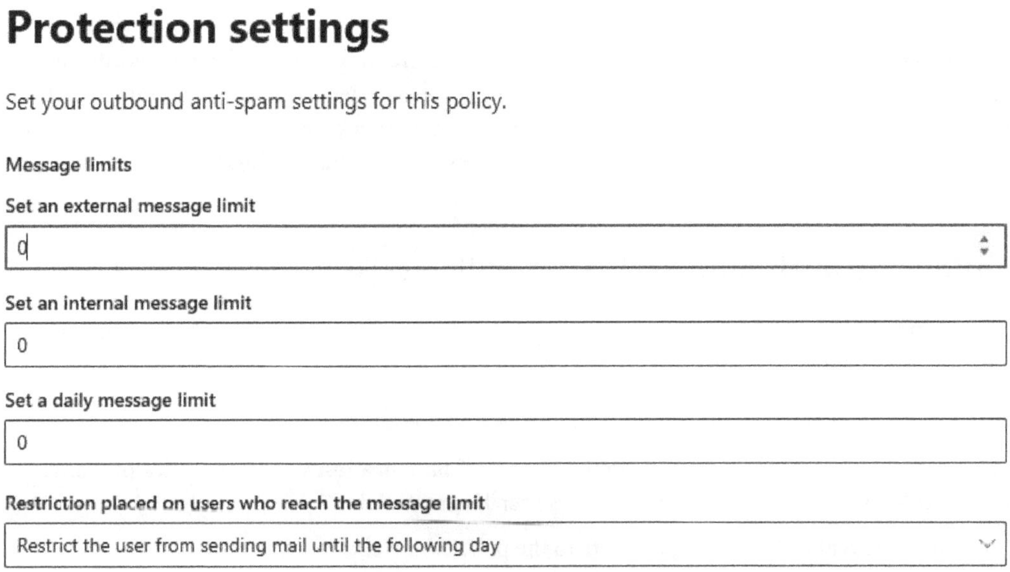

Protection settings

Set your outbound anti-spam settings for this policy.

Message limits

Set an external message limit

| 0 |

Set an internal message limit

| 0 |

Set a daily message limit

| 0 |

Restriction placed on users who reach the message limit

| Restrict the user from sending mail until the following day |

Figure 7.9 – The configurable limits in the outbound anti-spam policy

For users who exceed these specified thresholds, various enforcement actions can be applied. Options range from a complete restriction on email dispatches until further notice or a temporary suspension of sending privileges for the remainder of the day (aligned with UTC time), to simply issuing an alert, with the default action being the day-long restriction on sending. In terms of managing emails that infringe upon these limitations, the configuration for forwarding can be set to **Automatic - System-controlled** (functionally equivalent to **Off**), **Off** (the advocated setting), or **On** for situations necessitating further examination of the message, although this could potentially weaken the security posture in the absence of robust security measures. We can see these settings in the following figure:

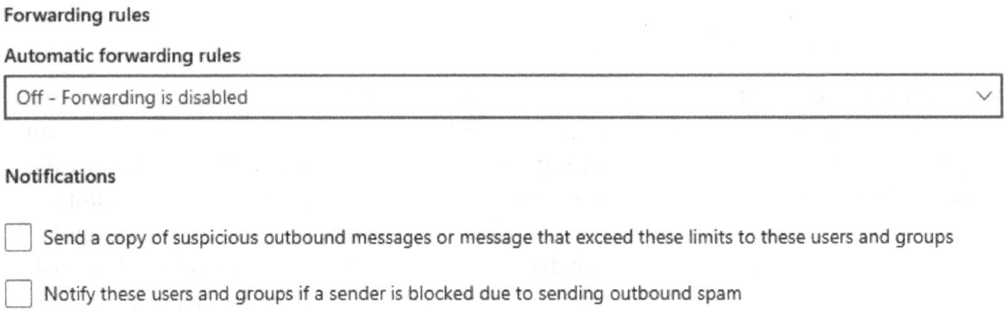

Forwarding rules

Automatic forwarding rules

Off - Forwarding is disabled   ∨

Notifications

☐ Send a copy of suspicious outbound messages or message that exceed these limits to these users and groups

☐ Notify these users and groups if a sender is blocked due to sending outbound spam

Figure 7.10 – The automatic forwarding rules and notifications sections

Lastly, the approach to notifications entails options for redirecting a copy of the offending email to an alternate address, a practice akin to establishing forward rules for such correspondence that is discouraged, and alerting users and groups upon a sender's blockage. The latter setting is highly recommended for security teams, ensuring that they receive immediate alerts and can initiate prompt investigative action.

Configuring outbound anti-spam policies

The configuration of these policies is a simple process that mimics how other threat policies are configured in Defender for Office 365:

1. Access the Microsoft Defender portal (`security.microsoft.com`).

2. On the left-hand menu, navigate to the **Email & Collaboration** section and click on **Policies & Rules**. On the **Policies & Rules** page, select **Threat policies**.

3. On the **Threat policies** page, navigate to the policies and click on **Anti-spam**.

4. On the **Anti-spam policies** page, click on **+ Create** in the top menu and select **Outbound** to start the custom policy creation.

5. Once the policy creation is completed, verify that it has the correct priority level.

As with other Defender for Office 365 and EOP policies, always remember that policies with higher priority will be the ones to take effect on a user, and that policy processing stops after the first policy, so ensure that the custom policy that you wish to affect the user has the highest priority (lower number).

Summary

To wrap up, establishing anti-phishing rules, enhancing email verification, and applying strict outbound spam guidelines in EOP and Defender for Office 365 are indispensable measures for safeguarding a company's communication channels. This thorough strategy spans from crafting detailed anti-phishing measures that use cutting-edge machine learning to spot and stop fraudulent emails, to setting up solid email verification protocols such as SPF, DKIM, and DMARC, significantly reducing vulnerabilities. Moreover, the introduction of outbound spam rules fortifies this protective framework, ensuring the company does not inadvertently serve as a spam distributor, thus preserving its standing. The combination of these measures forms a robust defense system crucial for securing confidential data and upholding the trust and dependability of the company's email interactions.

As we move forward, the subsequent chapter will investigate the domain of malware, introducing a comprehensive set of defenses designed to thwart such harmful threats. Recognizing the complex and constantly changing landscape of malware risks is critical for implementing defenses that can prevent, identify, and neutralize these dangers. We'll explore the newest developments and tactics in combating malware, focusing on how they can be seamlessly incorporated into EOP and Defender for Office 365 for all-encompassing security. This next chapter aims to arm you with the insight and resources required to bolster your defense mechanisms against the diverse and evolving nature of malware threats, significantly broadening the scope of security beyond merely countering outbound spam and phishing.

References

- Microsoft. (2023a, November 2). *Configure anti-phishing policies in EOP*. Microsoft Learn. https://learn.microsoft.com/en-us/microsoft-365/security/office-365-security/anti-phishing-policies-eop-configure?view=o365-worldwide

- Microsoft. (2023b, November 2). *Configure anti-phishing policies in Microsoft Defender for Office 365*. Microsoft Learn. https://learn.microsoft.com/en-us/microsoft-365/security/office-365-security/anti-phishing-policies-mdo-configure?view=o365-worldwide

- Microsoft. (2023c, November 15). *Configure outbound spam policies*. Microsoft Learn. https://learn.microsoft.com/en-us/microsoft-365/security/office-365-security/outbound-spam-policies-configure?view=o365-worldwide

- Microsoft. (2024a, February 12). *Configure trusted ARC sealers*. Microsoft Learn. https://learn.microsoft.com/en-us/microsoft-365/security/office-365-security/email-authentication-arc-configure?view=o365-worldwide

- Microsoft. (2024b, February 12). *How to use DKIM for email in your custom domain.* Microsoft Learn. `https://learn.microsoft.com/en-us/microsoft-365/security/office-365-security/email-authentication-dkim-configure?view=o365-worldwide`

- Microsoft. (2024c, February 12). *Set up SPF to identify valid email sources for your Microsoft 365 domain.* Microsoft Learn. `https://learn.microsoft.com/en-us/microsoft-365/security/office-365-security/email-authentication-spf-configure?view=o365-worldwide`

- Microsoft. (2024d, February 12). *Set up DMARC to validate the From address domain for senders in Microsoft 365.* Microsoft Learn. `https://learn.microsoft.com/en-us/microsoft-365/security/office-365-security/email-authentication-dmarc-configure?view=o365-worldwide`

8

Catching What Passed the Initial Controls

By now, we should have a good security foundation in the areas of spam and phishing protection, and while that lays a significant starting point, there are many other factors at play that we must look at, such as malware and what happens when our environment is the target. As with any security strategy, it is all about identifying gaps and implementing further layers that do not disrupt business. We will look at gaining an understanding of how mail travels through our organization and we will see what actions we can take to further fortify our defenses. We will also look at the controls Defender for Office 365 offers for things beyond emails, such as documents.

This chapter will cover the following topics:

- How to configure and use mail flow rules and trace messages
- How to prevent malware propagation via anti-malware and Safe Links policies and zero-hour purge
- What order do policies from different categories follow when being applied?
- How to configure alert policies

Let's continue our journey!

Understanding email flow

Understanding the journey of emails, both incoming and outgoing, is crucial for strengthening our defenses against email threats. This knowledge enables us to identify vulnerabilities, such as phishing attempts, malware infiltrations, and unwanted spam. By comprehending the route our emails take, we can implement security measures effectively at each stage. This involves setting up defense mechanisms such as anti-malware scans and spam filters to ensure that legitimate emails are delivered smoothly. We can map the path of emails to identify security breaches and take appropriate corrective actions. It can also help us pinpoint where a problem may have occurred, and help us troubleshoot any delivery issues, thus ensuring messages reach their intended recipients without fail.

We also cannot forget the need to comply with regulations and ensure our email practices are auditable and meet the required standards and protections while striking the right balance between security and user convenience. Finally, email flow understanding can help us further refine our security and operational procedures to maintain a proactive posture and allow for end user education and feedback. Familiarizing ourselves with both the inbound and outbound paths of email communication not only improves our ability to defend against security threats but also streamlines the email exchange process and instills confidence in our protective measures.

Inbound mail flow

When the sender clicks **Send** in their email, the first step along the way is the **Message Submission Agent (MSA)**, which acts as a *post office* for your email client, composing and submitting the email to the **Message Transfer Agent (MTA)**. The MTA, often the sender's email provider's server, is a crucial component that handles email relaying. It queries **Domain Name System (DNS)** servers to find the recipient's mail record. This record identifies the **Mail Exchange (MX)** server responsible for receiving emails in that domain. Optionally, organizations may use third-party inbound connectors for tasks such as spam filtering or content inspection. If employed, the email might pass through this intermediate step before reaching the recipient's server. The sending MTA uses the **Simple Mail Transfer Protocol (SMTP)** to communicate with the recipient's MTA. This establishes a secure connection and transmits the email message.

The email now enters the recipient's environment (in our case, Microsoft 365) and lands on **Exchange Online Protection (EOP)**, our cloud-based filtering system serving as a security gateway for incoming emails in Microsoft's infrastructure. It performs a series of checks to ensure the safety of the emails, including verifying the sender's IP address against block lists, scanning for malware using configured policies, and either quarantining or deleting infected emails, applying predefined actions such as adding disclaimers or redirecting messages based on specific criteria using transport rules, analyzing URLs and attachments for suspicious content to protect users from phishing and malware attacks using **Safe Links** and **Safe Attachments policies**, and leveraging advanced algorithms to filter out spam based on sender reputation, content analysis, and user reports with inbound anti-spam policies (*Microsoft, 2023f*).

The email now lands in the recipient's mailbox and is further processed by mailbox rules, which can automatically move or organize emails within the recipient's mailbox based on specific criteria. Now, the email is on its intended destination (the recipient's email), unless forwarding rules are configured, which automatically sends the message to another address. As an optional safeguard provided by Defender for Office 365, **Zero-Hour Auto Purge (ZAP)** scans emails even after they have been delivered and deletes any emails that are deemed malicious, even if missed by the initial filters.

What about the outbound mail flow?

Just as messages have a busy journey when they arrive at our mailbox, the same can be said for the journey they take when leaving. The journey starts from your mailbox, where your email client (such as Outlook) composes the message and submits it to the mailbox server within Microsoft 365 and onto

EOP. From there, predefined transport rules within your mailbox or organization can automatically modify the email, adding disclaimers or encrypting sensitive content. The email is then scanned for malware based on configured settings by anti-malware policies. If any threats are detected, the email may be quarantined or deleted. Additionally, outbound anti-spam policies utilize advanced algorithms to evaluate email for potential spam characteristics, blocking or filtering suspicious messages. Our message then leaves EOP and experiences a similar journey to inbound emails experienced in our environment.

Mail flow rules

Mail flow rules, or **transport rules**, are used to identify specific emails based on criteria such as sender address, keyword presence, attachment type, or recipient domain. These rules can perform actions on these emails, such as delivering them to specific recipients, forwarding them to alternative addresses, adding disclaimers or custom text, preventing delivery, holding suspicious emails for review, or notifying administrators.

From a security perspective, we can block emails with specific keywords to ensure compliance and prevent information leaks. We can configure rules to redirect risky emails, such as those with attachments from external senders, to a quarantine for review. We can optimize internal communication by adding disclaimers, routing replies to specific teams, and using a **Data Loss Prevention** (**DLP**) control by identifying and blocking the sending of sensitive emails (in a limited manner; for more efficient DLP, look at Microsoft Purview DLP). Properly configured rules can even supplement other security controls by filtering out specific spam patterns and malicious emails, all via centralized management through a single console.

To use mail flow rules effectively, it is important to define conditions and actions to avoid unintended consequences. An administrator can either discuss conditions and actions with the mail administrators or review any written documentation, which will allow for testing the rule in a logical and quantifiable manner before deploying. Finally, because of the major impact these rules can have, always inform users about any changes to rules, including the impact their email might experience.

Rule components

All mail flow rules are broken into multiple common components, which allow for a wide variety of combinations. Conditions for message rules can be specified based on various factors, such as the following:

- Sender/recipient address
- Keywords in the subject/body
- Message size
- Attachment type

We can use logical operators such as AND, OR, and NOT to combine these conditions. Exceptions are messages that meet the conditions but should not be affected by the rule. You can use the same options available for conditions to identify these exceptions. Actions define the outcome for messages that meet the conditions and are not exempted by exceptions. Among the actions to take are the following:

- **Deliver**: This sends to specific recipients or groups

- **Redirect**: This forwards the message to another address

- **Modify**: This adds disclaimers, prefixes, or alter headers

- **Block**: This option stops delivery, holds the message in quarantine for review, and sends a notification to administrators

- **Properties**: These allow for changing settings such as priority, enforcement date, logging, and scope

We can see an example of a mail flow rule in the following figure:

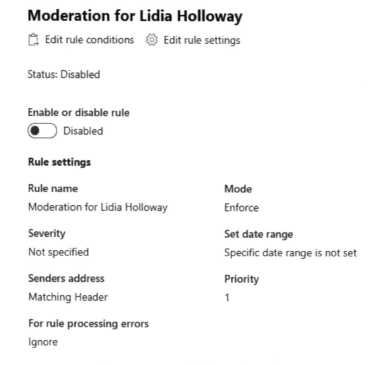

Figure 8.1 – Example of a mail flow rule

For example, let's assume we want to create a mail flow rule to block messages that use common phishing attempt keywords such as **urgent payment** or **you have 24 hours to reply**. Our mail flow rule would look like this:

- *Conditions:*

  - *Subject contains "you have 24 hours to reply" OR Subject contains "urgent payment" OR Body contains "you have 24 hours to reply" OR Body contains "urgent payment".*

- *Actions:*

  - *Block*

- *Optional:*

  - *You can add more specific keywords as needed.*

We could exclude trusted senders (for example, internal senders) from this rule.

For another example, we would like to add a disclaimer to all outbound emails sent outside our organization. Our mail flow rule would look like this:

- *Conditions:*

  - *Any message.*

- *Actions:*

  - *Append disclaimer with company contact information (e.g., website, phone number).*

- *Optional:*

  - *You can use different disclaimers for different users or departments.*

  - *You can exclude specific recipients from this rule.*

Now that we have clarity on how mail flow rules work, we can create our first set of mail flow rules.

Configuration steps

At the time of this writing, mail flow rules are a component of Exchange Online rather than a component of Defender for Office 365, as such configuration will be performed in a different location from where Defender for Office 365 components are configured. The following steps are presented:

1. Log in to the Exchange admin center: `https://admin.exchange.microsoft.com`.
2. Navigate to **Mail Flow | Rules**.

3. Click **Add a rule** and choose **Create a new rule** or **Select a template** based on your preference.

    If using a template, choose one closest to your desired functionality and customize it later. If creating from scratch, select **Create a new rule**.

4. Configure the rule conditions.

    We can see an example of the rule conditions section in the following figure:

Figure 8.2 – Mail flow rule conditions

The rule conditions are broken into the following fields:

- **Name**: Name your rule. Use descriptive names for rules to ease identification and management.

- **Apply this rule if**: Select the condition in the message body that triggers the rule. Use the + icon on the right of the condition to add other conditions to use alongside the first one. Examples of conditions include a specific recipient, a specific sender, specific headers, and message body contents.

- **Do the following**: Define the action to take on the message when the rule is triggered. Use the + icon on the right of the action to add other actions to perform alongside the first one. Examples of actions include blocking the message, changing the message properties, and adding words to the subject.

- **Except if**: Defines the conditions that would exempt the message from the rule. Use the + icon on the right of the condition to add other conditions to use alongside the first one. Examples of exemption conditions include emails sent from a VIP person and messages sent to a specific person.

1. Set the rule settings.

   We can see an example of the rule settings section in the following figure:

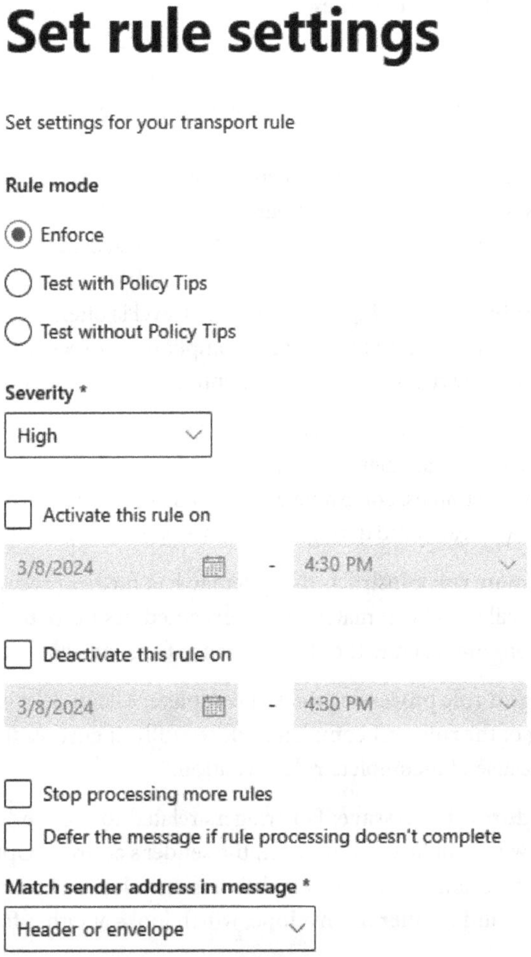

Set rule settings

Set settings for your transport rule

Rule mode

(●) Enforce

( ) Test with Policy Tips

( ) Test without Policy Tips

Severity *

| High   ∨ |

[ ] Activate this rule on

| 3/8/2024   🗓 | - | 4:30 PM   ∨ |

[ ] Deactivate this rule on

| 3/8/2024   🗓 | - | 4:30 PM   ∨ |

[ ] Stop processing more rules

[ ] Defer the message if rule processing doesn't complete

Match sender address in message *

| Header or envelope   ∨ |

Figure 8.3 – Mail flow rule settings section

Configuration of a rule is broken down into the following areas:

- **Rule mode**: Determines how a rule works in Exchange Online. There are three modes available: **Enforce**, **Test with Policy Tips**, and **Test without Policy Tips**. The **Enforce** mode option actively applies the rule to incoming messages, enforcing the specified actions or conditions. **Test with Policy Tips** allows testing of the rule's effects and provides policy tips to users, informing them of potential policy violations. **Test without Policy Tips** is like the previous mode, but it doesn't display policy tips to users during testing.

- **Severity**: Used for activity reports or rule management, and having no impact on how the rule behaves. **Severity** indicates the importance or criticality level of the rule and helps prioritize actions based on the potential impact or risk associated with the rule. Severity levels are divided into **Low**, **Medium**, **High**, **Not audit**, and **Not specified**. A **Low** severity implies issues that have minimal impact or urgency. These could be informational messages or events that don't require immediate attention. **Medium** signifies problems that warrant attention but aren't critical emergencies. They may require action or monitoring but are not as urgent as **High** severity issues, such as the detection of potentially harmful attachments in emails, triggering a **Medium** severity alert to prompt further investigation without causing immediate alarm. **High** denotes critical issues demanding immediate action or response. These could include serious security breaches, system failures, or compliance violations. For instance, a **High** severity alert might be raised when detecting a significant data leak of sensitive information that requires swift intervention to mitigate potential damage. **Not audit** typically indicates that an event or condition is not subject to auditing or logging. This level is often used for non-critical events or activities that do not require tracking for compliance or security purposes. Finally, **Not specified** is used for any rule that does not fall into any of the previous categories.

- **Activate this rule on/Deactivate this rule on**: Determines when the rule becomes active and inactive, respectively. These settings allow administrators to schedule the rule's operation based on specific criteria or timeframes. For example, a rule can be activated on a certain date to coincide with policy changes or deactivated during maintenance periods to avoid unintended consequences.

- **Stop processing more rules**: Instructs the system to halt further processing of additional rules once the current rule has been matched and executed. It's particularly useful for ensuring that specific actions are taken without interference from subsequent rules.

- **Defer the message if rule processing doesn't complete**: Allows delaying the message delivery if the processing of the rule isn't completed successfully. It ensures that messages aren't lost or misrouted because of incomplete rule execution.

- **Match sender address in message**: For triggers related to the sender's email address, this setting specifies what will be used to match the sender's address. Options available include **Header**, which only examines the message headers; **Envelope**, which only looks at the SMTP message envelope; and **Header or envelope**, which looks at either for a match.

2. Click **Finish** to create the rule.

> **Note**
> Remember to always begin with a few well-defined rules and gradually add complexity. The possible combinations make misconfiguration likely if rules are not discussed and defined beforehand. Once implemented, rules should be reviewed and changed periodically to ensure their relevance and effectiveness. It is recommended that these rules be classified according to the attack they aim at mitigating, along with procedures to trigger updates when new related security threats emerge or company policies are revised.

Message tracing – unveiling the mystery of your emails

Message tracing allows you to track the path an email takes within your Microsoft 365 environment, from sender to recipient. This valuable tool comes in handy for diverse scenarios, offering crucial insights into email delivery, troubleshooting problems, and investigating incidents (*Microsoft, 2023c*). To use this tool, follow these steps:

1. Visit the Exchange admin center (`admin.exchange.microsoft.com`).

2. On the left-hand menu, inside the **Mail flow** section, you will see **Message trace**. Click on it.

   The **Message trace** page will open with the following options: **Default queries**, **Custom queries**, **Autosaved queries**, and **Downloadable reports**.

   By default, the tool includes some queries for the most common scenarios where message tracing is needed. These default queries are variations of values in the following fields:

   - **Senders**: The originating address or domain of the email message.

   - **Recipients**: The address or domain that received the email message.

   - **Time range**: The time range of the messages, which can be configured using a slider that goes between 6 hours and 90 days from now, or a custom time frame between 2 dates.

   - **Delivery status**: The status of the message with options including **All**, **Delivered**, and **Expanded**. Other available statuses that only show when generating enhanced or extended summaries include **Failed**, **Pending**, **Quarantined**, **Filtered as spam**, and **Getting Status**.

   - **Message ID**: For investigation of specific messages, the message ID can be found in the message header inside the **Message-ID:** section.

   - **Direction**: Allows you to specify **Inbound**, **Outbound**, or **All**.

   - **Original client IP address**: Allows for investigations regarding a specific sender; be careful as IP addresses can be spoofed.

When looking at the default queries included, there will just be prefilled variations of these fields to get you started. You can also create custom queries for many different scenarios as needed by your organization. For both default and custom queries, you will have the option to view the results in either a summary report online or an enhanced or extended report, both of which can be downloaded as a CSV file. We can see an example of creating a message trace in the following figure:

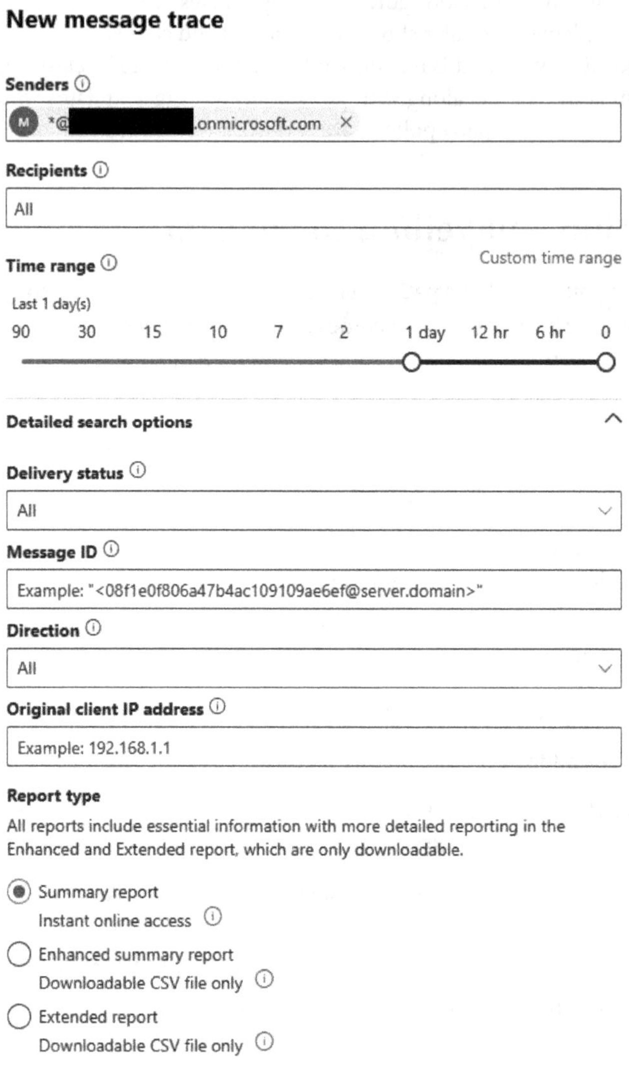

New message trace

Senders ⓘ

Ⓜ *@████████.onmicrosoft.com ✕

Recipients ⓘ

All

Time range ⓘ   Custom time range

Last 1 day(s)

90   30   15   10   7   2   1 day   12 hr   6 hr   0

Detailed search options   ∧

Delivery status ⓘ

All   ⌄

Message ID ⓘ

Example: "<08f1e0f806a47b4ac109109ae6ef@server.domain>"

Direction ⓘ

All   ⌄

Original client IP address ⓘ

Example: 192.168.1.1

Report type

All reports include essential information with more detailed reporting in the Enhanced and Extended report, which are only downloadable.

◉ Summary report
  Instant online access ⓘ

◯ Enhanced summary report
  Downloadable CSV file only ⓘ

◯ Extended report
  Downloadable CSV file only ⓘ

Figure 8.4 – New message trace page

> **Note**
>
> Remember that for the enhanced and extended reports, these reports will take some time to generate. When requesting either of these two reports, you will need to provide a name for the report and indicate who to email the report to once generated. You can then track the status of the report on the **Downloadable reports** tab of the **Message trace** page or wait for the email to arrive.

Rules of engagement when tracing a message

While an administrator can just start searching for emails, it doesn't mean they should. Any security team ready to use this tool should implement some rules of engagement to ensure no legal or company rules are violated. Common recommendations include the following:

- **Define the criteria**: Before initiating a search, clearly define what information you're looking for. Specifying sender/recipient addresses, time frames, message types, and keywords can significantly refine your results and save time.

- **Start with broad searches and refine progressively**: Begin with broader search parameters and gradually add filters to pinpoint specific messages. This approach helps avoid missing crucial information due to overly restrictive initial criteria.

- **Create a library of common searches**: If you frequently need to track similar types of messages, save your search criteria for easy future access. This saves time and ensures consistency in your search patterns. The **Autosaved queries** tab offers a list of the recently run queries in case you forgot to save them.

- **Respect privacy**: Exercise caution when searching for messages containing sensitive information. Only access message trace data when necessary and with proper authorization. If generating downloadable reports, ensure these are saved in an access-restricted location.

- **Document findings**: Record key details from the trace for future reference and reporting.

- **Maintain transparency**: Organizations should have clear policies outlining acceptable uses of message tracing and establish processes for authorization and oversight. This promotes transparency and accountability in user communication monitoring.

- **Respect compliance requirements**: Various regulations, such as the **General Data Protection Regulation** (**GDPR**) and the **California Consumer Privacy Act** (**CCPA**), govern the collection, use, and disclosure of personal data. Following established rules helps ensure compliance with these legal requirements.

Keep in mind that tracing messages offers a potent method for diagnosing issues and conducting investigations. However, it's essential to handle this tool with care and integrity. Following these guidelines, organizations can effectively utilize message tracing for problem-solving and inquiries, all while protecting individual privacy, complying with regulations, and fostering an environment of trust and ethical handling of information.

Diving deep into Safe Links policies – protecting your users from malicious URLs

Implementing Safe Links policies through Microsoft Defender for Office 365 plays a pivotal role in shielding your organization from phishing threats. These policies actively scan and authenticate hyperlinks in real time, deterring users from visiting harmful sites that could result in malware infections, data compromises, or phishing schemes. Moreover, Safe Links provide protection the moment a link is clicked, maintaining defense against threats, even if a website's risk level changes over time. Additionally, hyperlinks are modified before being sent to ensure continuous protection under the established Safe Links configurations, even when messages or links are forwarded. Organizations have the flexibility to tailor their security strategies, selecting specific URLs to either restrict or permit and adjusting protection intensity based on different user segments.

User warnings

Organizations can also use Safe Links to educate users by displaying warning pages or notifications when a potentially harmful link is clicked. When Safe Links identifies a dangerous link, the user may encounter different types of warnings, depending on the situation:

- **Malicious website or phishing attempt warning**: This is a clear message indicating that the link is known to be malicious and advising against proceeding

- **Unknown or suspicious link**: Sometimes, if a link cannot be definitively categorized as safe or malicious, users might be warned that the link is suspicious, prompting caution

Depending on the organization's configuration, these warnings will have the option for the user to continue to the website if they desire. While this option is not recommended, it is useful during the fine-tuning stages for misclassified URLs or to minimize user impact when the organization wants to provide some liberty for users to decide. If this is configured, it is recommended that other protections, such as endpoint detection and response and multifactor authentication, are employed to decrease the risk of a successful attack.

Safe Links policies

The security features provided by Safe Links can be fine-tuned by customizing policies. Depending on the previous configuration, you will find up to three policies already configured:

- **The built-in protection policy** covers the most common attacks out of the box for Defender for Office 365 customers

- **The Standard and Strict preset policies**, depending on what preset configuration has been deployed

Be aware that changes to Safe Links policies can only be performed on custom policies and include modifying the policies, enabling them, disabling them, deleting them, and changing their priority. Default policies cannot be changed; the built-in policy has the lowest priority (is applied last) and the preset policies have the highest priority (are applied first). This approach ensures that the organization always has a level of protection in place and an erroneously configured policy does not lower the organization's security when preset policies are in use (*Microsoft, 2023b*).

Policy fields

Before configuring these policies, it is important to understand what each section, field, and checkbox means to include the user impact. This will help us configure policies with a rationale that aligns with your organization's policies.

Section 1: Name your policy

This section is all about providing an easy way to identify the policy in the future. The **Name** field should have a descriptive name that aligns with the organization's naming standards; it should ideally include the impact, target audience, and version, for example, `SafeLinks_standard_testusers_20240130`. The **Description** field should briefly explain the purpose and scope of this policy. For example, `This policy applies the standard Safe links protection to the test users groups`.

Section 2: Users and domains

This section looks at the target of the policy. This is important not only during testing and fine-tuning efforts but also to allow for custom-tailored security. The policy can be targeted down to the individual user level via the **Users** field, to groups via the **Groups** field, and to entire domains via the **Domains** field, and you can exclude specific users, groups, and domains. You can even use these fields in combination to allow for very specific target audiences.

Section 3: URL & click protection settings

This section is where the main configuration for the policy occurs, and it is further divided into what product is protected, which includes email, Teams, and Office 365 apps. A final section, called **Click protection settings**, focuses on data gathered, the flexibility allowed to users, and how notifications are customized. We can see an example of this section in the following figure:

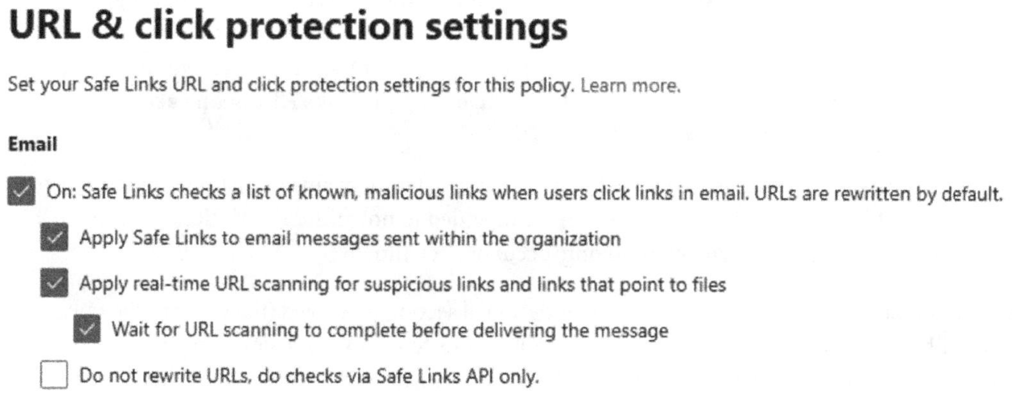

URL & click protection settings

Set your Safe Links URL and click protection settings for this policy. Learn more.

Email

☑ On: Safe Links checks a list of known, malicious links when users click links in email. URLs are rewritten by default.

☑ Apply Safe Links to email messages sent within the organization

☑ Apply real-time URL scanning for suspicious links and links that point to files

☑ Wait for URL scanning to complete before delivering the message

☐ Do not rewrite URLs, do checks via Safe Links API only.

Figure 8.5 – Upper part of Section 3 in the Safe Links policy configuration

We will look at all the parts of this section, starting with **Email**.

Email

These settings only apply to links in email messages and the level of protection offered:

- **On**: This checkbox turns on Safe Links for email to include URLs rewriting on inbound messages by default. When turning on Safe Links, any email with malicious links will be inspected during mail flow to the user's mailbox and not delivered if it is identified to be malicious. The rewriting and no delivery behaviors can be changed below, but it is recommended to leave these enabled to minimize user risk.

- **Apply Safe Links to email messages sent within the organization**: This extends Safe Links protection to intra-organization messages. By default, Safe Links only covers inbound emails originating from external locations. It is recommended to enable this setting to protect from internal attacks due to compromised users.

- **Apply real-time URL scanning for suspicious links and links that point to files**: This option enables scanning of the links on emails when the user clicks on the link. As it is possible that a link was not initially malicious at delivery time, but later changed by an attacker; this setting ensures links are scanned all the time.

- **Wait for URL scanning to complete before delivering the message**: An optional checkbox if real-time URL scanning is configured. It allows emails to be delivered right away to the user's inbox without waiting for a scan of the links during mail flow. This checkbox depends on real-time configuration as the intent is that the user will still have a level of protection, even if a malicious message is delivered, and is intended for messages that must be delivered urgently. It is recommended that this checkbox is not disabled, and if needed, should only be in rare circumstances.

- **Do not rewrite URLs, do checks via Safe Links API only**: Disables URL rewriting, but uses the Safe Links API for checks. It is recommended to leave URL rewriting on to ensure all links are always checked even if they are copied and pasted to other messages. Not rewriting the URL leaves the Safe Links protection to leverage the Safe Links API upon each click and is only supported on Outlook clients.

- **Do not rewrite the following URLs in email**: Creates an exclusion list for specific trusted URLs. It is important to note that it is recommended to not exclude any URLs or keep the list very small as it limits protection to only occur at click time.

For most organizations, it is recommended that the email section is enabled (the **On** checkbox ticked), along with a checkbox on the options to apply Safe Links to email messages, apply real-time URL scanning, and wait for URL scanning to complete before delivering the message, as this will provide the most protection.

Teams and Office 365 apps

These sections cover enabling URL scanning protection in other Office 365 applications such as Teams, and Office documents when using the supported desktop, mobile, and web apps. At the time of writing this, protection is limited to real-time scanning, and URLs are not rewritten:

- **Click protection settings**: This section customizes Safe Links behavior and allows for extra functionality.

- **Track user clicks**: This enables the tracking of user clicks on links scanned by Safe Links for reporting.

- **Let users click through to the original URL**: This option is available only if **Track user clicks** is enabled and impacts the behavior on warning pages when a malicious link is encountered. It allows users to continue to the link anyway, even if malicious. If it is recommended to leave this option disabled unless multiple links are being falsely identified as malicious.

- **Display the organization branding on notification and warning pages**: This option is available only if **Track user clicks** is enabled and allows for customization of the warning pages' appearance instead of using the default theme. If enabled, customization will follow the Microsoft 365 theme settings set at the organization level.

Users also need to be notified when Safe Links protection is triggered, which we see in the next section.

Section 4: Notification

This section allows using the out-of-the-box notification or customizing it to manage the user experience when a malicious link is encountered. If **Customize the notification text** is selected, the field will appear to enter a custom message, and a checkbox is provided for translation of this message via Microsoft Translator.

Configuring Safe Link policies

Configuration of these policies is a simple process that mimics how other threat policies are configured in Defender for Office 365:

1. Access the Microsoft Defender portal (`security.microsoft.com`).

2. On the left-hand menu, navigate to the **Email & Collaboration** section and click on **Policies & Rules**. On the **Policies & Rules** page, select **Threat policies**.

3. On the **Threat policies** page, navigate to the policies and click on **Safe Links**.

4. On the **Safe Links** page, click on **+ Create** on the top menu to start the custom policy creation.

5. Once the policy creation is completed, verify that it has the correct priority level.

> **Note**
>
> Always remember that preset policies are always applied first, and processing stops after the first policy, so if custom policies need to be used, the target user will need to be excluded from preset configuration settings, which will exclude them from other preset policies and require custom policies in other sections of Defender for Office 365.

Keeping malicious files out

Since the introduction of personal computers, criminals have been using malware as a popular tool to target systems. This malicious software is often delivered through email. To combat these threats, Microsoft offers two security solutions: EOP anti-malware and Defender for Office 365 Safe Attachments. These two solutions work together to provide multiple layers of protection.

EOP's anti-malware policies form the foundation of this protection. They offer a comprehensive defense against known malware on Windows, Linux, and Mac systems that enter or leave your organization. These policies utilize multiple anti-malware scan engines and advanced heuristic detection to safeguard against malware outbreaks, even in their early stages. Microsoft Teams also can promptly detect and address emerging threats through real-time threat response and definition deployment. In fact, they can release patches and rules within as little as two hours, thanks to their close partnerships with associates. EOP's anti-malware policies allow you to configure recipient and attachment filters for precise protection. You can implement ZAP when malware is detected after delivery and determine the notification settings for users. This level of customization ensures that your organization has control over its security measures.

Defender for Office 365 takes an extra step to enhance security by offering Safe Attachments. This feature adds a layer of protection to email attachments that have already undergone scanning by EOP's anti-malware protection. Safe Attachments employs a virtual environment to inspect attachments in email messages before they are sent to recipients. This process, known as **detonation**, provides an added layer of security. The protection provided by Safe Attachments is governed by Safe Attachments policies. Although there is no default policy, the built-in protection preset security policy ensures that all recipients receive Safe Attachments protection, unless they are defined in the Standard or Strict preset security policies, or in custom Safe Attachments policies. This flexibility allows organizations to tailor their security measures to meet their specific needs.

Anti-malware policies

As with other security policies in EOP and Defender for Office, the anti-malware policies available will depend on whatever preset security configuration has been implemented. Upon your first visit to the page for **Anti-malware** policies, you will find up to three policies already configured: the default (Default) policy, which covers the most common anti-malware attacks, and the Standard and Strict preset policies, which align with common Microsoft best practice recommendations. As with the anti-phishing default policy, you can make changes to the default policy (in contrast to the Standard And Strict preset policies), allowing for easy implementation of a catch-all policy as the default policy will have the lowest priority and will only come into effect if the user is not impacted by other policies. As before, priority cannot be changed on default and preset policies, with preset policies always having higher priority and the default policy the lowest, and processing of policies stopping only after one policy is effective on a user. Custom policies allow for the entire range of options, including enabling, disabling, deleting, and priority changing (*Microsoft, 2023e*).

Policy fields

As with other policies, let's look at the policy sections and configuration available before attempting to deploy our first custom policies. This will ensure we are making rational choices based on organizational requirements.

Section 1: Name your policy

This section focuses on easy identification and management of policies. A **Name** and **Description** field are provided, which should be completed following organizational vetted approaches that ensure commonly used descriptive names and descriptions apply to the policy. Always remember to include information on the policy type, impact, target audience, and version, as it will make for a simpler policy organization and easier setting conflict troubleshooting and resolution.

Section 2: Users and domains

As with other policies, the intent of this section is to define the policy target to either individual users, groups, domains, or a combination of all these, including exclusions in a similar manner.

Section 3: Protection settings

This is the main section of the policy where you define what the policy will do. We can see an example of this section in the following figure:

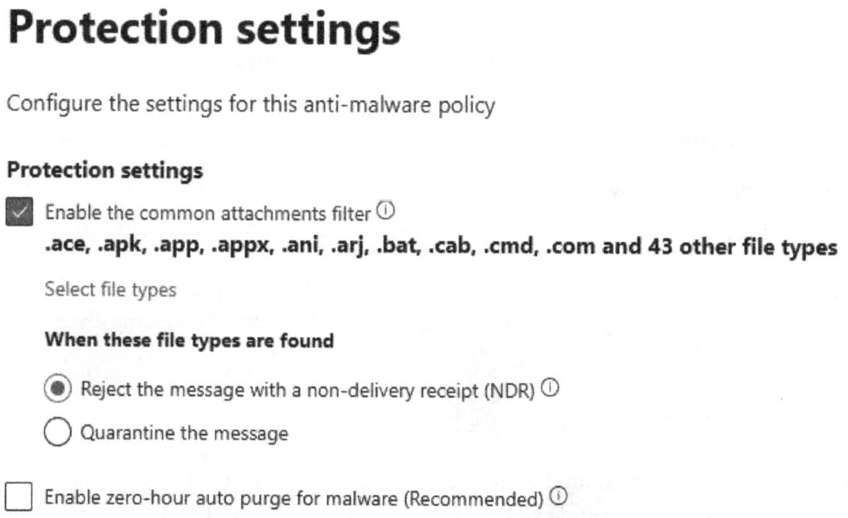

Protection settings

Configure the settings for this anti-malware policy

Protection settings

☑ Enable the common attachments filter ⓘ

    .ace, .apk, .app, .appx, .ani, .arj, .bat, .cab, .cmd, .com and 43 other file types

Select file types

When these file types are found

◉ Reject the message with a non-delivery receipt (NDR) ⓘ

○ Quarantine the message

☐ Enable zero-hour auto purge for malware (Recommended) ⓘ

Figure 8.6 – The Protection settings section of the anti-malware policy

The options included are as follows:

- **Enable the common attachments filter**: Use this option when you want to block emails containing certain file type attachments from a customizable list, even if the file is not clean. This is an approach used by many organizations to force users to use other more controlled forms of file sharing. When a message is encountered with this file, you can configure the policy to either reject the file or quarantine it for later inspection.

- **Enable zero-hour auto purge for malware**: Recommended to be enabled, it leverages Microsoft's network of information on malicious files to monitor messages with attachments already delivered, no matter whether the message has been read or not. An attachment that is identified as malicious prompts the automatic quarantine of any message carrying this attachment from all mailboxes in the organization targeted in the policy.

- **Quarantine policy**: Here, you select the quarantine policy to use for all emails quarantined under this anti-malware policy to include what users may see and do, and how users are notified of quarantined messages. You can go back to *Chapter 6* for more information.

The **Notification** area of this section has multiple options to change how administrators are notified of undelivered messages because of this policy, including which email address will receive notifications when either internal or external sender messages go undelivered. An optional **Customize notifications** checkbox enables fields for customizing the email and message used for notifications to administrators.

Configuring anti-malware policies

As with other threat policies, configuration is also as simple as visiting the page for **Anti-malware** page and following a simple wizard:

1. Access the Microsoft Defender portal (`security.microsoft.com`).

2. On the left-hand menu, navigate to the **Email & Collaboration** section and click on **Policies & Rules**. On the **Policies & Rules** page, select **Threat policies**.

3. On the **Threat policies** page, navigate to the **Policies** section and click on **Anti-malware**.

4. On the **Anti-malware** page, click on **+ Create** on the top menu to start the custom policy creation.

5. Once the policy creation is completed, verify that it has the correct priority level.

As previously mentioned, priority is important when creating multiple custom policies. Always remember that for all policies targeting a user, group, or domain, policy processing always stops after the first policy is applied.

Safe Attachment policies

This component provides the next level in malware protection by improving the protection provided to email via anti-malware policies and extending this protection to SharePoint, OneDrive, and Teams files. The initial policies present will depend on the previous configuration of preset security protection with up to three policies available, which include the built-in protection (Microsoft) policy, which sets foundational Safe Attachments coverage, and the Standard and Strict preset policies, which align with common Microsoft best practice recommendations. In contrast to the anti-malware policies, you can't change the settings or priority of the built-in protection, standard, or strict preset policies. As with other policies, custom policies allow for the entire range of options, including enabling, disabling, deleting, and priority changing (*Microsoft, 2023a*).

Policy fields

Before proceeding with implementing these policies, let's review the policy sections. For Safe Attachments policies, the configuration defined on these will apply to the email attachments of the target users by default and will be extended to Teams, OneDrive, and SharePoint in the **Global settings** section. As always, review these fields to understand what options you can implement with these policies and align these changes to organizational requirements.

Section 1: Name your policy, and Section 2: Users and domains

As with the anti-malware policies, Section 1 focuses on the name and description of the policy, to support making management and identification easier. Section 2 focuses on defining who the policy will target, which can be a combination of the inclusion and exclusion of individual users, groups, and domains.

Section 3: Settings

In this section, you can define the policy for handling attachments. While the focus of the anti-malware policy is on taking action, Safe Attachments behaves in a more proactive manner. It carefully examines all attachments suspiciously and uses a sandbox to detonate them, identifying any malicious behavior before delivering them.

We can see this section in the following figure:

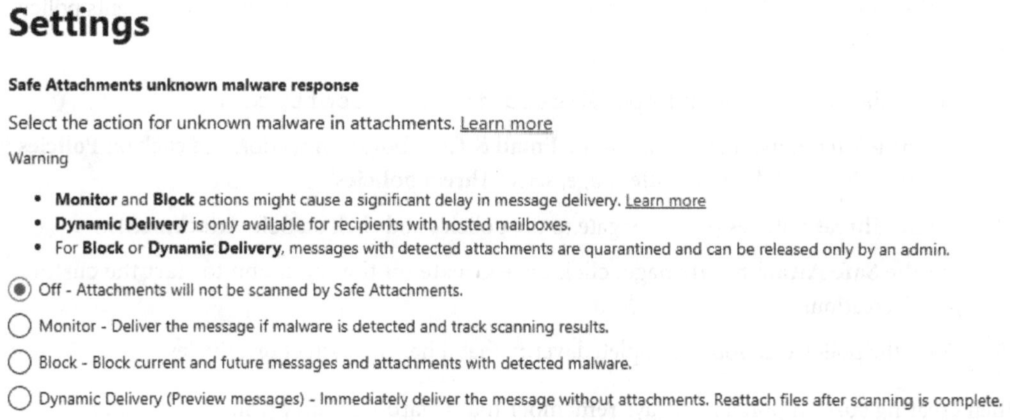

Settings

Safe Attachments unknown malware response

Select the action for unknown malware in attachments. Learn more

Warning

- **Monitor** and **Block** actions might cause a significant delay in message delivery. Learn more
- **Dynamic Delivery** is only available for recipients with hosted mailboxes.
- For **Block** or **Dynamic Delivery**, messages with detected attachments are quarantined and can be released only by an admin.

◉ Off - Attachments will not be scanned by Safe Attachments.

◯ Monitor - Deliver the message if malware is detected and track scanning results.

◯ Block - Block current and future messages and attachments with detected malware.

◯ Dynamic Delivery (Preview messages) - Immediately deliver the message without attachments. Reattach files after scanning is complete.

Figure 8.7 – The Settings section of the Safe Attachments policy

The first part of this section allows you to choose how to handle attachments with the following options:

- **Off** (not recommended): This option disables the scanning of attachments by Safe Attachments. However, they will still be subject to any anti-malware policy configurations.

- **Monitor**: Attachments are scanned, but even if they are detected as malicious, they will still be delivered. Additionally, the scan findings will be tracked. This option is typically used when establishing a separate monitoring mailbox for malware analysis and should only be configured in organizations with highly mature security procedures and an experienced security team, as mishandling could cause a major impact on the environment.

- **Block**: As the name suggests, any attachments identified as malicious will be blocked, and messages containing these attachments will not be delivered.

- **Dynamic Delivery**: This option is useful in scenarios where message delivery should not be delayed. By temporarily removing access to the attachment and adding a note that indicates that the attachment is being scanned, the message can be delivered with no delays. Scanning is performed in the background, providing the same level of protection as blocking malicious messages, but without the delays caused by waiting for the scans to finish before delivery.

For most scenarios, it is recommended to either use the **Block** or **Dynamic Delivery** options, with **Dynamic Delivery** being the option that causes the least disruption to users. The **Quarantine policy** section allows the selection of what quarantine policy to use when detecting malicious attachments in the **Block** or **Dynamic Delivery** configurations. Finally, the **Enable redirect** checkbox allows for specifying an email address to be used when the **Monitor** option is selected. As previously stated in the **Monitor** option, extreme caution should be used when enabling this option because of the increased risk of unintentionally spreading malware.

Configuring Safe Attachments policies

As with other threat policies, configuration is also as simple as visiting the Safe Attachments policy page and following a simple wizard:

1. Access the Microsoft Defender portal (`security.microsoft.com`).

2. On the left-hand menu, navigate to the **Email & Collaboration** section and click on **Policies & Rules**. On the **Policies & Rules** page, select **Threat policies**.

3. On the **Threat policies** page, navigate to the policies and click on **Safe Attachments**.

4. On the **Safe Attachments** page, click on **+ Create** on the top menu to start the custom policy creation.

5. Once the policy creation is completed, verify that it has the correct priority level.

When creating custom policies, always remember that regarding policy priorities, for all policies targeting a user, group, or domain, policy processing always stops after the first policy is applied, with the preset policies always having higher priority.

Extending beyond email

As mentioned previously, Safe Attachments can also protect from malicious files in Teams, SharePoint, and OneDrive. Configuration is not enabled at the policy level and, instead, is turned on via the **Global settings** menu option on the **Safe Attachments** page. When opening this page, the options available will be as follows:

- **Turn on Defender for Office 365 for SharePoint, OneDrive, and Microsoft Teams**: This option enables Safe Attachments for these applications when using them from the web browser

- **Turn on Safe Documents for Office clients**: This option extends the Safe Attachments protection to Office clients beyond the protection that **Protected View** provides for users with the proper Microsoft 365 security license

- **Allow people to click through Protected View even if Safe Documents identified the file as malicious**: Not recommended to be enabled, this option allows users to proceed with opening malicious files

For **Global settings**, it is key to remember that these options only extend the capability of the Safe Attachments policies to applications other than email. The protection's behavior will still be defined at the policy level, which will target a specific user.

Policy priorities

In organizations using Microsoft 365 with Exchange Online mailboxes, or those under standalone EOP without such mailboxes, incoming emails may encounter several protective measures. These include anti-spoofing safeguards, which are accessible to all Microsoft 365 users, and impersonation defense, exclusive to Microsoft Defender for Office 365 subscribers. Furthermore, each message undergoes a series of detection processes aimed at identifying malware, spam, phishing, and more. Amid these processes, it's easy to get puzzled over which policy takes precedence. Although it's feasible to prioritize policies within the same category, it's crucial to note that EOP and Defender for Office 365 adhere to a fixed sequence of protection processing. This sequence, which cannot be altered, establishes the order in which different types of defenses are applied. It's important to understand that when examining this order, protections labeled with high confidence, such as high-confidence phishing, belong under the broader umbrella of phishing defense. This indicates that the combined insights from Defender for Office 365, EOP deployments, and Microsoft's collected threat intelligence are sufficient to classify a message as likely phishing without needing further verification by other rules. A similar principle applies to anti-spam measures. The protection processing order is the following:

1. Anti-malware policies
2. High-confidence phishing
3. Phishing
4. High-confidence spam

5.   Spoofing

6.   User impersonation (protected users)

7.   Domain impersonation (protected domains)

8.   Mailbox intelligence

9.   Spam and bulk (*Microsoft, 2023d*).

When setting up policies, it's crucial to follow these key practices to make sure they provide the desired level of protection for your recipients:

- Allocate higher priority policies to fewer users, while reserving lower priority policies for a broader audience. Keep in mind that default policies are always enforced at the end.

- Ensure that your higher priority policies are configured with tighter or more specific settings compared to those of lower priority. While you can fully customize settings in your own created policies and adjust default policies, remember you lack control over the granular settings of preset security policies.

- It's advisable to minimize the use of custom policies; reserve them for users needing more particular settings than what's offered by the Standard or Strict preset security policies or the defaults.

The proper priority of policies can improve not only security operations by lowering noise but also the user experience by speeding delivery due to more effective controls. In the next section, we discuss alerts, which will also impact user experience and can help us detect issues early on.

Alert policies

Within Microsoft 365 environments with Exchange Online mailboxes, alert policies play a pivotal role by triggering notifications on the alert dashboard based on user actions that meet the policy's predefined rules and conditions. Each alert policy encapsulates a series of rules and criteria outlining the specific actions by users or administrators that lead to an alert, identifies who among the users activates the alert upon engaging in the specified activity, and sets a threshold for the frequency of the activity required to initiate an alert. Actions such as granting administrative rights in Exchange Online, encountering malware threats, phishing maneuvers, or unusual patterns of file deletion and external sharing can all prompt alerts. These policies empower you to organize the generated alerts, implement the policy across all organizational users, establish a threshold for alert activation, and choose whether to activate email notifications upon alerts. With several default alert policies available out of the box to aid in monitoring activities, Microsoft 365 also offers an **Alerts** page. This page allows you to view and sort alerts, assign an alert status for easier management, and dismiss alerts post investigating or resolving the event that triggered them (*Microsoft, 2024*).

Alert policy fields

Like any other policies, we should examine the details of the alert policy and identify potential modifications to ensure that these policies are in line with our organizational objectives and security strategy.

Section 1: Name your alert

In this section, we can configure the **Name** and **Description** fields that play a crucial role in policy organization. Similar to other policies, using these fields effectively ensures that your policies are systematically categorized, comply with organizational naming conventions, and offer a clear and succinct method for delineating the policy's nature, its impact, the intended recipients, and the policy version. The **Severity** field helps to determine the urgency of the alert, offering choices of **Low**, **Medium**, and **High**. It's critical for the **Severity** setting to sync with your organization's service-level objectives, reflecting the prioritization process akin to the classification of security incidents. This includes considerations such as the timeline for initiating an investigation, notification protocols, and the resolution time frame to prevent adverse effects on the operational environment. Adopting a logical and measurable approach to setting severity levels is key to mitigating alert fatigue among your team. We can see these fields in the following figure:

Name your alert, categorize it, and choose a severity.

Assign a category and severity level to help you manage the policy and any alerts it triggers. You'll be able to filter on these settings from both the 'Alert policies' and 'View alerts' pages.

Name *

Example Alert policy

Description

Enter a friendly description for your policy

Severity * ⓘ

Medium ⌄

Category *

Permissions ⌄

Figure 8.8 – Initial section of the alert policy

Furthermore, the **Category** field enhances alert identification, specifying the most equipped team to manage the alert based on options such as **Threat management**, **Information governance**, **Permissions**, **Others**, and **Mail flow** (which is typically managed within the Exchange admin center). As with setting the severity level, aligning category processes with the team possessing the highest competency in the relevant area ensures efficient and effective alert management. Also, be aware that different alert categories will require different roles to view and act on the alert.

Section 2: Create alert settings

In this section, you can establish the criteria that trigger an alert. The process begins with defining a filter to sift through the vast amount of logs Microsoft 365 processes. Setting up this filter is straightforward: start by choosing the specific activity you wish to monitor. Options include malware detection, phishing attempts, file or folder activities, external file sharing, and many others. Following this, you must select a condition to refine your filter further. This could be specifying a particular object, user, or location, such as a unique IP address, specific user, or filename, to name a few. Interestingly, conditions can also be defined inversely, such as excluding certain users, IP addresses, filenames, and so on. While adding a condition is optional, it's highly beneficial for reducing unnecessary alerts and concentrating your team's efforts on particular areas of interest. You have the flexibility to define multiple conditions as you progressively refine your alert system. We can see an example of this in the following figure:

Figure 8.9 – Alert policy trigger configuration

Subsequently, the decision on when to trigger the alert is made. This step is another strategy to curb alert fatigue among your team members, considering that some alerts might produce a lot of noise despite adjustments, and actual issues may only become evident after repeated occurrences. For trigger settings, you may opt for the alert to be activated with every instance of the event, upon reaching a certain threshold, or when the event frequency surpasses a baseline established for your environment. Should you choose the baseline option, keep in mind that you'll need to allow a week for the system to establish a baseline specific to your operational environment before any alerts are triggered.

Section 3: Set your recipients

In this section, the focus shifts toward specifying who gets notified via email upon the triggering of an alert, in addition to seeing the alert within the Defender portal. To action this, it's essential to activate the **Opt-In for email notifications** option and designate the precise users or groups slated for email notifications. With this feature enabled, there's the added ability to set a cap on daily notifications, which can vary from as few as 1 to as many as 200 notifications, or even opting for no limit at all.

However, it's crucial to underscore that while setting a daily notification limit is a useful tool in managing the volume of notifications, the primary strategy for reducing the occurrence of *noisy* alerts should ideally focus on adjusting the conditions under which an alert is activated. Tailoring the event frequency necessary for an alert to kick in is a more effective approach to streamlining alert management.

Configuring alert policies

Alert policies, like other policies in Defender for Office 365, follow a similar approach to modification and implementation:

1. Access the Microsoft Defender portal (`security.microsoft.com`).

2. On the left-hand menu, navigate to the **Email & Collaboration** section and click on **Policies & Rules**. On the **Policies & Rules** page, select **Alert policy**.

3. On the **Alert policy** page, click on **+ New Alert Policy** on the top menu to start the custom policy creation.

4. Once the alert policy is implemented, you can either turn it on or off by toggling the switch on the **Status** column of the **Alert policy** page. You can also view any triggered alerts by clicking on **Manage Activity Alerts** on the top menu of this page.

After finalizing the policy setup, it's important to note that access to view or interact with the policy is not universal. The visibility and actionability of a specific alert policy are largely determined by its category. This delineation ensures that individuals with the appropriate roles are the ones managing and responding to the alerts. For instance, any alert tied to data will necessitate roles associated with data management, such as those found in Purview, to effectively view and handle the alerts. Conversely, policies related to permissions and threat management are best suited for individuals with security-oriented roles, such as *Security Administrator* or *Security Reader*. This segmentation by role underscores the need for assigning the right responsibilities to the right team members for efficient and accurate policy enforcement.

Summary

In this chapter, we journeyed through the complex world of maintaining both secure and efficient email flow inside the Exchange ecosystem. We kicked things off by taking a close look at how Exchange handles email flow and quickly dove into the world of message tracing. This invaluable tool helps administrators untangle the web of email delivery, ensuring that messages not only reach their destination but do so reliably within an organization's communication network. We then explored the array of security defenses set up to protect email inboxes from the likes of malware and phishing – diving into how anti-malware policies, Safe Links policies, and the cutting-edge ZAP function work together to create a formidable barrier against cyber threats. This comprehensive set of security strategies, alongside our discussion on how protection policies are prioritized and the importance of alert policies, forms a solid frontline defense, ensuring smooth and secure email transmission across the board.

As we concluded this exploration, the critical nature of these features working in tandem became clear – they're the key to keeping an organization's email communications safe and sound. We're setting our sights next on expanding our knowledge of the reporting and data visualization capabilities within Exchange Online Protection and Defender for Office 365. The next chapter will arm us with valuable insights, allowing us to leverage data analytics to sharpen our security practices and keep our communication channels robust in the face of a digital landscape that's always changing. So, stay with us as we continue to dive into these advanced tools and strategies, all meant to bolster our defenses against the dynamic threats of the modern digital era.

References

- Microsoft. (2023a, June 19). *Safe Attachments*. Microsoft Learn. `https://learn.microsoft.com/en-us/microsoft-365/security/office-365-security/safe-attachments-about?view=o365-worldwide`

- Microsoft. (2023b, September 19). *Complete Safe Links overview for Microsoft Defender for Office 365*. Microsoft Learn. `https://learn.microsoft.com/en-us/microsoft-365/security/office-365-security/safe-links-about?view=o365-worldwide`

- Microsoft. (2023c, September 21). *Message trace in the new EAC in Exchange Online*. Microsoft Learn. `https://learn.microsoft.com/en-us/exchange/monitoring/trace-an-email-message/message-trace-modern-eac`

- Microsoft. (2023d, October 20). *Order and precedence of email protection*. Microsoft Learn. `https://learn.microsoft.com/en-us/microsoft-365/security/office-365-security/how-policies-and-protections-are-combined?view=o365-worldwide`

- Microsoft. (2023e, October 24). *Anti-malware protection*. Microsoft Learn. `https://learn.microsoft.com/en-us/microsoft-365/security/office-365-security/anti-malware-protection-about?view=o365-worldwide`

- Microsoft. (2023f, October 24). *Mail flow in EOP*. Microsoft Learn. `https://learn.microsoft.com/en-us/microsoft-365/security/office-365-security/mail-flow-about?view=o365-worldwide`

- Microsoft. (2024, February 9). *Microsoft 365 alert policies*. Microsoft Learn. `https://learn.microsoft.com/en-us/purview/alert-policies`

9

Incidents and Security Operations

In this chapter, we'll explore the incidents that might happen even with strong safeguards in place and the valuable insights they can reveal. We will delve into the important elements of automated remediation, examining these procedures and understanding the immediate responses. Our focus in this chapter will be establishing processes to investigate and manage all incidents generated by Defender for Office 365, leveraging automation to ease our workload, and fine-tuning our tools by pinpointing and correcting false detections.

Our processes will also be enriched by learning to break down incidents to understand the root cause, as well as how to enlist Microsoft to help with more advanced changes and improvement requirements. By the end of this chapter, you'll be well-equipped to maneuver through the intricacies of incident management and ready to polish your automated defenses, ensuring they serve as a more potent and dependable shield for your digital safety.

This chapter will cover the following topics:

- Incidents, the information they provide, and how to manage them

- Leveraging automated investigation and remediations

- Identifying and correcting false detections

- Some recommended security operations approaches

Let's continue our journey!

A holistic view of Microsoft Defender XDR

Addressing security incidents requires full visibility of the entire digital environment and relevant data streams to enforce robust protection. This comprehensive observation is pivotal in discerning whether an alert signifies a genuine threat or is merely a false positive, stemming from insufficient tuning. It illuminates multifaceted, targeted threats that, when viewed through the lens of a solitary tool, may be mistakenly de-prioritized. Microsoft Defender XDR excels in this arena, interlacing proactive and reactive security measures, before and after a cybersecurity incident. It adeptly coordinates a spectrum of security functionalities, detection, prevention, investigation, and response, across key vectors – endpoints, user identities, email systems, and applications. With this integrative defense strategy, it delivers a fortified wall against sophisticated cyber-attacks.

Leveraging data from Microsoft's extensive security ecosystem, encompassing Defender solutions for Endpoint, Office 365, Identity, Cloud Apps, and so on, Microsoft Defender XDR provides security personnel with the tools to construct an exhaustive view of potential cyber threats. In essence, it combines different **indicators of compromise (IoCs)** from separate sources, outlining a threat's life cycle from incursion to its lasting impact. It then initiates prompt, automated countermeasures to thwart the assault and rehabilitate the integrity of compromised sectors. Picture Microsoft Defender XDR as the quintessential electronic guardian for your corporation, a cohesive shield securing every aspect of your organization's digital sphere (*Microsoft, 2024b*). Let's examine the layers of security that Microsoft Defender XDR extends across various domains:

- **Vigilance in the cloud sector**: Defender for Cloud Apps operates akin to a keen sentinel, conducting relentless surveillance over shadow IT, reinforcing data governance, and forming an impenetrable line of defense against cyber threats. We can see an example of this in the following figure, showing the cloud discovery page:

Figure 9.1 – Defender for Cloud Apps' Cloud Discovery page

- **Communication defense mechanisms**: Defender for Office 365 serves as a sentry, scanning emails and digital collaborative environments to preempt and neutralize potential cyber threats, preserving the security of your organization's communications. We can observe this in the following figure, which shows the Defender for Office 365 **Explorer** page:

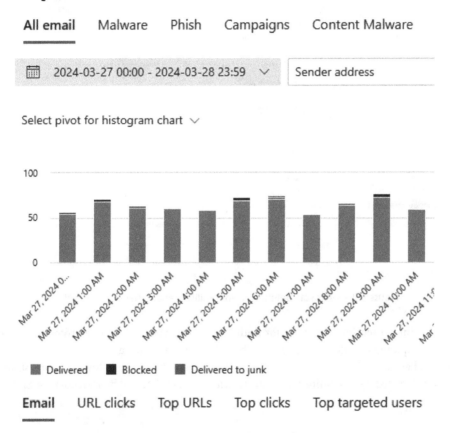

Figure 9.2 – The threat explorer page in Defender for Office 365

- **Proactive vulnerability management**: With Defender Vulnerability Management, your organization benefits from a modern vulnerability management approach that uncovers and promptly addresses any weaknesses within your digital infrastructure. We can see this in the following figure, which shows the Microsoft Defender Vulnerability Management dashboard:

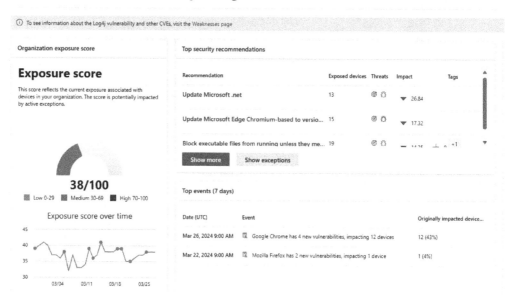

Figure 9.3 – Microsoft Defender Vulnerability Management dashboard

- **Endpoint protection**: Defender for Endpoint functions as a proactive protector, proficient in anticipating and deflecting security breaches with a robust arsenal for threat detection, immediate analysis, and agile intervention. We can see this in the following figure, showing a device isolated by an administrator due to critical issues detected by Defender for Endpoint:

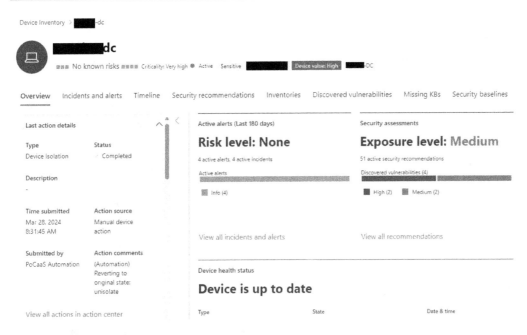

Figure 9.4 – An isolated device due to security concerns

- **The intersection of Identity and Access Management (IAM)**: The synergy between Defender for Identity and Microsoft Entra ID Protection erects a formidable barrier, shielding both cloud and on-premises environments from sophisticated threats, account breaches, and insider threats. We can see this in the following figure, where multiple identity-related attacks were detected by a mix of Defender for Identity and Entra ID Protection:

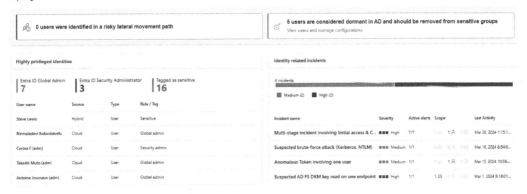

Figure 9.5 – An identity security dashboard

Beneath the overarching umbrella of Microsoft Defender XDR, these essential security components interlock smoothly, creating a vigilant and adaptable cyber defense network to ensure the digital integrity of your organization's complex systems. Beyond its individual preventative measures, Microsoft Defender XDR amplifies the efficacy of each security facet through the following actions:

- **Enhanced threat intelligence**: It equips cyber security specialists with advanced tools to dig deep into endpoint and office data, enabling tailored searches for traces of cyber breaches. We can see this on the **Threat analytics** page in the following figure:

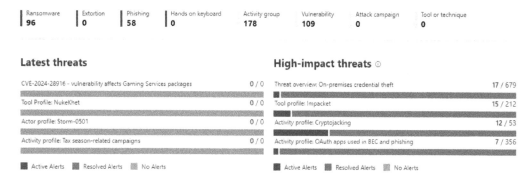

Figure 9.6 – The threat analytics page

- **A narrative of network intrusions**: By synthesizing data from diverse alerts and abnormal activities, it furnishes security teams with a comprehensive narrative of cyber intrusion activities. We can see this in the **Attack story** section of an incident shown in the following figure:

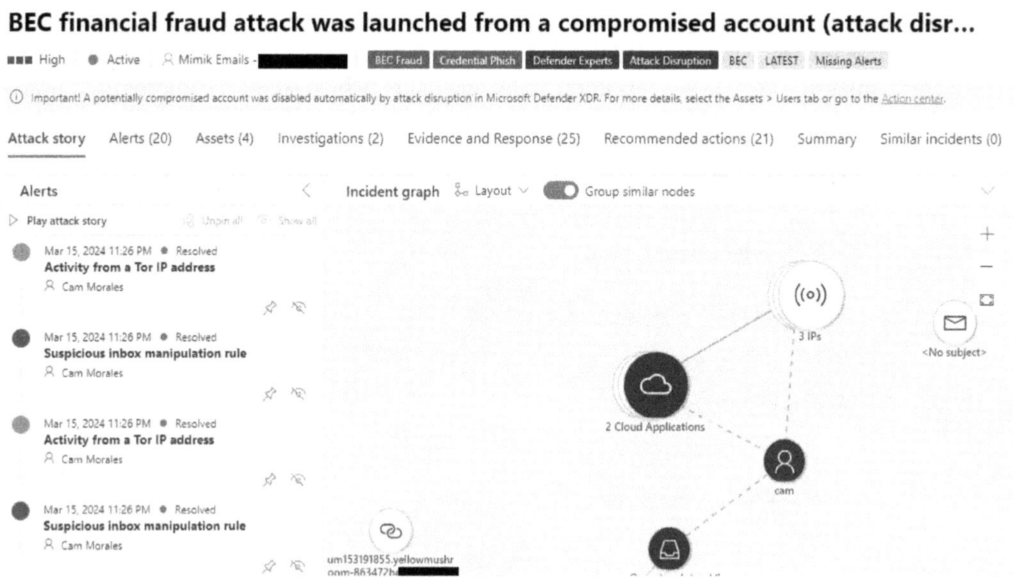

BEC financial fraud attack was launched from a compromised account (attack disr...

■■■ High   ● Active   ◯ Mimik Emails - ███████   `BEC Fraud` `Credential Phish` `Defender Experts` `Attack Disruption` BEC LATEST Missing Alerts

ⓘ Important! A potentially compromised account was disabled automatically by attack disruption in Microsoft Defender XDR. For more details, select the Assets > Users tab or go to the Action center.

Attack story   Alerts (20)   Assets (4)   Investigations (2)   Evidence and Response (25)   Recommended actions (21)   Summary   Similar incidents (0)

Alerts   <   Incident graph ⅗ Layout ⌄   ◉ Group similar nodes   ⌄

▷ Play attack story   ⤢ Unpin all   ⌖ Show all

● Mar 15, 2024 11:26 PM ● Resolved
  Activity from a Tor IP address
  ◯ Cam Morales

● Mar 15, 2024 11:26 PM ● Resolved
  Suspicious inbox manipulation rule
  ◯ Cam Morales

● Mar 15, 2024 11:26 PM ● Resolved
  Activity from a Tor IP address
  ◯ Cam Morales

● Mar 15, 2024 11:26 PM ● Resolved
  Suspicious inbox manipulation rule
  ◯ Cam Morales

um153191855.yellowmushr
oom-863472h...███████

Figure 9.7 – The attack story of an incident

- **A coordinated security strategy**: Microsoft Defender XDR's integrated approach to defense is not limited to the attack narrative; it also uses data gathered from all service vectors, configured with Defender security tools to coordinate the best defense across an entire security event. This approach is key when protecting against the highly complex attacks that are observed more frequently. We can see this by looking at the **Alerts** tab of the incident in the previous example in the following figure, which shows the multiple sources used to enrich the incident:

Alert name ⌄	Tags ⌄	Severity ⌄	Investigation state ⌄	Status ⌄	Category ⌄	Detection source ⌄
☐ Suspicious inbox manipul...		■■■ High		● Resolved	Defense evasion	Microsoft Defender for Cloud Apps
☐ Suspicious inbox manipul...		■■■ High		● Resolved	Defense evasion	Microsoft Defender for Cloud Apps
☐ BEC financial fraud		■■■ High		● Resolved	Collection	Defender XDR
☐ Suspicious inbox manipul...		■■■ High		● Resolved	Defense evasion	Defender XDR
☐ BEC financial fraud	Attack Disruption	■■■ High		● Resolved	Defense evasion	Defender XDR
☐ Anonymous IP address		■■◻ Medium		● Resolved	Initial access	AAD Identity Protection
☐ Anonymous IP address		■■◻ Medium		● Resolved	Initial access	AAD Identity Protection
☐ Anonymous IP address		■■◻ Medium		● Resolved	Initial access	AAD Identity Protection
☐ Anonymous IP address		■■◻ Medium		● Resolved	Initial access	AAD Identity Protection

Figure 9.8 – The alerts section of an incident, showing the multiple sources used

- **Restoration and resilience post-breach**: Microsoft Defender XDR proactively launches recovery procedures for any compromised entities, using automated systems to rapidly reestablish security integrity. We can see this in the following figure, which shows a compromised user account disabled automatically:

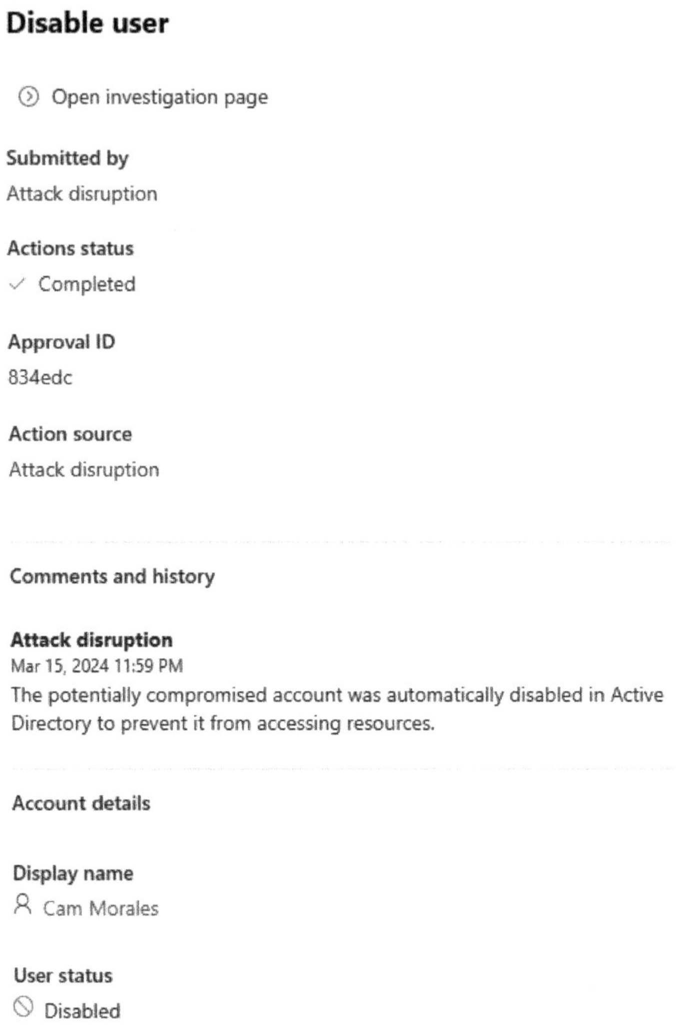

Disable user

⊙ Open investigation page

Submitted by
Attack disruption

Actions status
✓ Completed

Approval ID
834edc

Action source
Attack disruption

Comments and history

Attack disruption
Mar 15, 2024 11:59 PM
The potentially compromised account was automatically disabled in Active Directory to prevent it from accessing resources.

Account details

Display name
   Ⓐ Cam Morales

User status
⊘ Disabled

Figure 9.9 – A compromised user account disabled

Microsoft Defender XDR is not just a conglomerate of security tools; it is also a strategic ensemble of protective measures, each component orchestrated to enhance the collective strength, resulting in a vigilant, robust, and resilient security orchestra for your enterprise's extensive digital environment.

Incidents and more incidents

A Microsoft Defender XDR incident joins all related alerts and data, presenting a comprehensive narrative of a cyber assault. The **Incidents** page within Microsoft Defender XDR, also referred to as the **Incidents** queue, serves as a centralized hub, integrating Defender for Office 365 alerts, instrumental **Automated Investigation and Response (AIR)** initiatives, and the results of such in-depth investigations. These alerts are triggered by malicious or questionable activities bearing upon various entities, including emails, user accounts, and mailboxes, and help to provide intelligence on attacks in progress or those that have finished. It is commonplace for a multi-vector attack to generate an assortment of alerts across different points of origin. We can see this in the encapsulation of an aggregation of correlated alerts, in which many alerts that are triggered help create the story of how a complex attack occurred:

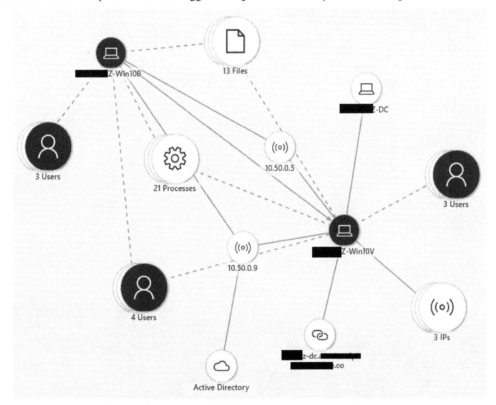

Figure 9.10 – The attack story of an incident

Certain predefined alerts automatically set off AIR playbooks, a set of preconfigured actions that meticulously carry out investigative procedures designed to unearth additional affected entities or suspicious conduct. Integrating Defender for Office 365 alerts, investigation results, and corresponding data points is an automatic process. An incident starts the moment a valid relationship between these elements is identified, serving as a beacon for security teams to perceive the full scope of an onslaught.

Within the **Incidents** queue of Defender for Office 365, incident management is streamlined, allowing functionalities such as filtration, categorization, tagging, and setting priorities. Incidents from the queue can be dealt with directly or allocated to team members. Incorporating comments and an archive of comment histories serves as a chronicle of progression and critical decisions.

Should an attack traverse additional areas protected by Microsoft Defender, any additional related alerts, investigative findings, and their data are combined into a common incident. The complexity of developing correlation logic is neatly sidestepped, as the system inherently provides this framework. Should there be nuances in needs, alerts can be manually appended to existing incidents or used to spawn new ones. We can observe the appending of an alert to an incident in the following figure:

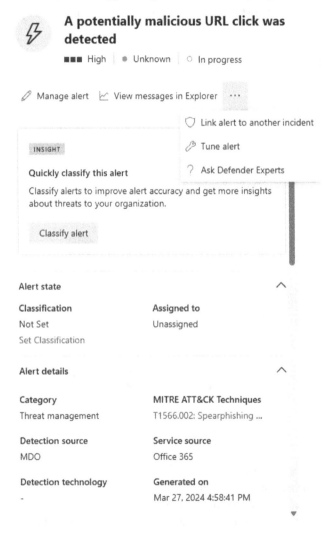

Figure 9.11 – Appending an alert to an incident

In the realm of Defender for Office 365, alerts, AIR investigation conclusions, and recommended courses of action are seamlessly integrated into incidents. In events where the AIR scrutiny discounts any threat, linked alerts are categorically resolved. An incident's status graduates to **Resolved** upon the resolution of all associated alerts. The **Evidence and Response** tab dutifully collates related evidence and enumerates response actions, which security personnel can execute directly from the incident narrative, such as the soft deletion of compromised emails or the dismantlement of intrusive inbox policies. We can see the **Evidence and Response** tab in the following figure with some of the data, such as IP addresses and user accounts (blacked out):

	First seen ↑ ⌄	Entity ⌄	Entity type ⌄	Verdict ⌄
☐	Mar 25, 2024 9:53 PM		Cloud logon session	Suspicious
☐	Mar 25, 2024 9:53 PM	1███████7	IP Address	Suspicious
☐	Mar 25, 2024 9:58 PM	it@c█████.c...		Suspicious
☐	Mar 25, 2024 10:00 PM	sonia@a███████...	Mailbox configuration	Suspicious
☐	Mar 25, 2024 10:00 PM	1█████5	IP Address	Suspicious
☐	Mar 25, 2024 10:00 PM	sonia@█████...		Suspicious
☐	Mar 25, 2024 10:00 PM	automation	Blob container	Suspicious
☐	Mar 25, 2024 10:00 PM	8█████8	IP Address	Suspicious

Figure 9.12 – The Evidence and Response tab

Recommended actions are systematically devised solely for emails deemed malicious and residing within cloud-based mailboxes. Action items pending execution are constantly refined to reflect the most up-to-date delivery locales. When emails have been preemptively addressed, this is aptly reflected in their status. Suggested countermeasures prioritize the direst of threats, encompassing malware, unequivocal phishing attempts, malicious URLs, and nefarious files.

Knowing how to prioritize

Microsoft Defender XDR leverages complex correlation analytics to synthesize alerts and outcomes from automated investigations across its diverse security suite into a single incident narrative. Malicious activity that would be difficult to see is easily to identitied thanks to alerts that provide a holistic view by leveraging signals from all the Microsoft Defender product suite products deployed in the environment. By delivering a panoramic view of the cyber battlefield, it equips security analysts with the context required to decode the narrative of attacks, enhancing the precision of the response to intricate threats permeating your organizational fabric.

The **Incidents** queue serves as a repository for synthesized incidents, amassing those spawned from interactions with various devices, user identities, and mailboxes. This queue is an invaluable tool in the incident triage process, empowering your cyber defense teams to sift through and duly prioritize incidents, which are key steps in sculpting a knowledge-driven cyber response tactic. Navigate to this repository via **Incidents & alerts | Incidents**, which can be found on the quick launch toolbar of the Microsoft Defender Security Center. We can see this in the following figure, with a queue filtered by high severity and Defender for Office 365 incidents:

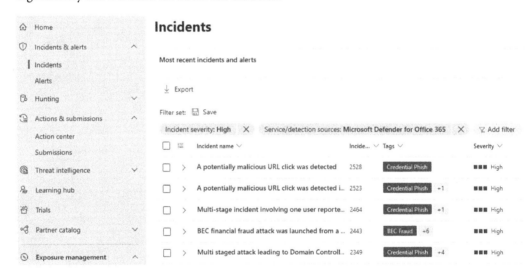

Figure 9.13 – The incidents page

Within the **Most recent incidents and alerts** dashboard, graphical representations illustrate the volume of alerts received, alongside the number of incidents generated within the last 24 hours. By design, the **Incidents** queue curated in the Microsoft Defender portal presents a six-month tableau of incidents, showcasing the freshest incident atop the roster, thus ensuring the immediacy of visibility (*Microsoft, 2023b*).

Customization of the **Incidents** queue is readily available (with a **Choose columns** feature) to discern the intricate aspects of each incident or affected entity. Such filtering granularity makes prioritization a simpler process. To augment visibility with minimal effort, incident names are automatically created that are derived from key attributes and observations in the related alerts, such as the tally of compromised endpoints, the impacted users, the sources of detection, or the nature of the threats, thereby enabling a quick understanding of an incident's breadth. For example, an incident could be named "a potential malicious URL was clicked by a group of users" to indicate an incident with alerts from multiple users who clicked on the same malicious URLs.

Filters

The **Incidents** queue within Microsoft Defender XDR presents an array of filtering tools to streamline your focus across the cybersecurity landscape of your organization. You can opt for an exhaustive review or pinpoint specific scenarios or adversarial patterns. Proper utilization of these filters can rapidly escalate critical incidents to the forefront of your attention. Filtering options include the following:

- **Alert policies**: Reference incidents by alert policy titles

- **Associated threat**: Identify incidents tied to known threats

- **Categories**: Refine your focus on the specific nature of the attacks

- **Classification**: Narrow down the incident view to specific alert classifications

- **Data sensitivity**: Urgently highlight incidents threatening to compromise highly sensitive information, using sensitivity labels from Microsoft Purview Information Protection

- **Device groups**: Focus on incidents associated with named groups of devices

- **Entities**: Focuses on the affected assets such as users, devices, or applications

- **Incident assignment**: Find incidents allocated to specific team members

- **Multiple categories**: For cross-cutting incidents that involve various tactics or techniques

- **Multiple service sources**: If incidents span across multiple service sources, specify so here

- **OS platform**: Differentiate incidents based on the operating systems affected

- **Severity**: Gauge the potential impact on your assets with the **High**, **Medium**, **Low**, or **Informational** severity options

- **Service Sources**: Drill down into incidents stemming from services such as app governance, the suite of Microsoft Defender solutions, and cloud apps

- **Status**: Choose from **New**, **In Progress**, or **Resolved** to track incidents through their life cycle

- **Tags**: Apply specific tag filters to streamline incident categorization

We can see some of these filters in the following figure:

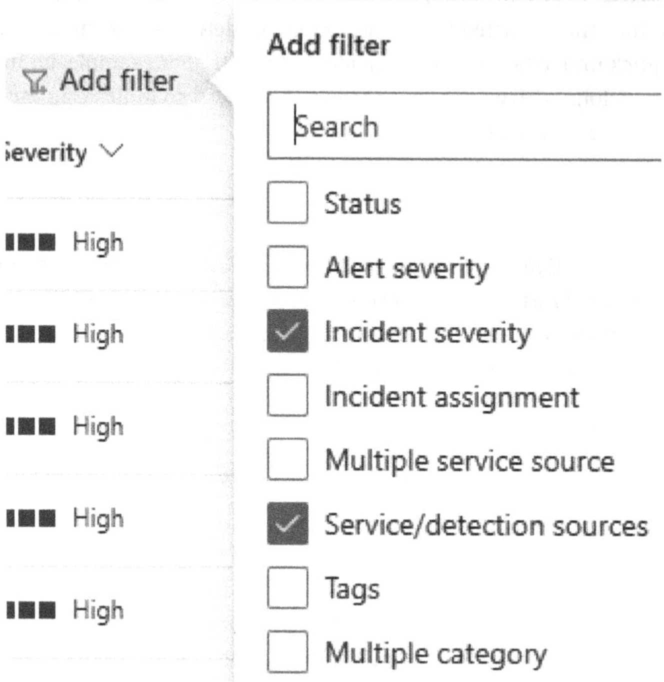

Figure 9.14 – The Incidents queue filters

By default, incidents and alerts are shown with a status of **New** and **In Progress**, respectively, alongside **Low**, **Medium**, or **High** severity levels. Remove specific filters with a simple click on the corresponding **X** in the filter pane, and preserve a customized set by using the **Create Filter Sets** feature. For convenience, you can save specific filtered views for recurrent access. You can also use bookmarked URLs for easier navigation to sections such as **High-severity incidents**, **High-severity, unassigned incidents**, **Incidents earmarked with particular tags**, **Incidents related to a named threat or actor**, **Incidents tagged to a particular analyst**, **Incidents tied to specific Defender services**, **Incidents within designated threat categories**, **New incidents**, and **Unassigned incidents**. Compiling these bookmarked filters allows for a practical and rapid approach to processing and prioritizing incidents. This repository, once constructed, becomes a cornerstone in the incident management ecosystem, bolstering you with swift access and strategic control over threat response and analysis.

Managing incidents

Efficient incident management is a cornerstone of a solid cybersecurity strategy, ensuring rapid containment and resolution. Within the Microsoft Defender portal (`security.microsoft.com`), navigate through the quick launch via **Incidents & alerts** | **Incidents** to manage incidents effectively.

After selecting an incident, you'll find the **Manage incident** option to tailor the incident's details. We can see the **Manage incident** window in the following figure:

Manage incident

Incident name

A potentially malicious URL click was detected

Severity

High

Incident tags

Type to find or create tags

Assign to

Unassigned

Status

Active

Classification

Not set

Comment

Normal   B I U S ⦿   ≣ ≣ ≣ ≣ I_x </>

Add comment

Figure 9.15 – The Manage incident window

The key management options include the following:

- **Incident name**: Although Microsoft Defender XDR generates initial incident names by drawing on alert attributes, such as the affected endpoints or the sources of detection, you have the leverage to customize this to encapsulate incident breadth accurately.

- **Severity**: Change an incident's urgency level based on the highest severity alert it encompasses, selecting from the **High**, **Medium**, **Low**, or **Informational** categories. This crucial step ensures prioritization aligns with potential impact.

- **Incident tags**: Infuse custom tags into incidents for nuanced categorization, aiding in subsequent filtering for traits or patterns common to specific incident clusters.

- **Assign to**: Delegate incidents to specific users or teams by specifying accounts using this field. Reassignment and unwinding assignments are straightforward, supporting dynamic incident ownership.

- **Status**: Once an incident is mitigated, select **Resolved** to reflect its status, simultaneously resolving associated alerts, with unresolved incidents marked as active.

- **Classification**: Categorize incidents accurately to refine Microsoft Defender XDR's detection capabilities. Classifications include **True positive** for verified threats, **Informational** for benign anomalies, and **False positive** for non-threatening alerts.

- **Comment**: Within the comment field, stakeholders can chronicle insights, adding textual data, formatted content, and multimedia. These comprehensive comments are integral to an incident's historical dossier.

Leveraging these options at the outset of an incident investigation provides your response team with structure and clarity, ensuring swift and strategic incident triage and resolution.

Understanding the information on the incident

Joining disparate alerts, executing a thorough analysis, and developing a complete narrative of an incident are the cornerstones of Microsoft Defender XDR. It combines related alerts, investigations, and evidence across the entire environment to combine endpoints, user accounts, cloud applications accessed by users, and email systems in a singular incident profile. This allows cybersecurity teams to understand the scope and impact of an attack more comprehensively. By scrutinizing these generated incidents, you can delineate the significance of each alert in relation to the environment, amass supporting data, and subsequently, develop a robust counterstrategy.

Attack story

The preliminary step involves examining an incident's attributes and the overarching narrative of the attack as captured by the system. These narratives serve a strategic function, granting swift examination, investigation, and rectification capabilities, while simultaneously presenting a seamless account of the assault within a unified interface. Remedial actions, such as file deletion or device isolation, are thus facilitated within this ecosystem without fragmenting the incident's contextual flow. Within this narrative framework, the alert page and the accompanying incident graph are pivotal components. Let's examine the different sections found on the alert page:

- **Attack story** consists of three parts – the order of events, the actions taken, and the interconnected incidents. We can see how this section looks in the following figure:

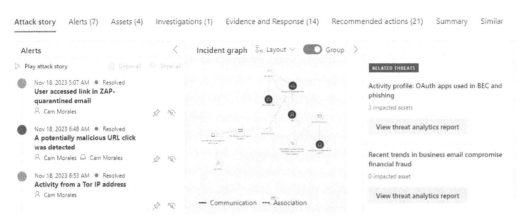

Figure 9.16 – The attack story section

- **Alert properties** can be seen when clicking on any of the alerts and provide a detailed breakdown of the status, details, story, and additional information. We can see these details in the following figure:

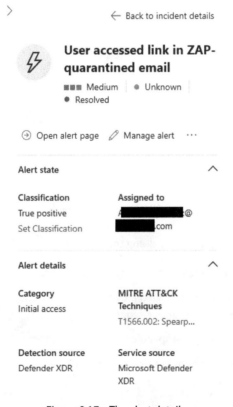

Figure 9.17 – The alert details

The incident graph explains the attack's progression, capturing its initiation, spread, and terminus within your network. It charts the trajectory of the assault, correlating suspect entities with corresponding assets, such as users, endpoints, and email accounts. We can see the incident graph in the following figure:

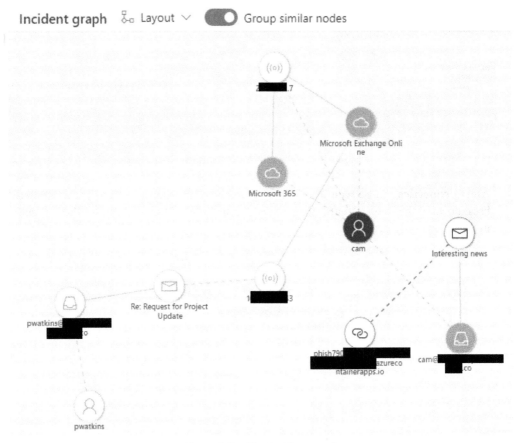

Figure 9.18 – The incident graph

Using the graph, insights can be gleaned by doing the following:

- Simulating the temporal sequence of alerts and nodes to find out an attack's chronology
- Inspecting an entity through a dedicated panel that offers a closer look and supports remedial measures, such as file erasure or device sequestration
- Prioritizing alerts as they relate to specific entities
- Pursuing detailed data on various entities, spanning devices, files, IP addresses, or URLs

By selecting an entity graphically represented within the incident graph, the **Go hunt** feature becomes accessible, powering the pursuit of detailed entity-related intelligence through the advanced hunting framework. This tool scours schema tables for any recorded events or alerts concerning the entity under scrutiny. We can see the **Go hunt** option in the following figure:

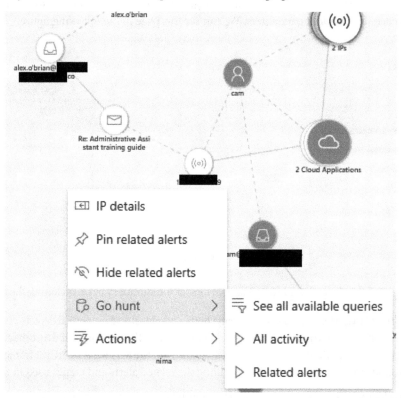

Figure 9.19 – The Go hunt feature

The options provided here include the following:

- **See all available queries**: Presents a repository of discoverable queries for your selected entity

- **All activity**: Details every recorded engagement of an entity, painting a full picture of the incident backdrop

- **Related alerts**: Identifying and collating all security alerts that are related to a particular entity, ensuring a comprehensive situational awareness

The Summary page

The **Summary** page within Microsoft Defender XDR serves as a pivotal tool to determine an incident's significance and to promptly navigate through related alerts and affected entities. This at-a-glance overview spotlights key aspects of the incident, segmented into alerts and classifications, scope, evidence, and detailed incident parameters. We can see this page in the following figure:

Figure 9.20 – The Summary page

A lot of information can be gathered from a simple look at the summary page, including the following:

- **Alerts and Categories**: This section presents both a graphical and numerical depiction of how far an incursion has penetrated the kill chain. Conforming to the structure used by other Microsoft security solutions, Microsoft Defender XDR aligns with the MITRE ATT&CK™ framework. An easy to understand view is provided, detailing the related alerts and tactics used in the attack.

- **Scope**: Here, you'll find a count of the compromised devices, users, and mailboxes, all organized according to their risk magnitude and urgency for investigation, allowing insights into which areas demand immediate attention.

- **Evidence**: Within this section, a compiled count of all affected entities by an incident is presented, offering a clear visual scope of the impact.

- **Incident Information**: Critical properties of an incident are neatly encapsulated here, including classification tags, the current state of the incident, and an assessment of its severity, equipping you with the essential details for an effective incident response strategy.

The **Summary** page can be a great source of training for your team on attacks, as it can provide information on the common approaches used by attackers, helping future threat hunters in your team develop a sense of these patterns.

Alerts

The **Alerts** tab in Microsoft Defender XDR is a central component that allows security analysts to scrutinize the catalog of warnings pertinent to a security incident. We can see the **Alerts** tab for an incident in the following figure:

☐ A potentially malici...	■■■ High	Remediated	● Resolved	Threat management	MDO	⋒ cam	◱ Cam Morales
☐ Email messages co...	■■■ Informational	Remediated	● Resolved	Threat management	MDO	⋒ 2 Accounts	◱ 2 Mailboxes
☐ Email messages co...	■■■ Informational	Remediated	● Resolved	Threat management	MDO	⋒ cam	◱ Cam Morales
☐ Suspicious inbox ...	■■■ High		● Resolved	Defense evasion	Microsoft Defender for...	⋒ cam	⬭ 2 Apps
☐ Suspicious emails s...	■■■ High	Queued	● Resolved	Suspicious activity	Defender XDR	⋒ cam	
☐ Suspicious inbox ...	■■■ High		● Resolved	Defense evasion	Defender XDR	⋒ cam	
☐ BEC financial fraud +1	■■■ High		● Resolved	Defense evasion	Defender XDR	⋒ cam	

Figure 9.21 – The Alerts tab for an incident

This tab provides a wealth of information, including the following:

- **Severity**: A vital indicator of the threat level associated with each alert

- **Engaged entities**: Details regarding which assets were implicated within an alert

- **Alert provenance**: Identification of the source generating an alert, whether it is Microsoft Defender for Identity, Microsoft Defender for Endpoint, Microsoft Defender for Office 365, Defender for Cloud Apps, or stemming from the app governance add-on

- **Correlation rationale**: Insights into the reasoning behind the grouping of alerts as part of the same incident narrative

To help you understand the sequence of events, alerts are organized in chronological order by default, offering a timeline that unveils how an assault progressed temporally. Selecting an individual alert within an incident pulls up pertinent details, aligning the information with the broader context of the incident at hand. Upon selection, the alert's narrative unfolds, detailing the events that made up the alert, identifying any subsidiary alerts that may have activated it, and outlining all entities and activities entangled in the breach, such as affected devices, files, users, and mailboxes. These insights are integral for cybersecurity teams as they construct a comprehensive picture of the attack's footprint and mobilize an informed response.

Assets

This centralized section displays a comprehensive list, including **Devices**, **Users**, **Mailboxes**, and **Apps**. Right at the top, you'll notice the total count of assets accompanied by their respective names. We can see the **Assets** tab for an incident in the following figure:

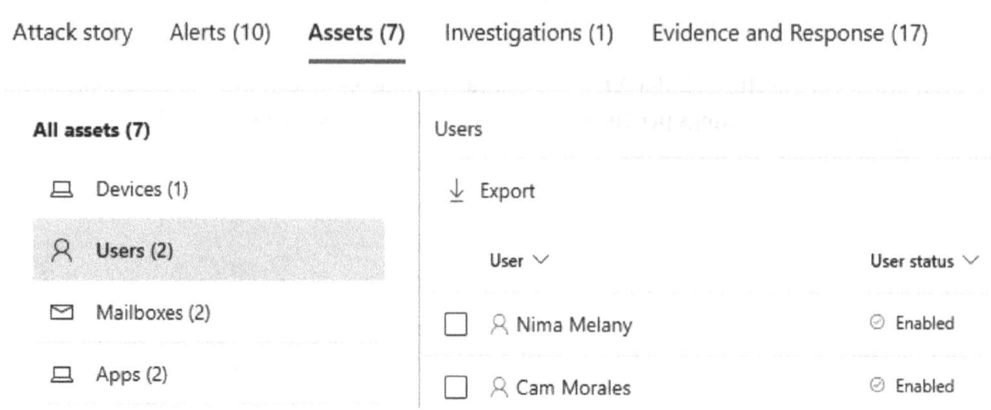

| Attack story | Alerts (10) | **Assets (7)** | Investigations (1) | Evidence and Response (17) |

All assets (7)		Users
🖥 Devices (1)		↓ Export
🧑 **Users (2)**		
✉ Mailboxes (2)		User ∨		User status ∨
🖥 Apps (2)		☐ 🧑 Nima Melany		⊘ Enabled
		☐ 🧑 Cam Morales		⊘ Enabled

Figure 9.22 – The Assets tab

The **Assets** tab allows you to explore various categories and the associated number of assets they contain, including the following:

- **Devices**: A detailed inventory of all devices linked to a security incident. Clicking on a device unveils a handy sidebar, equipped with options to handle your chosen device. You're able to swiftly execute exports, manage labels, and instigate an automated checkup, among other actions. Ticking a device's checkbox reveals its specific details, directory data, active notifications, and accounts that were recently active. By selecting the device's name, you unlock a full overview in the Defender for Endpoint inventory. Here, you're granted access to comprehensive insights – all alerts, a sequence of events, and expert security advice. For instance, within the **Timeline** tab, you can observe a chronological sequence of events and activities noted on a device peppered with alerts.

- **Users**: This displays a roster of individuals associated with a security incident. Upon marking a user's checkbox, you unveil insights into the user account's risks, vulnerabilities, and contact specifics. Clicking on a username pulls up extended data about the account.

- **Mailboxes**: Displays all email inboxes connected to a security issue. By selecting a mailbox's checkbox, you're treated to an overview of current alerts. Clicking the inbox's name takes you deeper into the mailbox's details on the **Explorer** pane within Defender for Office 365.

- **Apps**: Provides a listing of applications suspected in a security incident. Selecting an application's checkbox presents a tally of running alerts. You can click through on an app's title to glean more nuances on the **Explorer** view in Defender for Cloud Apps.

The **Assets** tab can help you better understand the impact of an incident and help fine-tune alerts for situations in which asset importance in the environment is misidentified.

Investigations

In the **Investigations** section, you'll find a comprehensive list of all the automated remediations started by an incident's alerts. We can see the **Investigations** tab for an incident in the following figure:

Attack story	Alerts (10)	Assets (7)	**Investigations (1)**	Evidence and Response (17)	Recommended actions (22)

☐	Triggering alert	ID	Investigation status	Service source
☐	Disabling a compromised user account suspected of financial fraud	243	Remediated	

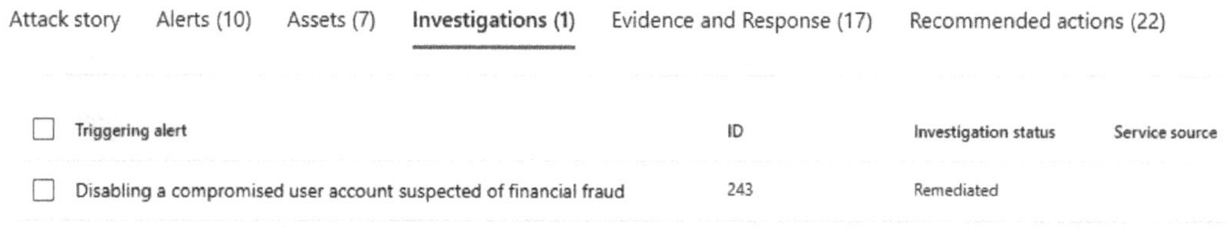

Figure 9.23 – The Investigations tab

These automated remediations might act straight away or wait for an analyst's green light, depending on the preferences set for your Defender for Endpoint and Defender for Office 365 systems. Click on any investigation to move to its details page, where you'll discover everything about the probe and its status in rectifying the issue. Should there be any decisions awaiting your authorization in the probe's process, they'll be shown under the **Pending actions** history section. Addressing these is a vital step in the process of managing an incident. The investigation graph segment provides a visual map, linking alerts to an organization's affected resources. This includes which entities connect to which alerts, their role in the attack's narrative, and the specific alerts tied to the incident. This graph is a tool to gain a swift and comprehensive understanding of an attack's breadth, linking various suspect entities to their associated resources, such as users, devices, and mailboxes.

Evidence and Response

The **Evidence and Response** tab displays all the relevant events and questionable entities linked to an incident's alerts. Microsoft Defender XDR takes the helm, automatically scrutinizing all such events and entities within the alerts. Your dashboard will highlight essential elements, such as emails, files, processes, services, and IP addresses, enabling you to swiftly identify and thwart potential threats. For each entity examined, a designation (such as **Malicious**, **Suspicious**, or **Clean**) alongside a resolution status is provided, offering you a clear picture of the incident's overall rectification progress and guiding you toward potential next steps. We can see the **Evidence and Response** tab for an incident in the following figure:

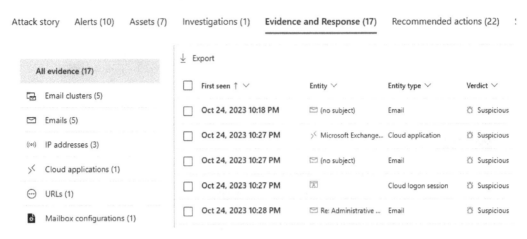

Figure 9.24 – The Evidence and Response tab

Let's remember the significance of the additional context provided by related alerts when you're delving into these investigations. It's advisable for an incident responder to deal with these incidents holistically at the overall incident level, rather than the individual alert level. This approach is not just more streamlined; it also delivers a broader context, ensuring that any alterations made on this scale are consistently applied to all relevant alerts. Tailoring settings at the individual alert level should ideally be reserved for precise environmental adjustments, such as alert classification.

AIR

Microsoft Defender for Office 365 is equipped with efficient AIR technology, aimed at streamlining tasks for your security operations team. When alerts are triggered, the team must diligently review and address them. With high volumes of alerts, this can be daunting, which is why automation proves beneficial.

AIR enhances your team's productivity by triggering automatic investigations for known threats. It proposes appropriate remedial actions, which are subject to your team's endorsement, enabling a swift reaction to threats. With AIR, your team can concentrate on urgent issues without overlooking critical alerts.

A high-level overview of AIR

Let's review at a high level how automated investigations work, from what triggers them to what occurs once these are active, as well as the benefits we can gain from them:

1. First, an automated investigation is prompted in two major ways:

    • A security analyst manually starts an automated investigation through the **Explorer** tool. We can see a user selecting the **Trigger investigation** option from the **Email** menu in the following figure:

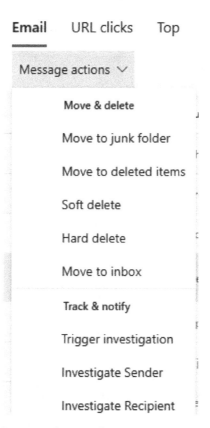

Figure 9.25 – The Email menu in Threat explorer to trigger an automated investigation

In the following screenshot, we can see the resultant confirmation window.

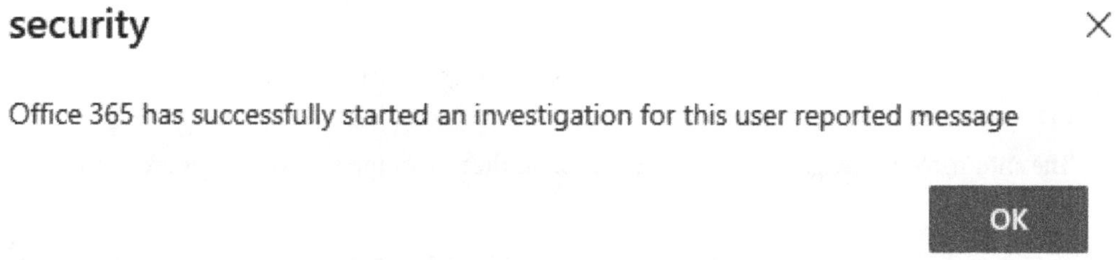

security   ✕

Office 365 has successfully started an investigation for this user reported message

OK

Figure 9.26 – The screen confirming the start of an automated investigation

- Suspicious activity such as attachments, URLs, or compromised accounts in emails set off an alert, thus beginning an investigation. We can see one of these automated investigations in the following figure:

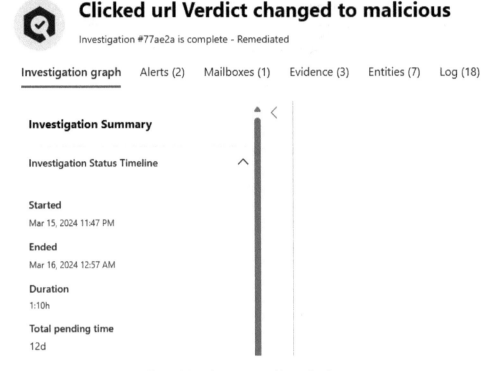

Clicked url Verdict changed to malicious

Investigation #77ae2a is complete - Remediated

Investigation graph   Alerts (2)   Mailboxes (1)   Evidence (3)   Entities (7)   Log (18)

Investigation Summary

Investigation Status Timeline

Started
Mar 15, 2024 11:47 PM

Ended
Mar 16, 2024 12:57 AM

Duration
1:10h

Total pending time
12d

Figure 9.27 – An automated investigation.

2. The investigation process compiles data about questionable emails and connected elements, such as files and URLs. Its scope might expand as it uncovers and integrates additional linked alerts.

3. Both during and after an investigation, detailed insights and suggested responses to identified threats are accessible for review.

4. The security operations team scrutinizes these findings, either validating or disapproving of the recommended remedial actions.

5. The automated investigation concludes as soon as these pending remedial steps are confirmed or denied.

It is important to remember that, in Microsoft Defender for Office 365, your team's explicit consent is required before any remedial steps are enacted. AIR significantly saves time by identifying and detailing actions, and facilitating decisions made with full awareness of the facts. Throughout the investigation, your team can do the following:

- Look at the details of an alert

- Review an investigation's detailed results

- Assess and allow subsequent actions based on the investigation's findings

AIR features are a fundamental component of Microsoft Defender for Office 365, automatically active with auditing enabled. For new setups in Microsoft 365, confirm your audit logging status with this Exchange Online PowerShell command:

```
Get-AdminAuditLogConfig | Format-List UnifiedAuditLogIngestionEnabled
```

A `True` value confirms that auditing is operational. A `False` value shows that it is not. Activate auditing through the compliance portal or with the following Exchange Online PowerShell command:

```
Set-AdminAuditLogConfig -UnifiedAuditLogIngestionEnabled $true
```

Note that if activation is required, it might take up to 60 minutes before the setting is fully effective.

Alerts that trigger AIR

Microsoft 365 features an array of default alert policies designed to detect misuse of Exchange admin permissions, malware, potential security threats, and compliance risks. Automated investigations can be started through several of these alerts, detailed in the following points:

- **High Severity Alerts:**

  - **A potentially malicious URL click was detected**: Triggered under the following circumstances:

    - A user covered by Safe Links clicks on a suspect link

    - Microsoft Defender for Office 365 identifies changes in verdicts for URLs

    - Your organization's Safe Links policy allows users to bypass warning pages

  - **A user is restricted from sending email**: Occurs when an individual in your organization is barred from sending out mail, usually indicative of a compromised account

- **Medium Severity Alerts:**

  - **Suspicious email sending patterns detected**: This alert flags unusual sending behaviors that may lead to a user being prevented from emailing. It hints at a potentially compromised account and requires verification, even though false alarms are infrequent.

  - **Admin triggered user compromise investigation**: Raised when an admin manually starts an investigation into a user from Threat Explorer, alerting an organization that this probing has begun.

- **Low Severity Alerts:**

  - **An email message is reported by a user as malware or a phishing attempt**: Arises when a user reports an email as phishing using the Report Message or Report Phishing add-ons

- **Informational Severity Alerts**:

  - **Email messages containing malicious files removed after delivery**: Generated when messages with a malicious file are initially delivered but later eradicated from mailboxes by Microsoft's **zero-hour auto purge (ZAP)**

  - **Email messages containing malware removed after delivery**: Like the previous alert, this is triggered when messages with malware are delivered and are subsequently removed via ZAP

  - **Email messages containing malicious URL removed after delivery**: Alerts when messages with a malicious URL are delivered and later removed by ZAP

  - **Email messages containing phish URLs removed after delivery**: Notifies when messages with phishing URLs are delivered and later cleaned up using ZAP

  - **Admin triggered manual investigation of email**: This alert communicates that an admin has begun a manual email investigation via Threat Explorer

If you need to review these policies, adjust them, or turn them off, from the Microsoft Defender portal (`security.microsoft.com`), select **Email & collaboration | Policies & rules | Alert policy** to open the alert policy page. Be aware that for default policies, the settings that can be modified will be limited.

Reviewing AIR pending actions

In Microsoft Defender for Office 365, when an automated investigation renders a finding, labeling content as **Malicious** or **Suspicious**, it proposes corresponding remedial actions. These can include quarantine of suspicious emails or disabling unauthorized external email forwarding. Importantly, these actions are conditional upon review and sanction by your security team. Prompt approval of these actions is advised to ensure investigations conclude efficiently.

To prevent redundancy, additional measures have been implemented to check for repetitive approvals of identical clusters within a short timeframe. Approved actions for the same cluster within the previous hour will not be processed again, eliminating unnecessary duplication and enhancing the speed of remediation. The details of these duplicated actions won't be displayed in the **Action center** side panel. There are three methods to identify and address auto-investigation actions.

The Incidents queue

Reviewing pending actions from the **Incidents** queue is a simple process.

1. Navigate to the **Incidents** page on the Defender portal at `https://security.microsoft.com`, and on the left-hand menu, select **Incidents & alerts | Incidents**.

2. You can optionally filter for **Pending action** under **Automated investigation state**, as shown in the following figure:

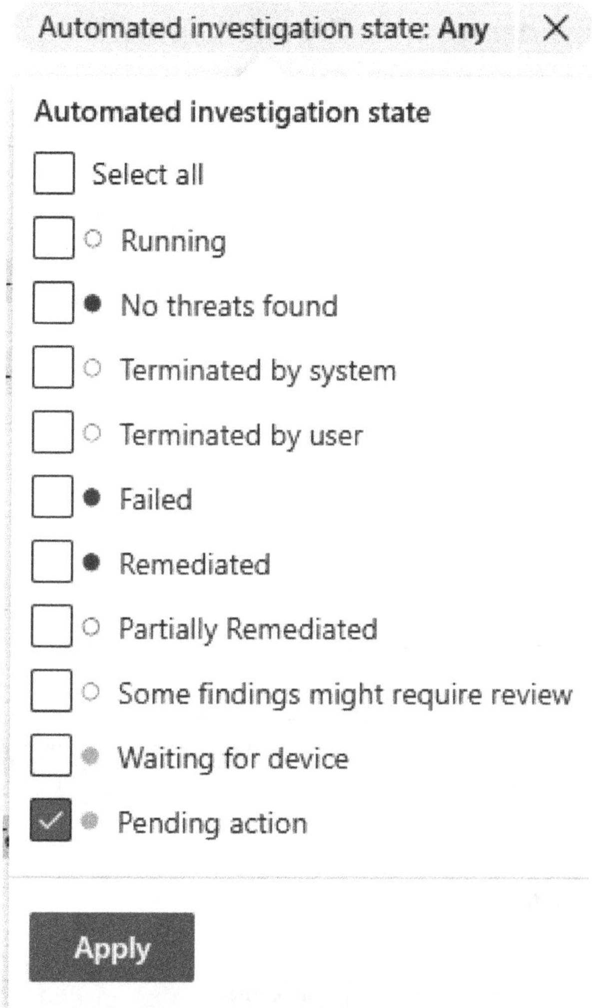

Automated investigation state: **Any**   X

Automated investigation state

☐ Select all

☐ ○ Running

☐ ● No threats found

☐ ○ Terminated by system

☐ ○ Terminated by user

☐ ● Failed

☐ ● Remediated

☐ ○ Partially Remediated

☐ ○ Some findings might require review

☐ ● Waiting for device

☑ ● Pending action

Apply

Figure 9.28 – Filtering by automated investigation state

3. Click an incident name to view its summary.

4. Go to the **Investigations** tab and select an item to open its details.

5. Go to the **Pending actions** tab, select one of the pending actions, and in the **Actions details** window, choose to approve or reject an action, as shown in the following figure:

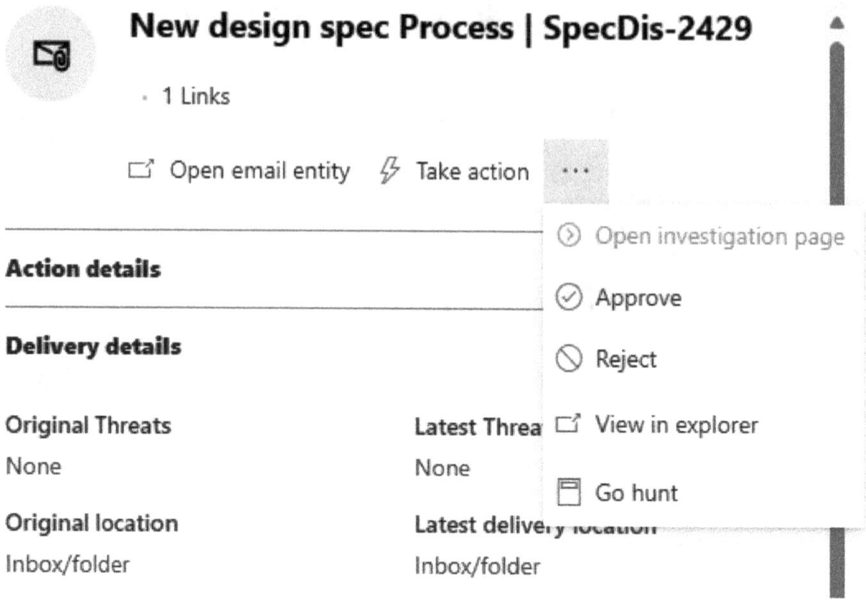

Figure 9.29 – Approving or rejecting pending actions

Before approving or rejecting a pending action, understand the impact of this action on an environment. A repeatable process of what points to look for when reviewing these actions should be part of any security team maturity roadmap.

The Action center

Reviewing pending actions from the Action center is just as simple:

1. Visit the Action center at `https://security.microsoft.com`, and on the left-hand menu, select **Actions & submissions | Action center**.

2. Make sure the **Pending** tab is chosen, and then browse the actions needing approval.

3. Click on any action, and the details panel will appear.

4. Click **Approve** to allow a pending action or **Reject** to cancel it. We can see this panel and options in the following figure:

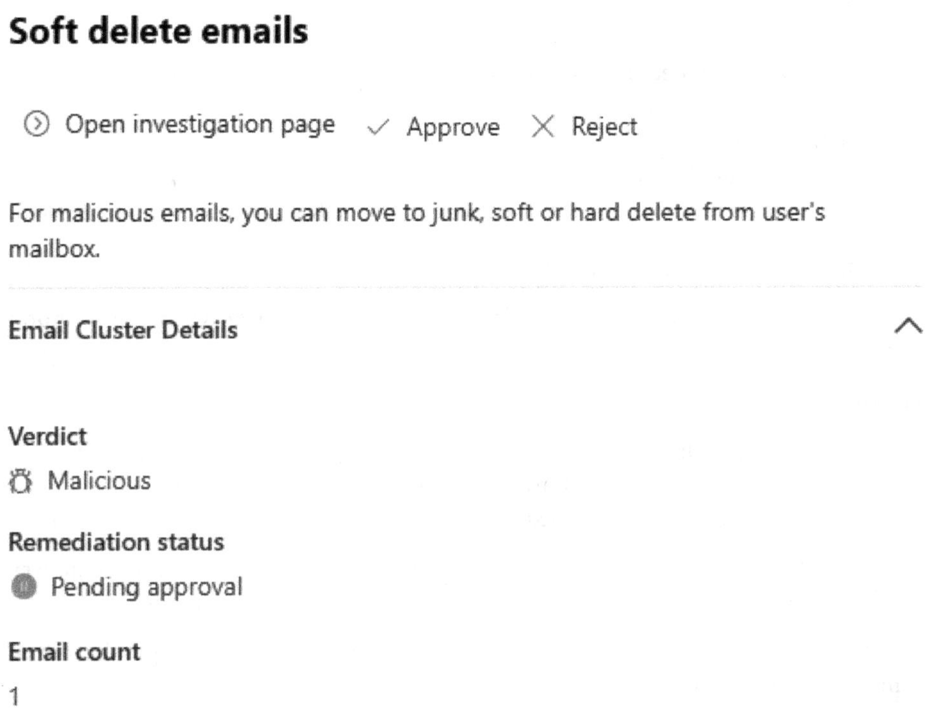

Soft delete emails

⊙ Open investigation page   ✓ Approve   ✕ Reject

For malicious emails, you can move to junk, soft or hard delete from user's mailbox.

Email Cluster Details   ∧

Verdict

🐞 Malicious

Remediation status

⬤ Pending approval

Email count

1

Figure 9.30 – The pending action details panel

If in doubt about what to do, open the investigation page in the **Details** panel to obtain more context on why an action is pending.

Investigation queue

Reviewing pending actions from the investigation queue is also a simple process:

1. Access the investigations page at `https://security.microsoft.com`, and on the left-hand menu, select **Email & collaboration | Investigations**.

2. Filter items by the **Pending action** status and open one of the items in a new window.

3. Go to the **Pending actions** tab, select one of the pending actions, and in the **Actions details** window, choose to approve or reject an action.

The investigation queue and incident queue processes are similar, as they use the same windows for pending actions.

Viewing AIR results

When you initiate an automated investigation with Microsoft Defender for Office 365, you get a comprehensive view of the process, both during and after an event. These details not only keep you informed about the status but also let you take charge of any necessary actions pending approval. The investigation status is regularly updated, signifying whether threats have been detected and whether any actions await confirmation. Here's a rundown of the various status updates you may encounter:

- **Starting**: This status means the investigation is queued to begin soon.

- **Running**: The investigation is active. This status will also appear when you've approved pending actions.

- **No Threats Found**: The investigation wraps up, detecting no threats.

- **Partially Investigated**: The investigation identified irregularities, but there aren't clear-cut actions to rectify them. This might involve scenarios such as data loss prevention incidents, odd email-sending patterns, or detected malware or phishing without a clear remediation route.

- **Terminated By System**: If an investigation halts, it's typically because pending actions weren't confirmed in time or the number of actions required exceeded system limits.

- **Pending Action**: Threats have been identified and are waiting for your go-ahead to start remediation. As the investigation continues, more items might join the pending list.

- **Remediated**: The investigation is complete, with all remedial actions given the go-ahead and recorded as resolved, although you may need to double-check for any execution errors, since the status won't reflect these.

- **Partially Remediated**: Some threats found have been dealt with, while other actions are still under review.

- **Failed**: Occasionally, an investigation might experience a glitch and not follow through as intended. Interestingly, approved actions could still have been successful, so a detailed check is recommended.

- **Queued By Throttling**: To prevent system overloads, investigations might be queued, kicking off as others finish.

- **Terminated By Throttling**: Investigations stuck in the queue too long are eventually terminated.

To dive into the details of an AIR case, follow these steps:

1. Log into the Microsoft Defender portal through `https://security.microsoft.com`.

2. In the left-hand menu, under **Actions & submissions**, select **Action center**.

3. Choose an item from either the **Pending** or **History** category. This will open the relevant details. We can see the **History** tab under **Action Center** in the following figure:

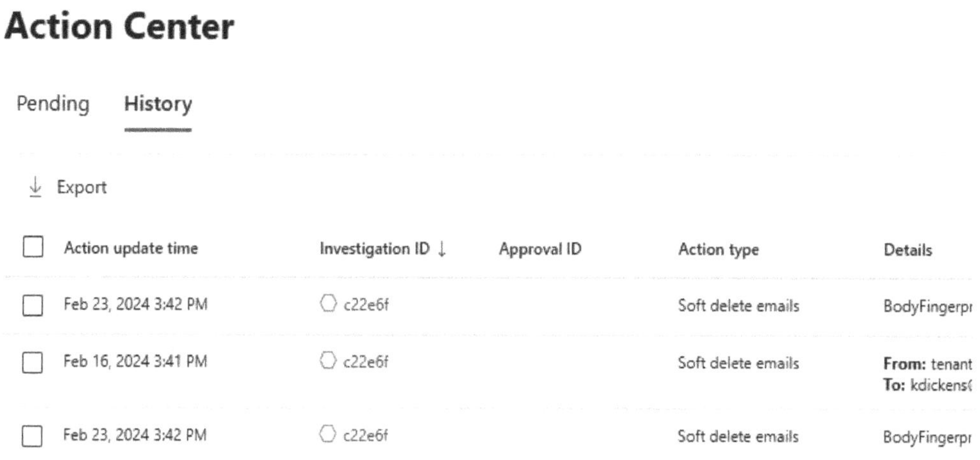

Action Center

Pending   **History**

⤓ Export

	Action update time	Investigation ID ↓	Approval ID	Action type	Details
☐	Feb 23, 2024 3:42 PM	◯ c22e6f		Soft delete emails	BodyFingerpr
☐	Feb 16, 2024 3:41 PM	◯ c22e6f		Soft delete emails	**From:** tenant **To:** kdickens(
☐	Feb 23, 2024 3:42 PM	◯ c22e6f		Soft delete emails	BodyFingerpr

Figure 9.31 – Action Center | the History tab

4. Within the **Details** pane, click on the **Open investigation** page to get a bigger picture.

5. Use the array of tabs to analyze the investigation further, and peruse through the **Alerts** section to assess individual alerts and gather more specifics.

This thorough approach arms you with the information to oversee and guide the automated investigation process effectively, keeping your organization's digital environment safe and sound.

Managing false positives and false negatives

False detections refer to alerts and signals that indicate the opposite of what is occurring in an environment. For example, if an alert appears indicating malware in an email attachment but the file is benign, this is a **false positive**. The opposite can also occur, with malware being identified as a benign file, which would be considered a **false negative**. If AIR capabilities in Office 365 miss or wrongly detect something, there are steps your security operations team can take to fix this. Such actions include some steps that will be covered in the following sections.

Reporting a false positive/negative to Microsoft

If AIR fails to identify or incorrectly identifies a threat within an email, an attachment, or a URL, your team has tools at hand within the Microsoft Defender portal. By utilizing the **Submissions** platform, messages, attachments, and URLs can be sent directly to Microsoft for further examination. There are two main pathways to submit these items:

*   **Admin submissions**: Should your team discover questionable entities; they can be flagged by selecting **Submit to Microsoft for analysis** from the **Submissions** page. To reach this page, go to the Defender portal (`security.microsoft.com`) left-hand menu and select **Actions &**

submissions | **Submissions**. During submission, you will be provided with options to explain why items are being submitted, as shown in the following figure:

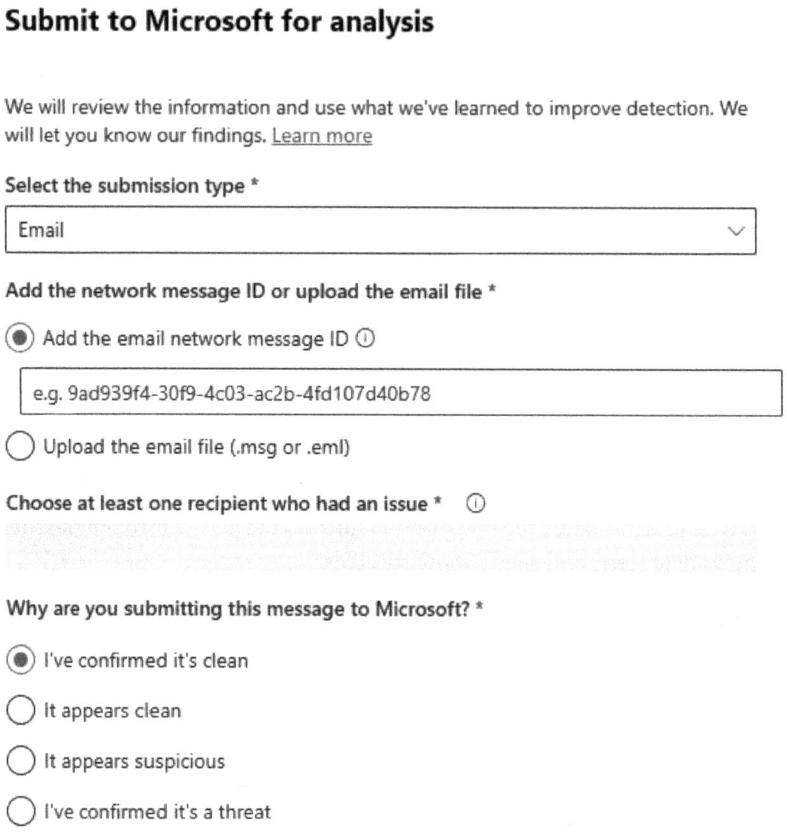

Submit to Microsoft for analysis

We will review the information and use what we've learned to improve detection. We will let you know our findings. Learn more

Select the submission type *

| Email | ⌄ |

Add the network message ID or upload the email file *

◉ Add the email network message ID ⓘ

| e.g. 9ad939f4-30f9-4c03-ac2b-4fd107d40b78 |

◯ Upload the email file (.msg or .eml)

Choose at least one recipient who had an issue * ⓘ

Why are you submitting this message to Microsoft? *

◉ I've confirmed it's clean

◯ It appears clean

◯ It appears suspicious

◯ I've confirmed it's a threat

Figure 9.32 – Submitting to Microsoft for analysis

- **User submissions**: With the user reporting function enabled in Outlook, users can flag messages through a dedicated report button. These user-reported messages pop up on the **User reported** tab, from which admins can then forward them to Microsoft.

Both the admin and user submissions can be tracked on the **Submissions** page upon submission. During admin submission, each entity is subjected to a thorough check, including the following:

- **Email authentication check**: This is especially relevant for emails and scrutinizes whether they passed authentication protocols at the delivery point.

- **Policy Hits**: The system searches for any relevant policies or exceptions that could have impacted an email's delivery; this includes both allowances and blocks that might have overruled standard filtering decisions.

- **Payload evaluation**: For a real-time inspection, the URLs and attachments housed in the message are analyzed anew.
- **Grader analysis**: A final layer of review involves human inspection to verify the nature of the messages, affirming whether they're indeed threats or false positives.

Reporting false detections will not only provide more fine-tuned detections and less noise but also make your security team more savvy when encountering the same threat again. There might be times when your team might need more experienced resources to determine with certainty whether a file is malicious; this is where submitting files for malware analysis might come into play.

Submitting a file to Microsoft for malware analysis

Organizations with a Microsoft Defender XDR plan or those equipped with Microsoft Defender XDR for Endpoint Plan 2 have the capability of submitting suspicious files for scrutiny. This process is streamlined via the **Submissions** section on the Microsoft Defender portal (*Microsoft, 2024a*). Alternatively, files can also be submitted through the Microsoft security intelligence file submission page (https://www.microsoft.com/wdsi/filesubmission), as shown in the following figure:

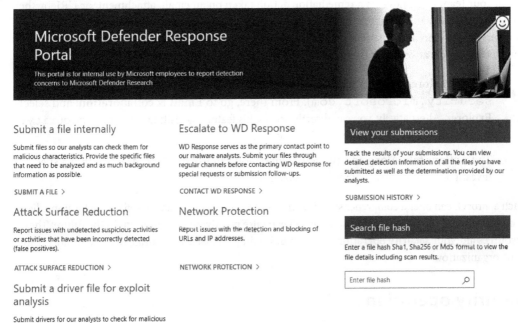

Figure 9.33 – The Microsoft malware analysis Enterprise customer page

To kick-start the analysis and ensure you can follow the process, follow these steps:

1. Log into the portal or enter your details; this can be a Microsoft work or school account. Select whether you are a home customer, enterprise customer, or software developer.

2. Upon uploading the file(s) in question, make a note of the unique submission ID assigned to your case.

3. Post-upload, Microsoft embarks on a thorough inspection to determine the nature of the file.

If Microsoft establishes that the submitted file harbors malware, measures will be taken to augment the system's capacities to spot and stop such malware in the future. In scenarios where you notice the recurrence of an infected message or attachment in user mailboxes following your submission, additional steps can be taken. Compile the malicious message's header and the previously recorded submission ID, and then forward these to Microsoft Customer Service and Support, who will then assist you with more advanced investigative measures.

Undoing remediation actions that were taken

When your team identifies that a remediation action taken on an email, attachment, or URL by the security system was unwarranted, meaning the item isn't malicious, there's a straightforward process to reverse the action and ensure such a false positive doesn't happen again. To correct such issues, you have two tools at your disposal:

- **Threat Explorer**: For this approach, navigate to the Microsoft Defender portal (`https://security.microsoft.com`). From there, go to **Email & collaboration**, and select **Explorer**. Alternatively, you can directly access this feature by visiting `https://security.microsoft.com/threatexplorer`.

- **The Actions tab**: Alongside Threat Explorer, you can use the **Actions** tab within a specific investigation to reverse any measures taken.

Each approach can offer a comprehensive view and control over security actions, allowing for immediate remediation and the adjustment of settings to keep false positives to a minimum in the future. These resources aid your security operations team in upholding robust and accurate defense protocols for your organization.

Security operations

In shaping your organization's security operations, it is wise to adhere initially to establish best practices that are tried and true. From there, you may tailor these foundations as necessary to fill the unique security gaps your organization may face. Aim to keep alterations to the essential structure of your security strategy to a minimum; such restraint will render your processes more straightforward to oversee and refine when the need arises. As your organization evolves, you'll likely begin with many tasks that are executed manually. However, your trajectory should aim toward increased automation as your team becomes more adept and familiar with your environmental intricacies. Beginning with AIR is a sound first step. This sets the stage to progress into integrations with **security information and event management** (**SIEM**) and eventually develop into sophisticated **security orchestration, automation, and response** (**SOAR**) capabilities. To set your team on the right track, here are some key pointers.

Start by doing the following:

- Implementing an AIR system to handle routine security threats with efficiency, decreasing response time and labor

- Considering integrations with SIEM to centralize security alerts and improve response strategies

- Evaluating SOAR solutions to orchestrate and automate complex workflows, freeing your team to tackle more strategic tasks

Once the initial measures are in place, focus on the following:

- Continual analysis and reassessment of standard procedures to align with evolving security landscapes and organizational needs

- Advanced training for your team to ensure a deep understanding and effective management of the technologies employed

Remember, the foundation of any robust security program lies in the balance between proven best practices and tailored solutions where necessary, leading to scalable and manageable operations. Next, we will focus on some recommended activities that any team can implement into their processes.

Daily activities

Every day, your security team should be actively involved in certain tasks to ensure the smooth running of your organization's cyber defenses. The upcoming sections cover what should be on your daily checklist.

Monitor the Incidents queue

Access the **Incidents** queue from the Microsoft Defender portal or use the direct URL (`https://security.microsoft.com/incidents-queue`). This page lets you oversee and manage events that come from alerts and AIR. Each day, you should do the following:

- **Prioritize and triage**: Start by addressing high- and medium-severity incidents.

- **Investigate incidents**: Examine every incident thoroughly, and either take action as the system recommends or perform manual interventions.

- **Resolve and classify**: Finish any remedial work for incidents and classify them accurately as true incidents or false positives. For true alerts, detail the nature of the threat to improve your system's understanding and pattern recognition.

Security teams not staying on top of their incident queues is a common trait of compromised environments. If your team can't keep up, it might mean fine-tuning is required, such as managing false detections.

Tracking and managing false detections

To keep your security system's accuracy sharp, you'll need to manage incorrect detections by doing the following:

- **Reporting errors**: Submit any **false positives** (erroneously identified as malicious) and **false negatives** (erroneously identified as clean) via `https://security.microsoft.com/reportsubmission`.

- **Diving into details**: Investigate each false detection to understand what triggered it and assess your current settings.

- **Updating an allow/block list**: Add confirmed false negatives (attacks that passed controls) to your tenant block list (`https://security.microsoft.com/tenantAllowBlockList`).

- **Releasing incorrect quarantines**: Free any items that were quarantined by mistake.

False detection management makes for a robust and effective security solution, but this is just part of what you need to do, and staying ahead of what attackers are doing, such as campaigns, is also necessary.

Reviewing phishing and malware campaigns

Keep an eye out for phishing and malware campaigns infiltrating your organization:

- **Assess campaigns**: Check the campaigns page (`https://security.microsoft.com/campaigns`) and look for patterns or repeat offenders. Be aware that this page will focus on campaigns impacting your environment, so there might be instances that appear blank. The following figure shows a campaign impacting an environment:

Phish.10D885D9 - 務署からのお知らせ【宛名の登録確認及び秘密の質問等の登録に関するお知らせ】

Campaign ID 10D885D9.B3374922.5157F0DE.CC3395D2.20152
● Inactive 2404 min(s) ago
-

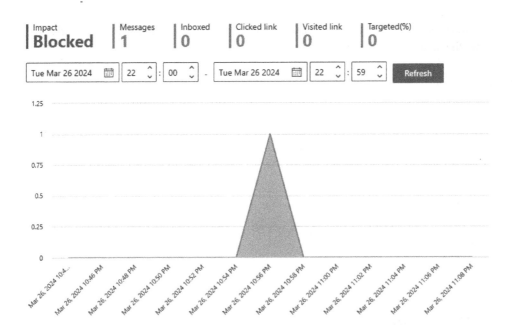

Figure 9.34 – A phishing campaign blocked

- **Verifying actions**: Confirm whether AIR has already dealt with these campaigns. If it hasn't, manual measures might be required.

By following these protocols daily, your team will ensure a proactive defense against threats, adapting quickly to any potential weaknesses in your security posture.

Weekly activities

Your security team's weekly to-do list is critical for maintaining a robust defense against email-based threats. Let's break down the tasks that should be taken regularly.

Email detection trends analysis

Use the following reports to assess and understand email detection trends:

- **Mailflow status**: The Mailflow status report (`https://security.microsoft.com/mailflowStatusReport?viewid=type`) helps you track email flow, including detections of malware, phishing, and spam. Comparing them against legitimate emails is essential. We can see the mailflow status report in the following figure:

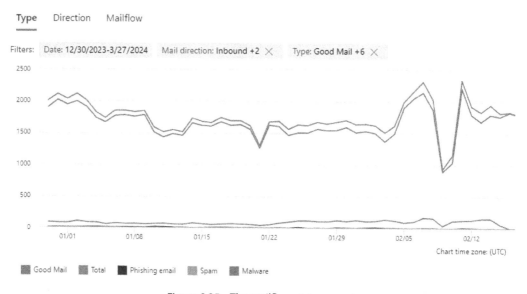

Figure 9.35 – The mailflow status report

- **Threat protection status**: Examine the threat protection status (`https://security.microsoft.com/reports/TPSAggregateReportATP`) to gauge the effectiveness of your current policies and whether adjustments are warranted. We can see the **Threat protection status** page in the following figure:

Threat protection status

The Threat protection status report provides information about threats found prior to email delivery, covering relevant detection technologies, policy

Filters: Date (UTC): 12/30/2023-3/28/2024   Detection: **Email Malware +3** ✕   Protected by: **MDO +1** ✕   Tag: All ✕   Direction: All ✕

Figure 9.36 – The threat protection status page

Monitoring these trends over time can reveal threat patterns and the necessity to tweak your Defender for Office 365 policies accordingly.

Threat analytics

Let's say that there is a new threat in the news; you can leverage threat analytics to quickly identify whether an environment has been impacted and report any findings to your team for further action. To track and act upon emerging threats, we can leverage the threat intelligence included with Defender for Office 365. The **Threat analytics** page can be viewed from the Defender portal (`security.microsoft.com`) and by selecting **threat intelligence | threat analytics**, or from `https://security.microsoft.com/threatanalytics3`. We can see this page in the following figure:

Threat analytics

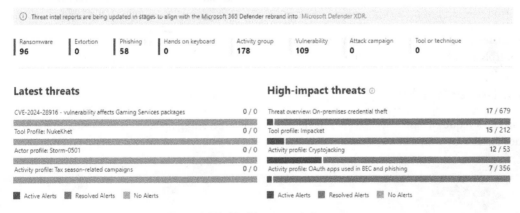

Figure 9.37 – The Threat analytics page

This page can help you focus on active threats that have a higher probability of impacting your environment and their characteristics. You can also obtain information to further enrich your threat hunting queries, such as IoCs, prepared hunting queries focused on the threat actors' strategies, recent attack methods, related vulnerabilities, and the malware used.

Top targeted user assessment

Investigate individuals most frequently targeted by cyber threats, as this might help identify users that need more education and controls. This assessment is divided into two activities:

- **Top targeted users**: Visit the **Top targeted users** tab (`https://security.microsoft.com/threatexplorer`) in Threat Explorer to identify who among your users is heavily targeted by malware and phishing attempts. Determine whether policy or protection adjustments are required for these individuals.

- **Priority accounts**: If a successful attack against a user can cause major damage to an environment, the user should be elevated to a priority account. This distinction allows for extra oversight and specialized safeguards, tailored to high-profile email patterns.

Don't forget the value of identifying highly targeted individuals and comparing this to previously identified individuals, which can shed light on how campaigns impact different users in your organization and help improve controls and education.

A malware and phishing campaigns review

Stay ahead of targeted attacks by with the help of **Campaign views**. They analyze campaign views (`https://security.microsoft.com/campaigns`) for insights into malware and phishing attacks directed at your organization. Adapt your defenses based on the insights gleaned from these campaigns.

Campaign views are also part of a mature threat intel approach, and its effectiveness might be directly impacted by how much a resource spends on analysis. Large-sized organizations will have at least one person focused just on threat intelligence efforts.

Implementing these essential weekly protocols facilitates a pre-emptive security stance, refining your controls against evolving cyber threats.

Ad hoc activities

On an as-required basis, let's consider the tasks integral to brushing up your security stance that are not dictated by a regular schedule.

Email threat management and investigation

During the investigation of alerts and incidents related to email threats, perform the following actions as necessary:

- **Email threat removal**: Use Threat Explorer (`https://security.microsoft.com/threatexplorer`) to investigate and eradicate harmful emails reported by users

- **Automated investigation initiation**: Trigger an automated investigation in Threat Explorer to aggregate data around an email's origin, potentially unearthing a larger threat context, including related risks and mitigation recommendations

While these ad hoc activities are necessary, they should be slowly automated to allow you to focus on proactive threat hunting efforts.

Proactive threat hunting

As your operational security maturity solidifies, your reliance on tools to detect and manage will increase; this will free resources to find the threats that the security controls miss in the form of proactive threat hunting. This activity can be accomplished by doing the following:

- **Hunting for threats**: Launch proactive threat searches via Threat Explorer or advanced hunting (`https://security.microsoft.com/v2/advanced-hunting`). Note that this requires considerable time and should be reserved for when your team isn't swamped with daily incidents. While prebuilt queries are provided, hunters will require some knowledge of the KQL to be effective. We can see the **Advanced hunting** page in the following figure:

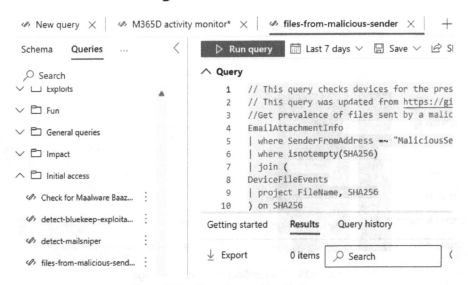

Figure 9.38 – The Advanced hunting page

- **Leveraging a hunting query repository**: Maintain and leverage a database of hunting queries, which can be converted into custom detection rules (`https://security.microsoft.com/custom_detection`) to enhance threat spotting efficiency.

From all the ad hoc activities identified, proactive hunting will be one that will continue to provide benefits even as an organization matures. While security tools do a great job detecting threats, sophisticated attackers can also spend a great deal of resources on circumventing protections. For organizations that might have a high chance of experiencing highly targeted attacks from sophisticated attack groups, setting some budget aside to maintain one or two proactive hunters for further fine-tuning protections might pay great dividends.

A policy review and adjustment

These activities are part of fine-tuning and should be performed regularly, and with a heightened frequency if false detections spike:

- **A Defender for Office 365 policy review**: Delve into the configuration analyzer tool (`https://security.microsoft.com/configurationAnalyzer`) to ensure policy settings align with Standard or Strict recommendations, or any other agreed-upon configuration, and to verify that no harmful configurations have slipped in (*Microsoft, 2023a*). We can see the history of configuration changes on the **Configuration analyzer** page, as shown in the following figure:

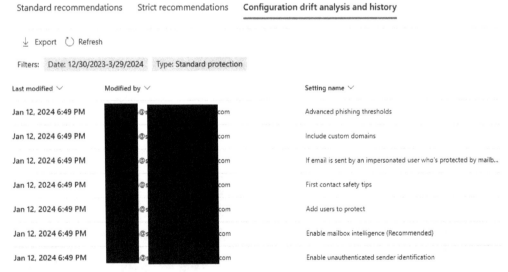

Figure 9.39 – The Configuration analyzer page

- **Detection override analysis**: Inspect the detection overrides on the **Threat protection status** page and change the filters to **View data by system override**. Alternatively, visit `https://security.microsoft.com/reports/TPSMessageOverrideReportATP` to understand why certain messages might be slipping through filters.

Not keeping an eye on configuration changes is an easy path to an ineffective security solution, and so is not looking at other signs of active attacks, such as spoofing and impersonation attempts.

Detecting spoofing and impersonation

Checking for spoofing attempts and identifying patterns can lead to identifying attack attempts that were missed. As part of effective security, you should also periodically perform the following checks:

- **Spoof intelligence**: Examine the spoof intelligence insights (`https://security.microsoft.com/spoofintelligence`) to ensure that your spoofing filters are up to the task

- **Impersonation insights**: Similarly, analyze the impersonation detection insights (`https://security.microsoft.com/impersonationinsight`) to keep your filters effective

Spoofing attempts might signal targeted attacks, so the effectiveness of these checks is also dependent on important accounts being tagged as a priority account.

Priority account checks

Properly tagging the accounts that can cause the most impact during an attack, as well as removing any accounts no longer under this classification, will ensure effective focused detection. We can accomplish this via two main activities:

- **Priority account audits**: Reassess the user list for priority account status (`https://security.microsoft.com/securitysettings/userTags`) to confirm that it includes only users whose compromise could significantly impact an organization. We can see the **Priority account** tag configuration in the following figure:

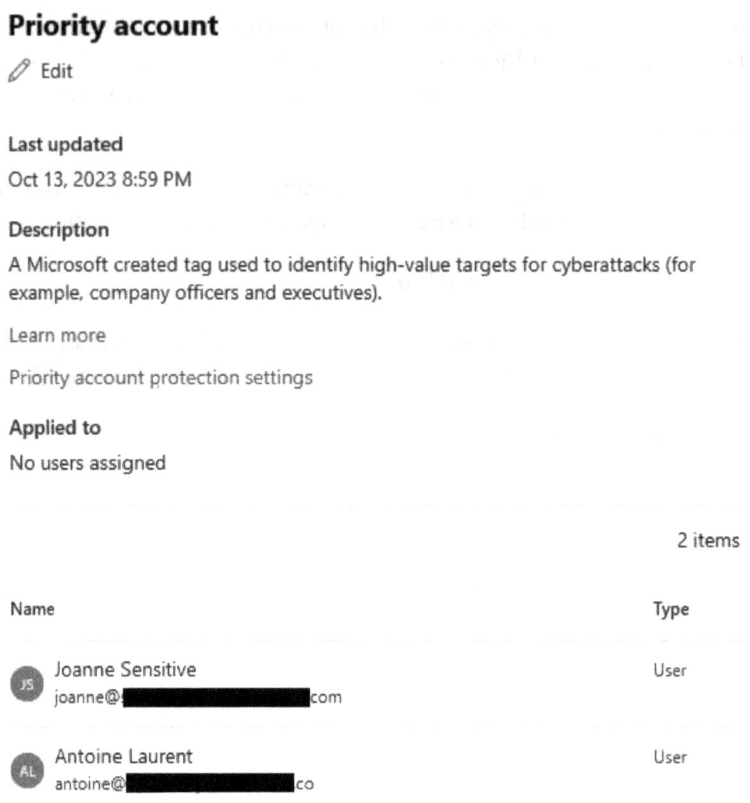

Priority account

🖉 Edit

Last updated

Oct 13, 2023 8:59 PM

Description

A Microsoft created tag used to identify high-value targets for cyberattacks (for example, company officers and executives).

Learn more

Priority account protection settings

Applied to

No users assigned

2 items

Name	Type
(JS) Joanne Sensitive joanne@███████████com	User
(AL) Antoine Laurent antoine@███████████co	User

Figure 9.40 – The Priority account user tag configuration

- **Custom user tags:** For other users that require enhanced reporting visibility or incident isolation, consider applying custom user tags.

Staying vigilant in these activities, despite their irregular nature, is pivotal in enhancing the security maturity of any organization.

Summary

In this chapter, we covered the concept of adopting a holistic view of environmental security by leveraging a suite of multi-layered defense tools. Recognizing and managing different security incident types is an intricate process, and we provided insightful methodologies to interpret the nuanced information they present. We also looked at the pivotal role of automated investigations and remediations, dissecting the processes behind these operations and the pertinent actions that trigger in response to identified threats. We also looked at false detections within the Defender for Office 365 automated systems and provided an approach for identification and rectification, including collaborating with Microsoft to fine-tune detection algorithms and remedy false positives effectively.

The next chapter will shift focus toward an analysis and utilization of reports, insights, and other pertinent data. It will aim to equip you with the knowledge to rapidly assess the health and efficiency of security operations through accelerated information access and visibility.

References

- Microsoft (2023a, March 1). *Optimize and correct security policies with configuration analyzer.* Microsoft Learn: `https://learn.microsoft.com/en-us/microsoft-365/security/office-365-security/step-by-step-guides/optimize-and-correct-security-policies-with-configuration-analyzer?view=o365-worldwide`

- Microsoft (2023b, November 15). *Incident response with Microsoft Defender XDR.* Microsoft Learn: `https://learn.microsoft.com/en-us/microsoft-365/security/defender/incidents-overview?view=o365-worldwide`

- Microsoft (2024a, February 8). *Submit files for analysis by Microsoft.* Microsoft Learn: `https://learn.microsoft.com/en-us/microsoft-365/security/defender/submission-guide?view=o365-worldwide`

- Microsoft (2024b, March 8). *What is Microsoft Defender XDR?* Microsoft Learn: `https://learn.microsoft.com/en-us/microsoft-365/security/defender/microsoft-365-defender?view=o365-worldwide`

Part 3 –
Making the Tool Work for
Your Organization

In this part, we visit the advanced topics that many organizations do not get a chance to implement, but which can make an environment much more robust to advanced threats. We will commence with how to use threat intelligence to enrich our signal and improve our visibility. We will also see how reports can be used beyond just an executive status document and as an additional tool to understand the organization's current security posture including how to best address identified security gaps. We will then move on to integrating the data in third-party tools into our Defender suite to further enhance our visibility. We will also discuss Copilot for Security and how it can remove the most tedious areas of security operations while speeding up our detection and issue resolution. Finally, we will touch upon how we can educate our end users using realistic simulations and how to identify what areas we should focus our training efforts on.

This part contains the following chapters:

- *Chapter 10, Magnifying the Unseen – Threat Intelligence and Reports*
- *Chapter 11, Integration and Artificial Intelligence*
- *Chapter 12, User Awareness and Education*

10

Magnifying the Unseen – Threat Intelligence and Reports

Embarking on a journey through **threat intelligence**, this chapter starts with the genesis of the concept and the meticulous methods utilized by Microsoft to gather solid threat data. We will probe into how this knowledge is effectively applied across the Defender product range, emphasizing the importance of real-time insights in shaping proactive defenses. The chapter unveils an assortment of Microsoft's available threat intelligence reports, scrutinizing their content and the role they play in shaping cybersecurity tactics. At the same time, we explore the comprehensive **Defender Threat Intelligence (Defender TI)** offering, which can further improve an organization's threat intelligence approach. Additionally, we examine the critical, uniquely crafted security reports for Defender for Office 365, clarifying how these documents act as guiding lights through the complex process of identifying hidden threats in your environment, ensuring that enterprises not only endure but excel in the face of cyber challenges.

This chapter will cover the following topics:

- An introduction to threat intelligence and what is available from Microsoft
- Threat intelligence's role in improving security operations
- Further enhancing our threat intelligence strategy with Defender TI
- Other security reports available to further improve our visibility

Let's continue our journey!

Threat intelligence comes to the rescue

Threat intelligence includes all the information we gather, study, and apply to understand the strategies, goals, and methods of those who pose digital threats. By harnessing threat intelligence, security teams can predict and obstruct cyberattacks as well as enhance the quick reactions needed during security breaches. Analysts build this intelligence, pulling from a variety of sources such as security tool data, the dark web, world events, and other organization's breach experiences. They then draw connections to get insights into both current and potential digital dangers. For threat intelligence to be truly beneficial for a company, it needs to be custom fit to the organization, detailed, measurable, actionable, and kept up to date (*Shweta, 2023*). When looking at threat intelligence, several types support different aspects of a security strategy:

- **Tactical intelligence**: This helps security teams during an attack, searching for hidden threats. Used primarily by **security operations centers (SOCs)**, it aids defense maneuvers.

- **Operational intelligence**: This focuses on tweaking security measures to prevent attacks and setting priorities. Executives use this to strategize and fund allocation.

- **Strategic intelligence**: This offers a broad view of threats at the global level, aiding all levels of the organization in assessing how their risks stack up against peers.

Integration of Defender for Office 365 with Microsoft Defender's other security tools means enhanced threat intelligence as it allows access to threat analytics from Microsoft's security experts as well as attacks occurring in the field. This empowers security teams against the latest threat tactics, dangerous actors, common vulnerabilities, and widespread malware.

You'll find threat analytics on the Microsoft Defender XDR's navigation bar or a dashboard card featuring prime threats to your organization. Visualizing ongoing campaigns and actionable tactics through these analytics provides security teams with the knowledge to make smart decisions. With the cyberthreat landscape expanding, it's vital to swiftly pinpoint new threats, figure out whether you're being targeted, understand threat impacts, and take the necessary steps to safeguard your networks against them. The analytics also includes an extensive strategy on how to counter these threats and is paired with data about your environment to show you if you're currently at risk and if you have the right defenses established.

The Threat analytics dashboard

The **Threat analytics** dashboard in the Defender portal is your go-to place to stay informed about potential cyber dangers. Find it either in the menu under **Threat intelligence | Threat analytics** or directly at `security.microsoft.com/threatanalytics3`. We can see this dashboard in the following figure.

Figure 10.1 – The Threat analytics dashboard

The dashboard prioritizes information beneficial to your organization and breaks down threat summaries into accessible sections:

- **Highest exposure threats**: This section displays the threats your organization is most vulnerable to, judging by the severity of associated vulnerabilities and the number of susceptible devices you have. Your team can use this information to prioritize the patching of vulnerabilities.

- **High-impact threats**: Here, threats posing the greatest risk to your organization are listed and prioritized by the quantity of active and resolved alerts they've generated. Your team can use this information to identify what security controls are lacking or missing in your environment.

- **Latest threats**: This area showcases the newest threat reports and counts of both active and resolved alerts. This area can quickly give you a glimpse of what is happening around the world and provide good topics for security team training.

On the lower part of this dashboard, you can see all the threats being reported that your environment might be susceptible to along with filters to help focus your efforts. We can see this section in the following figure.

Threat	Alerts ↓	Threat exposure level ⓘ	Misconfigured devices ⓘ	Vulnerable devices ⓘ	Report type
Threat overview: On-premises c...	49 active / 728	20 - Low	19	Not available	Tools & techniques
Tool profile: Mimikatz	32 active / 400	20 - Low	19	Not available	Tools & techniques
Technique profile: Kerberos atta...	32 active / 391	19 - Low	18	Not available	Tools & techniques
Storm-0300 ransomware activity	32 active / 389	20 - Low	19	Not available	Activity groups
Technique profile: Antivirus tam...	32 active / 234	20 - Low	19	Not available	Tools & techniques
Tool profile: Impacket	15 active / 209	20 - Low	19	Not available	Tools & techniques
Activity profile: Cryptojacking	12 active / 53	0 - Low	Not available	Not available	Attack campaigns
Activity profile: OAuth apps use...	7 active / 333	0 - Low	Not available	Not available	Tools & techniques

Figure 10.2 – View of current threats impacting the environment

This view can be modified when trying to perform a focused search by using the following filters:

- **Report type**: You can filter according to the report type, which includes attack campaigns, tools and techniques, activity groups, and other options.

- **Threat tags**: You can filter reports according to the threat variety they discuss with options including ransomware, extortion, phishing, hands-on keyboard, and many others.

The Microsoft Threat Intelligence team meticulously tags each report, which not only ensures easy visibility of threat types but also helps an organization align efforts with other security approaches such as cybersecurity frameworks and attack-related guidance such as the one provided by MITRE. A quick look on top of the **Threat analytics** dashboard allows a quick assessment of what the organization should be focusing on thanks to these tags.

The threat analytics report

Every report within **Threat analytics** is comprehensive and allows for a quick assessment of the impact of the threat on the environment. Each report is broken down into various sections:

- **Overview**: This is a summary presenting a quick description of the threat, published date, and any alert of exposure information in your environment as it relates to the threat.

- **Analyst report**: This is an in-depth exploration of the threat compiled by Microsoft security experts, which breaks down the threat's nature, tactics, potential objectives, and recommendations on how to hunt and protect against this threat.

- **Related incidents**: This is a record of incidents connected to the threat, providing context and scope. For major incidents, this is a good way to track the impact discovered and work ongoing.

- **Impacted assets**: This is a list of resources in your organization that have been or could be affected including devices, users, mailboxes, apps, and cloud resources.

- **Endpoints exposure**: For environments using Defender for Endpoint, this tab provides information on current vulnerabilities on endpoints that increase the exposure of devices to this threat. This information is key to ensure prompt containment of a threat.

- **Recommended actions**: This page provides security controls that can protect and prevent further advances of a threat in your environment. Depending on what defender security tools are deployed in your environment, some of these controls will be reported as "addressed."

We will look at these sections in greater detail in subsequent sections.

Overview

The **Overview** section in the threat analytics report offers a sneak peek at the exhaustive analysis provided in the full analyst report. It features graphical depictions that convey the threat's scale regarding your organization and unveils exposure because of misconfigurations and outstanding patches (*Microsoft, 2023b*). We can see this section in the following figure.

Figure 10.3 – The Overview section of a threat report

The information gathered in this section alone can help gauge the priority we should assign to mitigating this threat. The main sections related to the impact on our organization are as follows:

- **Impacted assets**: This graph tallies impacted assets to include users, devices, cloud resources, apps, and mailboxes along with which of these assets still have unresolved alerts. This panel is a great way to gauge progress in resolving an ongoing threat.

- **Alerts over time**: This graph shows active and resolved alerts over a period, revealing response times to threats. It can also quickly tell us whether our efforts are containing the threat or whether the threat impact is expanding.

- **Endpoints exposure**: This graph depends on our organization using Defender for Endpoint and provides a quick reference of what vulnerabilities are open in devices that could expose them to the threat at hand.

- **Related incidents**: This part looks at the threat's current influence within your company by showcasing the number of active alerts and the associated active incidents, along with their severity.

- **Recommended actions**: This graph shows recommended controls to help mitigate the threat and our progress in implementing these.

The **Overview** section can quickly give you a glimpse of the ongoing efforts and offers valuable data not only on the mitigation progress but any gaps in how your security operations processes align with real-life threats.

Analyst report

In the **Analyst report** section, you'll find a detailed account from cybersecurity experts. Most of these reports offer rich descriptions of the attack strategies employed, correlating them with tactics and techniques from the MITRE ATT&CK framework. We can see one of these reports in the following figure.

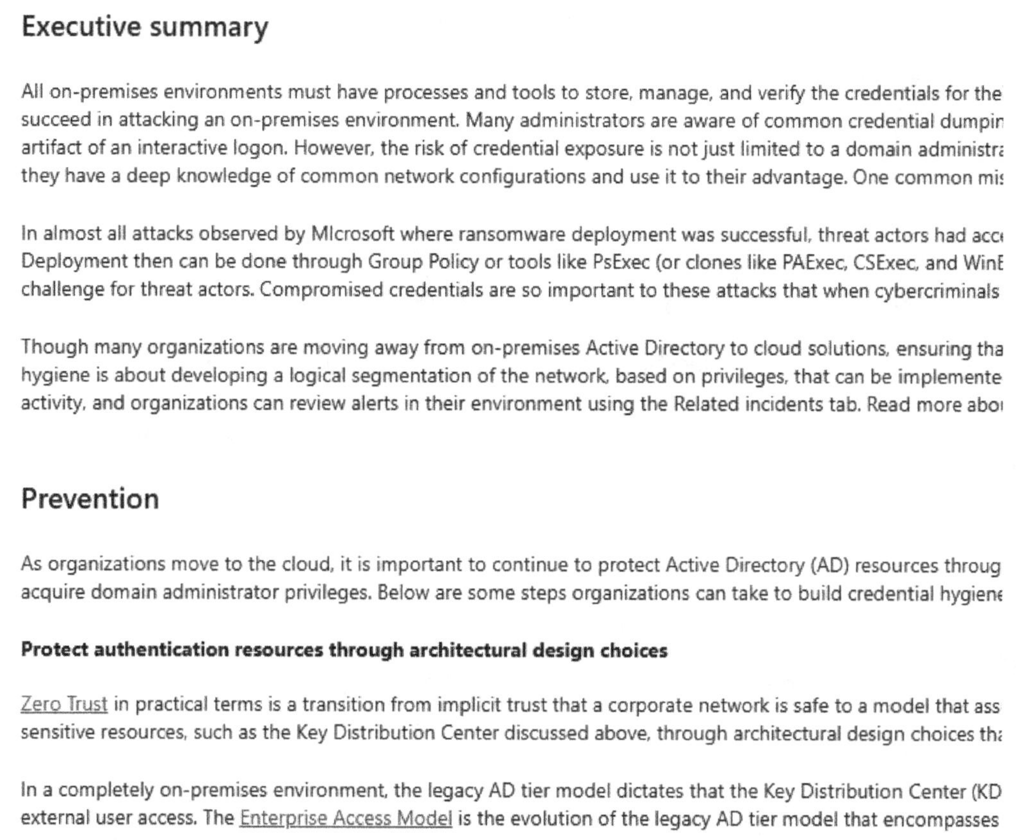

Executive summary

All on-premises environments must have processes and tools to store, manage, and verify the credentials for the
succeed in attacking an on-premises environment. Many administrators are aware of common credential dumpin
artifact of an interactive logon. However, the risk of credential exposure is not just limited to a domain administra
they have a deep knowledge of common network configurations and use it to their advantage. One common mis

In almost all attacks observed by MIcrosoft where ransomware deployment was successful, threat actors had acce
Deployment then can be done through Group Policy or tools like PsExec (or clones like PAExec, CSExec, and WinE
challenge for threat actors. Compromised credentials are so important to these attacks that when cybercriminals

Though many organizations are moving away from on-premises Active Directory to cloud solutions, ensuring tha
hygiene is about developing a logical segmentation of the network, based on privileges, that can be implemente
activity, and organizations can review alerts in their environment using the Related incidents tab. Read more abo

Prevention

As organizations move to the cloud, it is important to continue to protect Active Directory (AD) resources throug
acquire domain administrator privileges. Below are some steps organizations can take to build credential hygiene

Protect authentication resources through architectural design choices

Zero Trust in practical terms is a transition from implicit trust that a corporate network is safe to a model that ass
sensitive resources, such as the Key Distribution Center discussed above, through architectural design choices tha

In a completely on-premises environment, the legacy AD tier model dictates that the Key Distribution Center (KD
external user access. The Enterprise Access Model is the evolution of the legacy AD tier model that encompasses

Figure 10.4 – The Analyst report section

These reports come packed with comprehensive lists of ways to fortify your defenses and insightful
tips for threat hunting. This section will be split into an executive summary, prevention guidance,
detection guidance, recommended response in case the threat becomes active in the environment,
MITRE ATT&CK-related techniques and tactics, and any related articles published by Microsoft
Security on the threat.

Related incidents

The **Related incidents** tab catalogs all the incidents and alerts observed in your environment that are
linked to the threat. We can see this section in the following figure.

Figure 10.5 – The Related incidents section

Security teams can use this section to get more details on the response tasks, delegate appropriately, and review progress on threat mitigation and impact.

Impacted assets

This section provides a list of users, devices, mailboxes, apps, and cloud resources impacted by the threat. We can see this section in the following figure.

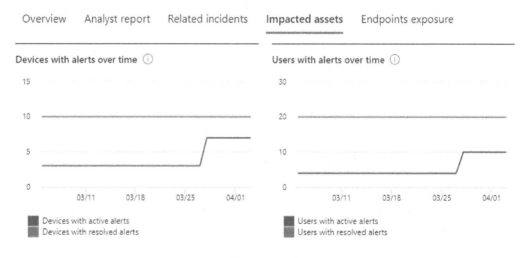

Figure 10.6 – The Impacted assets section

Graphs are provided for each one of the impacted asset types showing active and resolved alerts. A list is also provided at the bottom of this section for each of the impacted assets, as shown in the following figure.

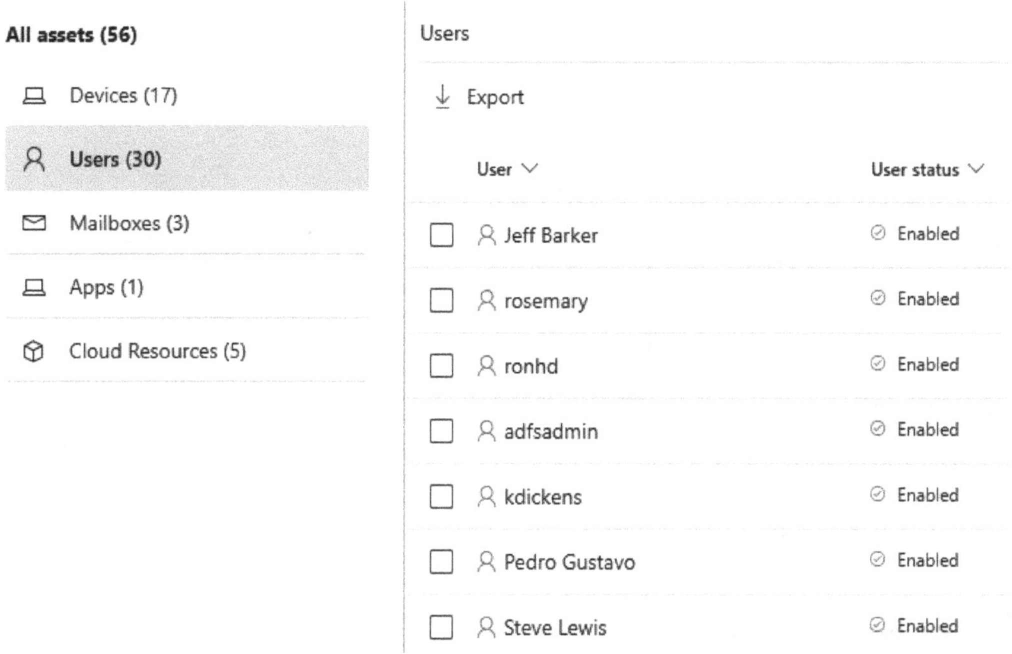

All assets (56)

| Devices (17) |
| Users (30) |
| Mailboxes (3) |
| Apps (1) |
| Cloud Resources (5) |

Users

↓ Export

User ∨	User status ∨
☐ Jeff Barker	⊘ Enabled
☐ rosemary	⊘ Enabled
☐ ronhd	⊘ Enabled
☐ adfsadmin	⊘ Enabled
☐ kdickens	⊘ Enabled
☐ Pedro Gustavo	⊘ Enabled
☐ Steve Lewis	⊘ Enabled

Figure 10.7 – A list of the impacted assets and details

It is important to keep in mind that this section's level of detail will depend on how many Defender security products are deployed in the environment. The products needed will depend on the asset, for example, devices will depend on data from Defender for Endpoint, users on Entra ID Protection, mailboxes will depend on Defender for Office 365, Apps will depend on Defender for Cloud Apps, and cloud resources will depend on a mix of all these applications along with Defender for Cloud (depending on the resource).

Endpoint exposure

This section looks at how exposed the endpoints in your organization are to the threat in the report. The information in this section is focused on trying to prevent damage from the threat and the focus on device vulnerability information gathered using Defender for Endpoint, along with any misconfiguration information detected. An exposure score is provided on the page to help provide context on how these misconfigurations and vulnerabilities increase the chance of impact.

Recommended actions

This section aligns with the security best practices and controls recommended for your environment and is heavily aligned with an organization's secure score. A **secure score** is an easy way for a security team to view the secure posture of an organization against best practices and other organizations of similar size, as seen in the following figure.

Microsoft Secure Score

Overview   Recommended actions   History   Metrics & trends

Microsoft Secure Score is a representation of your organization's security posture, and your opportunity to improve it.

Applied filters:

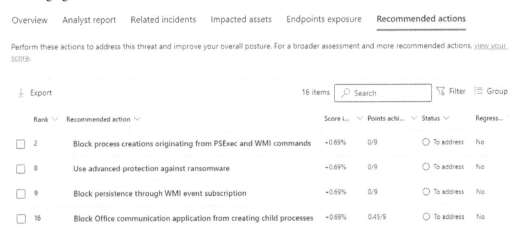

Your secure score   Include ∨

Secure Score: 55.15%

722.44/1310 points achieved

Actions to review				
Regressed	To address	Planned	Risk accepted	Recently added
11	133	0	0	0
Recently updated				
0				

Breakdown points by: Category   ∨

Identity   63.29%

Data   88.89%

Top recommended actions

Recommended action	Score impact	Status	Category
Block Office applications from creating executable content	+0.69%	○ To address	Device
Block process creations originating from PSExec and WMI co...	+0.69%	○ To address	Device
Block all Office applications from creating child processes	+0.69%	○ To address	Device
Block Win32 API calls from Office macros	+0.69%	○ To address	Device
Block execution of potentially obfuscated scripts	+0.69%	○ To address	Device

Figure 10.8 – An organization's secure score

The **Recommended actions** section of the threat report looks at controls that are closely related to the way the threat behaves and provides guidance on which controls have been addressed, which are pending, and security tools recommended to implement the control. We can see this section in the following figure.

Overview   Analyst report   Related incidents   Impacted assets   Endpoints exposure   **Recommended actions**

Perform these actions to address this threat and improve your overall posture. For a broader assessment and more recommended actions, view your score.

↓ Export   16 items 🔍 Search   ▽ Filter ≡ Group

Rank ∨	Recommended action ∨	Score i... ∨	Points achi... ∨	Status ∨	Regress...
□ 2	Block process creations originating from PSExec and WMI commands	+0.69%	0/9	○ To address	No
□ 8	Use advanced protection against ransomware	+0.69%	0/9	○ To address	No
□ 9	Block persistence through WMI event subscription	+0.69%	0/9	○ To address	No
□ 16	Block Office communication application from creating child processes	+0.69%	0.45/9	○ To address	No

Figure 10.9 – The Recommended actions tab

This tab plays an important role during the active containment of a breach as it gives concise guidance on what to do for each control. For scenarios in which an active breach is not occurring, it is recommended to focus on improving the organization's secure score as this will decrease the risk of any future breaches occurring.

Don't forget about email notifications!

To keep up with current security events, configuring **Email notifications** for any new or updated **Threat analytics** reports can be configured easily. Follow these steps to ensure you're always updated with the newest insights:

1. Visit the Defender portal (`security.microsoft.com`) and on the left-hand menu, click on **Settings**, then on the **Settings** page, select **Microsoft Defender XDR**.

2. Navigate to the **Email notifications** section and select **Threat analytics**. Click on the **Create a notification rule** button to start the wizard. We can see this in the following figure.

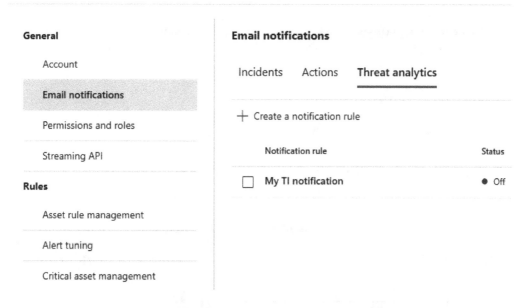

Figure 10.10 – The Threat analytics notification page

3. Type a unique name rule and optionally a description that will help your team quickly identify what the intent of the rule is. Ensure that the checkmark for **Turn rule on** is checked and click on the **Next** button. We can see this in the following figure.

Name your notification rule

Provide a name and a description for this notification rule to make it easier to identify and manage.

Name *

threat analytics notification rule

Description

This rule will notify the security team on important threat analytics data.

☑ Turn rule on

Figure 10.11 – The rule name section

4. Decide which reports you wish to receive alerts for, either all new/updated reports or only those that are tagged specifically. Once completed, click on the **Next** button. We can see this section in the following figure.

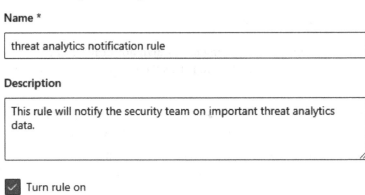

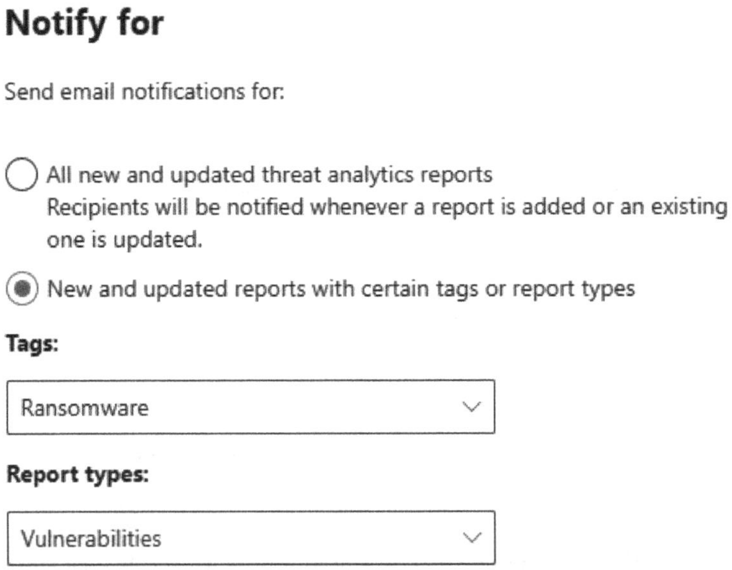

Notify for

Send email notifications for:

○ All new and updated threat analytics reports
   Recipients will be notified whenever a report is added or an existing one is updated.

◉ New and updated reports with certain tags or report types

Tags:

Ransomware ⌄

Report types:

Vulnerabilities ⌄

Figure 10.12 – Configuring the notification report scope

5. In the **Recipients** section, add the email addresses of those who should receive these updates and click the **Send test email** button to send a test email to confirm the format of the notifications. If everything is correct, click the **Next** button to continue. We can see this section in the following figure.

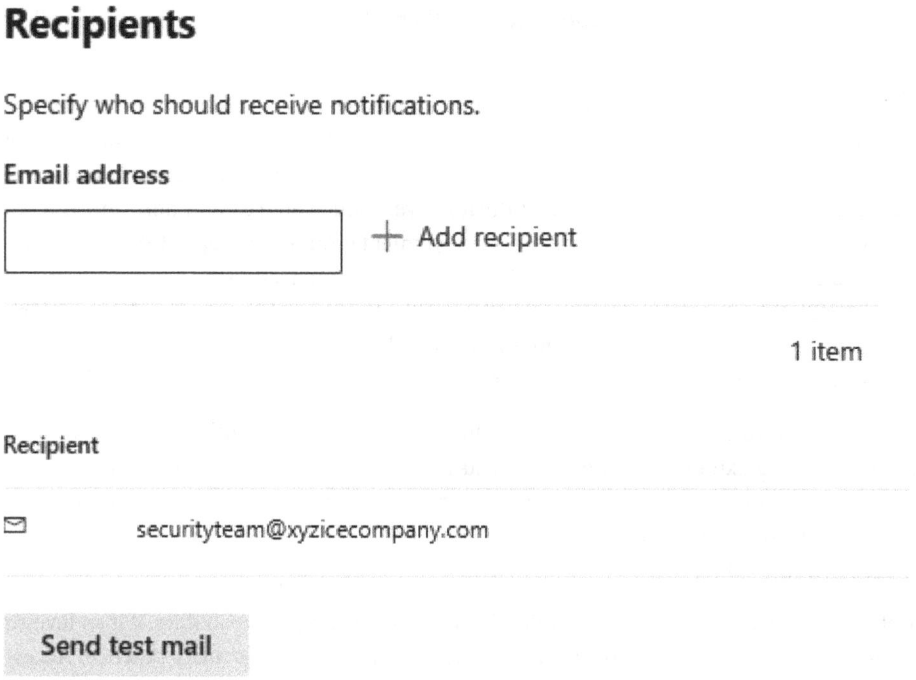

Recipients

Specify who should receive notifications.

Email address

[   ]   + Add recipient

1 item

Recipient

✉   securityteam@xyzicecompany.com

[Send test mail]

Figure 10.13 – The Recipients section

6. A new screen will show along with the review of the notification rule settings. If you need to edit anything, you can either click the **Back** button or click on **Edit** in any of the sections. If all the settings look correct, click the **Create rule** button to finish, and the **Done** button to exit the setup. Your new notification rule will now show up in your **Threat analytics** email notification list.

When creating notification rules, it is important to ensure the rule has a rationale for being implemented and its implementation is limited, as this ensures the security team does not suffer from notification fatigue and start ignoring these notifications. These notifications, along with the threat analytics report, are important tools used by security teams to quickly sort through the noise and find the signals that identify an active or previously active attack. These tools' efficacy is not only dependent on how well the security team aligns them to security operations processes but also on the quality of the threat intelligence used to enrich these signals. The following section will touch on ways to further improve this enrichment beyond what is offered out of the box by using Defender TI.

Going a step further with Defender TI

Defender TI serves as an integrated solution designed to optimize various aspects of cybersecurity operations encompassing incident response, threat analysis, and vulnerability management. By utilizing Defender TI, analysts are empowered to focus on generating actionable insights by examining threat actor tactics and coordinating data more efficiently than sifting through extensive data collection and parsing.

The gathering of essential data to evaluate potential threats such as suspicious domains, hosts, or IP addresses can become an arduous task. Analysts often navigate through disparate sources to collect DNS records, WHOIS details, SSL certificate information, and malware indicators. These data pieces are crucial for understanding compromise indicators, yet their scattered and non-uniform nature can hinder timely and comprehensive analyses. This fragmentation of data repositories proves to be a significant drain on cybersecurity teams, as it diverts their attention from proactive defense measures to time-intensive data management tasks. Add to this issue that it might be challenging to determine which intelligence is most relevant to the environment and threats experienced makes this a key task that impacts security operations efficiency before and during an incident (*Microsoft, 2023a*).

Defender TI goes beyond the initial threat intelligence data provided with the Defender security products by providing extra information on threats and helps speed the correlation of **indicators of compromise** (**IOC**) found in the environment with relevant content and vulnerabilities discovered. All of this is provided in a way that allows for easier collaboration amongst the licensed Defender TI users in the environment.

While Defender TI requires a license for its full suite of threat intelligence capabilities, it is an investment that organizations are encouraged to consider as they advance in their security practice. Accessible via the Defender portal, the Defender Threat Intelligence features, such as intel profiles, intel explorer, and intel projects, are neatly organized for user convenience. Even for those not holding a license, a free version with a basic set of profiles is available, ensuring a range of options suited to different operational needs.

Intel Profiles

Intel Profiles within Microsoft Defender TI offer a deep dive into the threat environment, spotlighting crucial knowledge on vulnerabilities, adversary strategies, and the malicious infrastructure powering cyber assaults. These profiles harness an eye-opening array of 65 trillion signals and meld them with the discernment of over 8,500 cybersecurity experts, distilling this global intelligence into immediate, actionable insights. By unraveling threat actor techniques, infrastructure, and playbooks, security squads can proactively fortify their defenses against would-be infiltrations. We can see the profile details for a threat actor called Aqua Blizzard in the following figure.

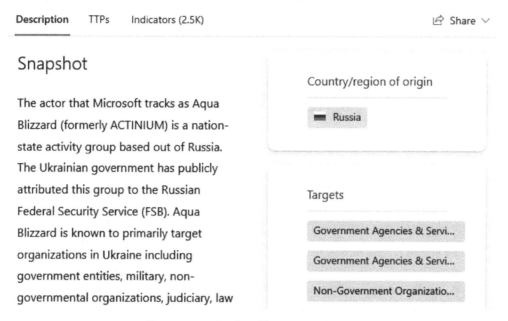

Figure 10.14 – Intel profile on Aqua Blizzard

In Defender TI, Intel Profiles systematically organize adversaries by geography, targeted sectors, and their preferred attack patterns. They consolidate a comprehensive list of IOCs linked to their infrastructure plus daily updates on the adversaries' **tactics, techniques, and procedures** (**TTPs**). Penetration testing products such as Cobalt Strike that are frequently utilized in cyber offensives are carefully tracked by Defender TI, allowing organizations to spot, and sidestep, potential threats proactively by identifying malignant domains and IP addresses.

The Intel Profiles are precisely divided into two categories: **threat actors**, which is another name for the attackers identified in the report, and **tools**, which are what these attackers use to perform their attacks. This demarcation provides an in-depth portrayal of internet threat actors and their modus operandi, covering their targets, attack mechanics, infrastructure, and digital toolsets. Unlike static datasets or intelligence feeds from other sources, Intel Profiles are perpetually refreshed to ensure organizations have the latest insights for informed security strategies.

Defender TI empowers organizations with the capability to detect adversary infrastructure on any scale, no matter whether it's a singular antagonist or a multitude, ensuring swift and effective threat response. Intel in Defender TI encapsulates detailed dossiers on threat actor collectives, inclusive of aliases, preferred targets, **Common Vulnerabilities and Exposure** (CVE), any associated state sponsorship, their **TTPs**, and **IOCs**. We can see an example of the **TTPs** reported for Aqua Blizzard in the following figure.

Aqua Blizzard phishing email

Aqua Blizzard has inserted web bugs within the body of a phishing message. Web bugs are small external image references that enable the actor to track when a message has been opened and rendered. While not malicious themselves, web bugs might indicate that the email is intended for malicious use, as shown below.

```
<img src="http://eyeofra[.]ru/images/icons/312rz45d7/43oFI4b/cached.gif"
height="0" width="10" style="height:0px;width:10pxpx">
```

Figure 10.15 – Aqua Blizzard TTPs section

Tools in the threat actor's arsenal comprise software or scripts designed for exploiting weaknesses or methods for conducting attacks. Some tools are crafted by the offenders themselves, while others could be legitimate software designed for penetration testing, used by red teams, that threat actors misappropriate. Such commercial tools, known for their up-to-date exploit libraries and ease of deployment against targets, become a part of the threat actors' campaigns when repurposed for malicious intent. These tactics facilitate the concealment of their operations since the use of popular, non-custom software reduces the risk of being associated with a specific group.

Intel Profiles in Defender TI offer meticulous details on the implements of digital war, including their intended use, the typical users, the threat actors who employ them, and evidence of their existence online. Having access to this information primes security teams to predict adversary tactics and take effective countermeasures, safeguarding their organization against cyber threats.

Intel Explorer

The **Intel Explorer** page offers a central location to understand not only what the latest threats are via articles, but also to search for IOC, threat actors, and other information. This page serves as a key to making informed decisions when customizing the security strategy and security controls to ensure a more proactive stance.

Intel articles

Intel articles are well-crafted pieces penned by Microsoft that shed light on the nuances of cyber threats, spanning from the actors and their digital armaments to the assaults they perpetrate and the security loopholes they exploit. These articles, placed prominently within the Defender TI, transcend mere blog posts; they not only encapsulate a summary of diverse threats but also forge connections to pragmatic content and pivotal IOCs, empowering users to spring into protective action. By weaving such technical details into the narrative, Microsoft equips users with the resources to monitor the evolution of threat actors, their tools, their methods of attack, and any exploitable vulnerabilities adeptly and persistently. Each article on the **Intel Explorer** page is organized into distinct sections for ease of understanding and use.

Description

The **Description** part of the article detail interface presents a comprehensive view of the incident or the profile of the attacker in question. Depending on the nature of the source, such as **open-source intelligence (OSINT)** bulletins, the information could be concise. On the other hand, more exhaustive reports, especially those enriched by Microsoft with additional insights, tend to be lengthier. We can see the **Description** section in the following figure.

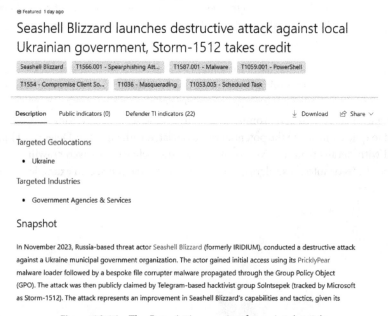

Figure 10.16 – The Description section for an intel article

These extended descriptions can incorporate various elements such as visual aids, hyperlinks to in-depth material, quick links for in-platform searches within Defender TI, excerpts of code used by attackers, and even specific firewall rules that could be implemented to thwart the reported attack.

Public indicators

The **Public indicators** segment of the display highlights the IOCs that have been previously released and are associated with the article. The list of IOCs will include anything from IP addresses, file hashes, and URLs, which can be clicked to direct you either to the pertinent information housed within the Defender TI ecosystem or to external resources that are relevant to the indicators listed.

Defender TI indicators

In the **Defender TI indicators** section, you'll find a curated list of IOCs identified by the Defender TI's research team and integrated into the articles. Clicking on these links will pivot to the related data within Defender TI itself or redirect to the appropriate external sources where these indicators are discussed or analyzed.

Indicator reputation scoring

Defender TI offers an insightful feature that assesses reputation scores for any host, domain, or IP address. You can retrieve these scores by clicking on indicators within an intel article or searching directly via the Intel Explorer. These scores are crucial for swiftly evaluating the trustworthiness of a given entity, clarifying its relation to potentially malicious or dubious activity. The information provided by Defender TI includes the timeframes in which these entities were first and last noticed, **autonomous system numbers** (**ASNs**), geographical locations, linked infrastructure, and any applicable rules affecting the reputation score.

The insight into IP reputation is an asset for gauging the security of your network's attack surface and is invaluable for probing unfamiliar hosts, domains, or IP addresses that emerge during security investigations. Reputation scores reveal past malevolent actions or suspicious engagements linked to the entity, along with correlated compromise indicators. These scores stem from complex algorithms that are crafted to quickly measure the potential risk associated with an entity. Defender TI proprietary data, enriched with details from the Microsoft threat researcher team's web crawling systems and external IP records, feeds into these algorithmic calculations. We can see an example of this score in the following figure.

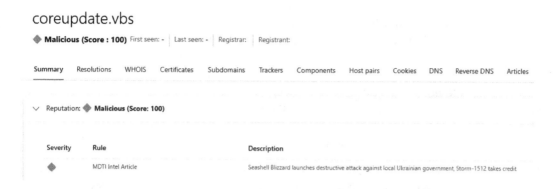

Figure 10.17 – The reputation score for an IOC

Displayed as a numerical value on a scale from 0 to 100, a score of 75 or more indicates a clearly malicious artifact. A score of 50 to 74 indicates a suspicious artifact in which three or more machine learning rules were triggered. A score of 25 to 49 indicates that there is a possibility of it being a malicious artifact as at least two machine learning rules were triggered, but not enough data is available to confirm at this time. Finally, a score of 0 to 24 indicates that not enough information is available to confirm or deny the artifact's malicious relation with certainty with a green color indicating one machine learning rule being triggered in the past and a grey color indicating no machine learning rules triggered.

Intel projects

Defender TI empowers users to curate private projects tailored for personal or team use, aimed at organizing and tracking indicators of interest and potential IOCs found during investigations. Each project is equipped with a compilation of related artifacts and maintains a comprehensive log that holds information regarding names, descriptions, involved collaborators, and monitoring profiles. We can see a project in the following figure.

Figure 10.18 – Intel project

For enhanced situational awareness, users can create projects in which multiple related suspicious artifacts such as IP address, domain, or host can be included, along with some context on why or how it was added. The Defender TI platform facilitates the creation of various Intel project types, allowing for the systematic organization of investigative findings, including both indicators of interest and IOCs. Other team members can then access the project from the Defender portal (`security. microsoft.com`) by using the left-hand menu, **Threat intelligence | Intel projects**. This approach offers immediate insight into the indicator's relevance within the broader context of the investigation, ahead of delving into more extensive datasets.

Private projects can also be created for team members who are still determining the validity of their findings. The privacy of these projects can be changed at any time when the project is ready for team-wide collaboration by clicking the **Edit project** button and changing the privacy. We can see this in the following figure.

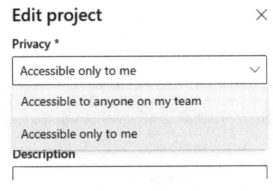

Figure 10.19 – Changing a project privacy setting

Furthermore, users have the option to download artifacts from a project with a simple click on the download icon. This feature proves invaluable for threat-hunting teams who can use these downloaded artifacts to initiate blocks on identified IOCs or to craft new detection rules for integration into their SIEM systems.

Creating a project

There are two convenient methods to initiate a project within the Defender TI sphere. Through the **Intel projects** page in the Defender portal, users can easily glimpse a dashboard populated with their own projects as well as those shared by other Defender TI users in the organization. Here, initiating a new project is as straightforward as clicking the **New project** button in the top-right corner. This approach is typically the one most people are familiar with as it is most intuitive and doesn't require any extra steps to perform. It will create a new blank project with no items on it, so you will still be required to associate an artifact or other intelligence items for the project to make sense to other security team members.

Another approach is during searches within the **Intel Explorer** page. When an artifact of interest has been identified, you have the option to incorporate it into a current project or new project by selecting **Add to project** and selecting either an existing or new project, as seen in the following figure. This approach will either allow for the creation of a new project and adding the artifact of interest to it or allow for adding the artifact to an existing project. Mature security teams will tend to use this approach during their investigations as it saves time and better aligns with typical daily security processes.

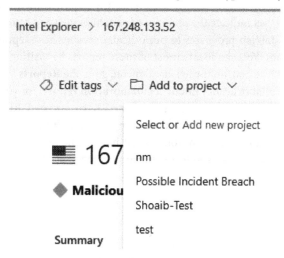

Figure 10.20 – Adding a malicious IP address to a project

Defender TI enables users to conduct a variety of searches, making it vital to adopt an intelligence-gathering strategy that first delivers wide-ranging results before delving into nuanced examinations of specific indicators. Take, for instance, conducting a file hash search through Defender TI's home page. Here, one can uncover what articles mention the file hash, which allows you to piece together a contextual narrative regarding the file hash that isn't immediately apparent from simply accessing its specific **Data** tab. Questions such as whether the file hash has been flagged as a potential ransomware file, the identity of the associated threat actor, additional linked IOCs, TTPs utilized, and the intended targets can all be addressed through this initial broad search.

Also, remember that Defender TI isn't just about independent searches, it is intended to be a collaborative platform where users can enhance the investigative process by creating projects, aggregating related indicators to these projects, and adding colleagues as collaborators when more than one person is addressing the same case. This collaborative feature is designed to streamline the analytical effort, avoiding duplication of work on the same IOCs and aiming to accelerate the overall workflow.

The value of threat intelligence cannot be undermined as the constantly changing nature of technology, attacker interest, world politics, and vulnerabilities discovered means that even the most mature security teams will miss suspicious activity. We can see this scenario play out in many high-profile breaches such as those observed in the SolarWinds and Equifax incidents. All security teams should have at least one person spending several hours daily verifying new threat intelligence information and building queries for other team members to effectively use the threat intelligence in their hunting. In the next section, we will look at another tool that can further help customize these queries along with identifying other gaps, security reports.

Security reports

Defender for Office 365 includes a suite of reports to showcase the efficacy of its security measures. These reports can provide visibility on changes or events in your environment that might not have been triggered by any alerts yet indicate the presence of a sophisticated bad actor in the environment. Mature security teams establish processes to periodically review these reports and validate the behavior reported on them. You can locate most of these reports by visiting the Defender portal (`security.microsoft.com`); on the left-hand menu, go to the **Reports** section and follow this path: **Reports | Email & collaboration | Email & collaboration reports**, or you can directly access it via the `security.microsoft.com/emailandcollabreport` URL. There are also some mail-specific reports included with the **Exchange admin center** (**EAC**) (`admin.exchange.microsoft.com`), which can be accessed from the left-hand menu by selecting **Reports** and either selecting **Mail flow** or **Migration** (*Microsoft, 2024*). Let's now look at the different kinds of security reports available in Defender.

Permissions required

These reports are not viewable to all users and a set of permissions are required for your team to access them. These permissions can be assigned via Microsoft Entra ID by adding the user to one of the following groups:

- **Global Administrator**: The highest privilege role in a Microsoft Entra environment, which allows the user to perform any task in the Entra tenant. This role should have limited membership along with both multifactor authentication and **Entra Privileged Identity Management just-in-time access** configured due to its high impact if a user account is compromised. This is the role the organization management role is required to allow users to use the **Create schedule** and **Request report** actions in reports.

- **Organization Management (on-premises only role)**: One of the highest privilege built-in exchange server roles that allows the user to perform almost any task in the exchange environment. This role will only be available when an environment also uses an on-premises installation of an Exchange server. This role should have limited membership due to its high impact if a user account is compromised. This role or the Global Administrator role is required to allow users to use the **Create schedule** and **Request report** actions in reports.

- **Security Administrator**: A high privilege Microsoft Entra built-in role that allows the user to read security information and reports along with managing the entire configuration for all the Defender security tools, and some aspects of the Entra ID tenant, and Office 365 environment configuration.

- **Security Reader**: A Microsoft Entra built-in role that allows the user to read security information and reports without permitting configuration changes. While this role cannot perform changes in an environment, it is recommended that only members of the security team are provided with access to this role due to the visibility it provides on the security configuration of the environment.

- **Global Reader**: A Microsoft Entra built-in role that is also considered the read-only version of the Global Administrator role. While this role cannot perform changes in an environment, membership should be controlled as this role is a common target for attackers when performing enumeration and reconnaissance of an environment.

Note that for the Microsoft Entra built-in roles, due to the tendency of attackers to target user accounts with this role, it is key to protect them as part of our multi-layer approach to security. Part of the measures to protect users with these roles should include as a minimum configuring both multifactor authentication (or **passwordless authentication**) and **Entra Privileged Identity Management just-in-time access** for the role, which ensures the account does not always have the privileges enabled all the time and decreases the chance of it being used during an attack. User groups should be used to easily manage and limit role membership to a need-to-use basis; we can also use automated access reviews on these user groups to prevent privilege creep. We can see access review being configured under Microsoft Entra ID Identity Governance in the following figure.

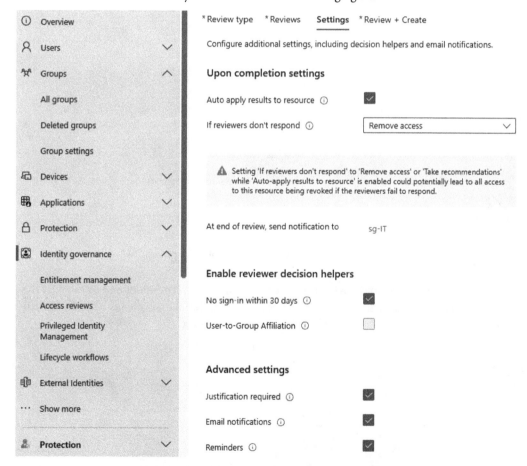

Figure 10.21 – Access review being configured

A second approach but limited to the Defender portal only (no capabilities via PowerShell) is by using the **Microsoft Defender XDR Unified RBAC**, an approach at single permissions management that all Defender security products are slowly being migrated to. At the time of writing this, only some parts of Defender for Office 365 use these permissions, and no granular permission option is available. While this approach seems more complicated at first sight, as Defender security products evolve, it is expected that this will be the way to manage permissions in the future, so it is important to familiarize ourselves with it. Let's learn more about this approach:

1. To configure these, we should first be global administrators in the tenant. We need to visit the Defender portal (`security.microsoft.com`) and on the left-hand menu, select **Settings**, and then select **Microsoft Defender XDR**.

Settings

Name ∨	Description ∨
⚙️Microsoft Defender portal	General settings for the Microsoft Defender portal
⛊Microsoft Defender XDR	General settings for Microsoft Defender XDR
🖥️Endpoints	General settings for endpoints
⬦Email & collaboration	General settings for email & collaboration
⚲Identities	General settings for identities

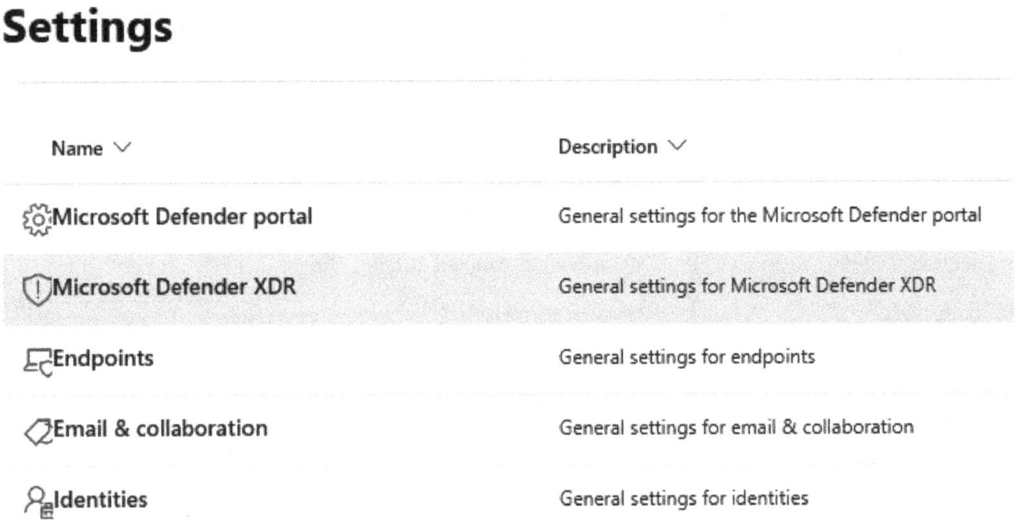

Figure 10.22 – The settings page in the Defender portal.

2. Once the **Microsoft Defender XDR** page opens, we can select **Permissions and roles** under the **General** section of the left menu. We will need to ensure that the workloads related to **Email & Collaboration** are enabled, as shown in the following figure.

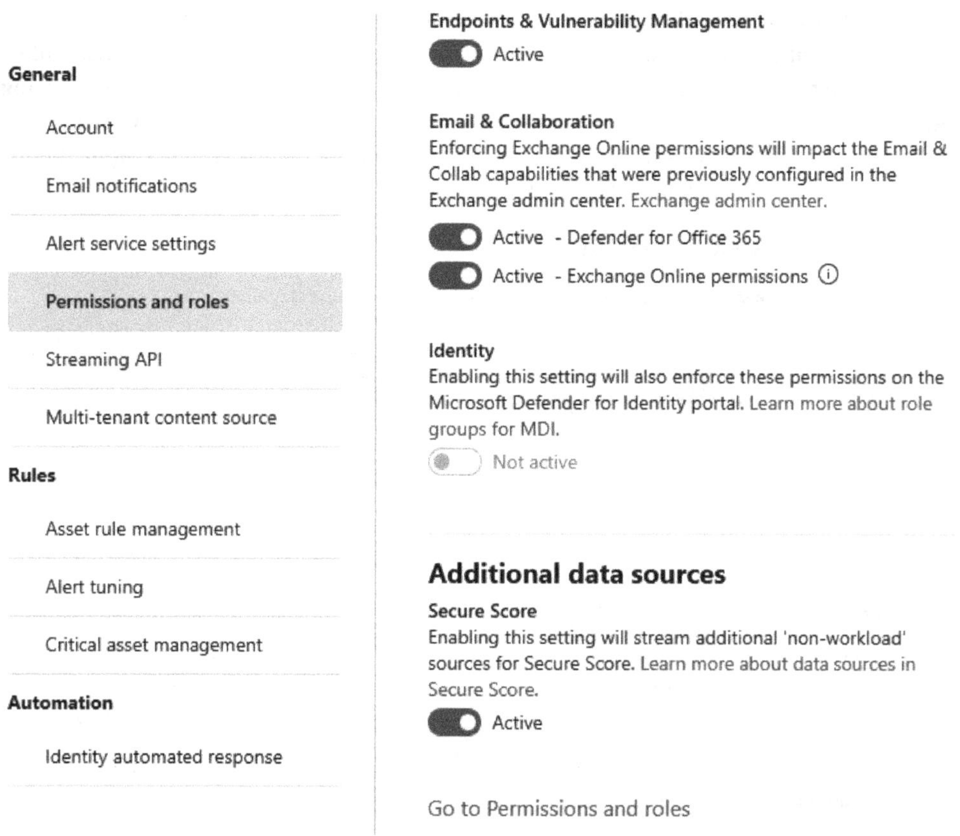

General

Account

Email notifications

Alert service settings

Permissions and roles

Streaming API

Multi-tenant content source

Rules

Asset rule management

Alert tuning

Critical asset management

Automation

Identity automated response

Endpoints & Vulnerability Management

🔘 Active

Email & Collaboration
Enforcing Exchange Online permissions will impact the Email &
Collab capabilities that were previously configured in the
Exchange admin center. Exchange admin center.

🔘 Active - Defender for Office 365

🔘 Active - Exchange Online permissions ⓘ

Identity
Enabling this setting will also enforce these permissions on the
Microsoft Defender for Identity portal. Learn more about role
groups for MDI.

⚪ Not active

Additional data sources

Secure Score
Enabling this setting will stream additional 'non-workload'
sources for Secure Score. Learn more about data sources in
Secure Score.

🔘 Active

Go to Permissions and roles

Figure 10.23 – The Permissions and roles section of the Microsoft Defender XDR settings

3.   Once the proper workloads are enabled, we can click on the **Go to Permissions and roles** link
     at the bottom of the page. The **Permissions and roles** page will open, and we will need to click
     on the **Create custom role** option on the top menu, as seen in the following figure.

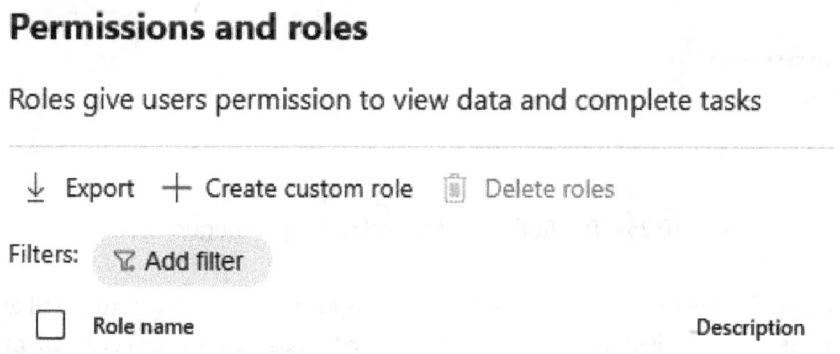

Permissions and roles

Roles give users permission to view data and complete tasks

↓ Export   + Create custom role   🗑 Delete roles

Filters: ▽ Add filter

☐   **Role name**   **Description**

Figure 10.24 – The Permissions and roles page

4.   The custom role wizard will start, and we will provide a descriptive role name and description to help our team easily identify this role in the future, then we can click on the **Next** button. The **Choose permissions** page will open, and you should select the **Authorization and settings** group of permissions, and in the side panel that appears, click on **Select custom permissions**. Under the **System settings** section, select **Read and manage** and click on the **Apply** button, as seen on the following figure.

Authorization and settings

Select the permissions for users who need to configure your security and system settings, and create and assign roles.

ⓘ If you select any permissions on this page, you will also assign the security data read permission under the Security operations permission group.

✕ Clear all permissions

◯ All read-only permissions

◯ All read and manage permissions

◉ Select custom permissions

Authorization ⓘ

◯ Read-only

◯ Read and manage

Security settings ⓘ

◯ Read-only

◯ Select all permissions

◯ Select custom permissions

☐ Detection tuning (manage) ⓘ

☐ Core security settings (read) ⓘ

☐ Core security settings (manage) ⓘ

System settings ⓘ

◯ Read-only (Defender for Office, Defender for Identity)

◉ Read and manage

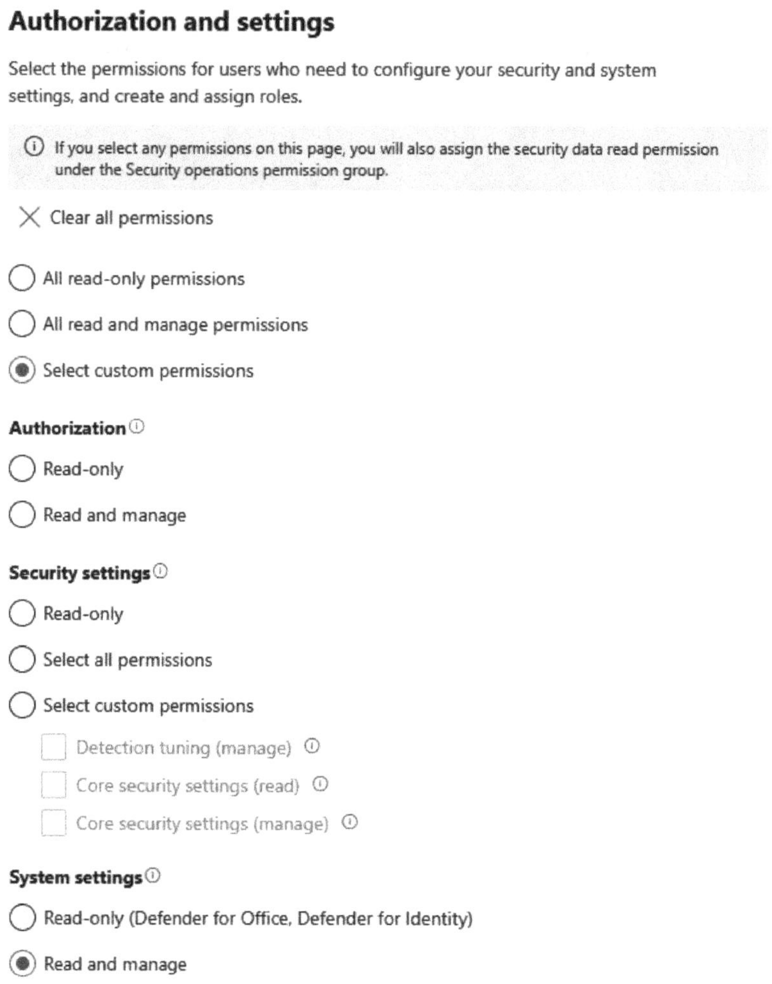

Figure 10.25 – The Authorization and settings configuration

5.   Next, click on the **Security operations** permissions group, and a side panel will appear. Click on **Select custom permissions** and under the **Security data** section, select **Read-only** followed by clicking on the **Apply** button at the bottom of the page, as seen on the following figure.

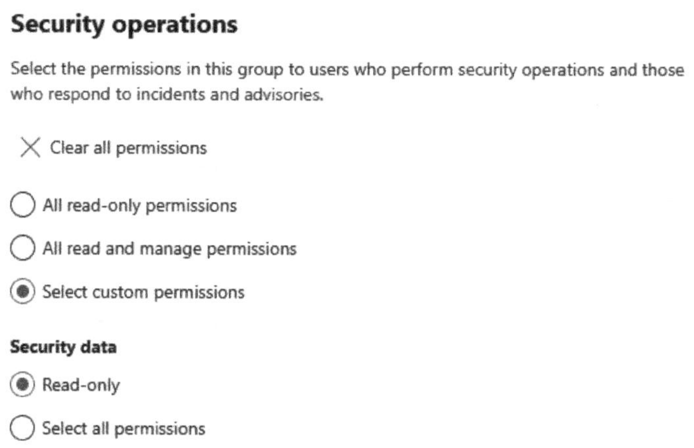

Security operations

Select the permissions in this group to users who perform security operations and those who respond to incidents and advisories.

× Clear all permissions

◯ All read-only permissions

◯ All read and manage permissions

◉ Select custom permissions

Security data

◉ Read-only

◯ Select all permissions

Figure 10.26 – The Security operations configuration.

6. Once we have confirmed the permissions are selected correctly, on the **Choose permissions** page, we can click the **Next** button to proceed with the permissions assignment. Click on the **Add assignment** button on the top of the page to open the assignment wizard.

7. In the wizard, we will provide a name for the role and assign it to a group of users. Also, ensure that all the data sources available are selected, then click on the **Add** button, as shown in the following figure.

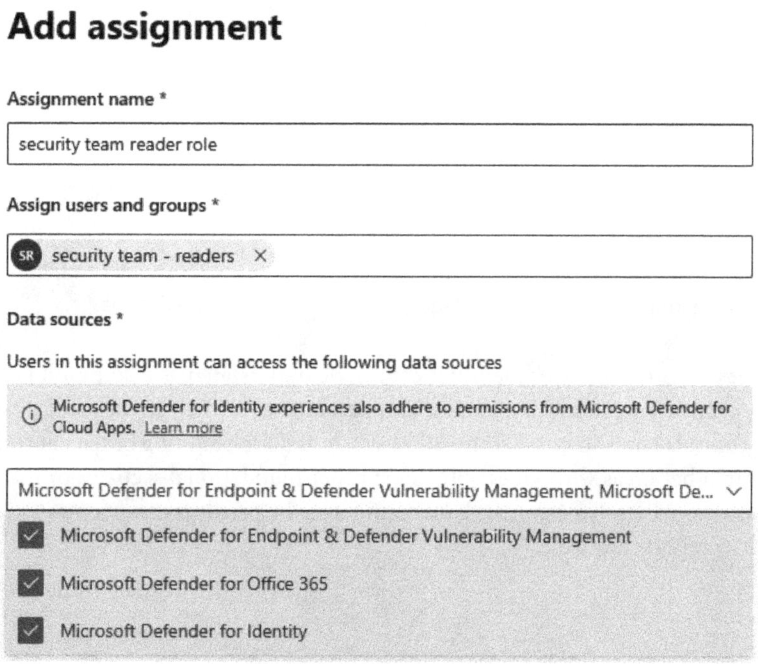

Add assignment

Assignment name *

security team reader role

Assign users and groups *

SR security team - readers   ×

Data sources *

Users in this assignment can access the following data sources

ⓘ Microsoft Defender for Identity experiences also adhere to permissions from Microsoft Defender for Cloud Apps. Learn more

Microsoft Defender for Endpoint & Defender Vulnerability Management, Microsoft De... ▾

☑ Microsoft Defender for Endpoint & Defender Vulnerability Management

☑ Microsoft Defender for Office 365

☑ Microsoft Defender for Identity

Figure 10.27 – Custom role assignment

8.   The **Assignment and data sources** page will show our recent selection, and we can click on the **Next** button at the bottom of the page and click on the **Submit** button to start the role creation. Once the role is created, we can click on the **Done** button to close the wizard.

The users should now be able to see the reports but be aware that this custom role cannot be directly assigned via Entra Privileged Identity Management, as such, it is recommended that the role be assigned to a custom user group and Entra Privileged Identity Management groups leveraged to enable just in time for these roles. We can see this assignment in the following figure.

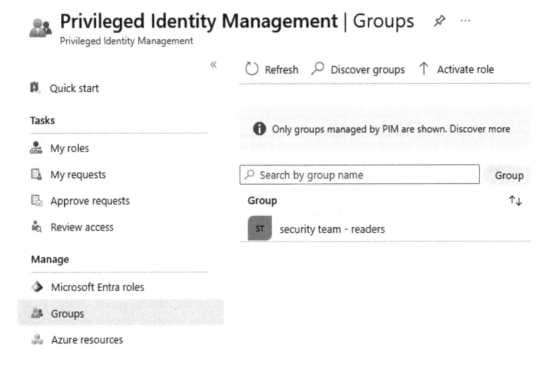

Figure 10.28 – Entra Privileged Identity Management | Groups

The impact of these controls on user identities cannot be understated as a common factor in most breaches is poor control and monitoring around user accounts with high-privilege roles assigned. The lack of security controls tends to come from either a lack of understanding of the controls required, too much focus in other areas without establishing proper foundational security, or thinking it will hinder administrative efforts. Now that we have configured the privileges and controls required, let's look at the security reports available.

The compromised users report

In Microsoft 365 environments, the safeguarding of mailboxes, data, and service access hinges on secure credentials, such as a username paired with a password or PIN. An account is deemed compromised if these credentials fall into the wrong hands. Once an attacker has the credentials, they can tap into the Microsoft 365 mailbox and access SharePoint folders or OneDrive files. Compromised accounts are often exploited to send emails from the affected user, aiming to reach colleagues and individuals beyond the organization. This act of emailing sensitive data outside the company is referred to as **data exfiltration**.

The **Compromised users** report in the Microsoft Defender portal highlights user accounts flagged as **Suspicious** or **Restricted** within the past week. These labels indicate potential or definite account compromises. Regular monitoring of this report can help identify patterns or sudden increases in such accounts. Any user accounts identified in the report as compromised will be grouped as either a **Suspicious** account, which indicates suspicious emails being sent from the account, or a **Restricted** account, which identifies an account restricted from sending emails due to a suspicious pattern being identified. We can see this report in the following figure.

Compromised users

Figure 10.29 – The Compromised users report

The report allows filtering according to a date range, activity type (restricted or suspicious), and tags such as **All** or **priority account**. During an incident, this report can be exported if the user has the proper permissions by selecting the **Export**, **Request report**, or **Create schedule** options. Finally, various other logs can be used to validate the findings such as the unified audit logs, and the Microsoft Entra sign-in logs, which can help identify any suspicious activities, related IP addresses, sign-in locations and times, and any success or failure of sign-ins.

For compromised account recovery, adherence to a predefined protocol is crucial to limit further harm and collect data on the extent of the breach. At the very least, changing the account password, setting up multifactor authentication, removing dubious email forwarding setups, and suspending any unknown inbox rules are actions recommended until the account is verified and not compromised. Check whether spam control prevented the account from sending emails, disable the sign-in feature until the incident is addressed, and temporarily replace the account if necessary. Remove the account from administrative roles and contemplate enforcing Microsoft Entra conditional access policies that factor in user sign-in and user behavior risk assessments.

The Exchange Transport Rule report

In the EAC, the **Exchange Transport Rule report** offers insights into how mail flow rules, also known as transport rules, are influencing message traffic. A line graph renders a day-by-day depiction of inbound and outbound messages impacted by mail flow rules. Doughnut charts are also provided that show the message volume according to direction as well as the severity level of the mail flow rule impacting the message. We can see this in the following figure.

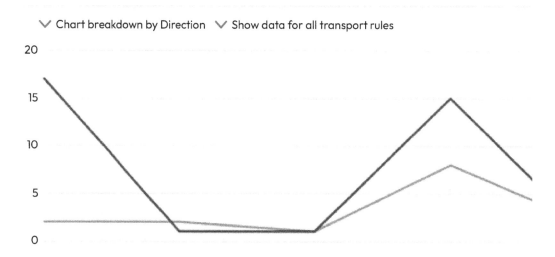

Figure 10.30 – The Exchange Transport Rule report

As with other reports discussed in this section, you can filter the report according to the categories shown in the data. For this report, filters include the direction of mail, the severity of the mail flow rule, and even the time range. The report can also be exported if the user has the proper permissions. For this report to be effective in detecting unexpected behavior, the mail flow rules severity must have been configured in a documented manner that aligns with agreed-upon security controls.

The Auto forwarded messages report

In the EAC, the **Auto forwarded messages** report provides insights on emails that are automatically redirected outside your organization to external domains. Configuring forwarding rules is a common tactic used by attackers to exfiltrate data once an account has been compromised, and as such, periodic monitoring of this report can be a proactive measure to detect potential data leaks or bad actors moving inside the environment. The report provides information in three areas: **Forwarding type**, **Recipient domain**, and **Forwarding users**. We can see this in the following figure.

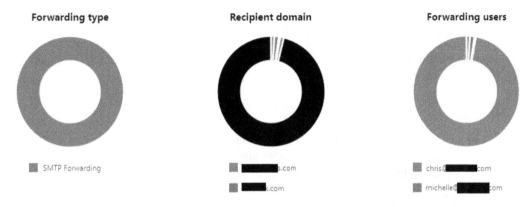

Figure 10.31 – The Auto forwarded messages report

Let us learn more about the entities visible in the figure:

- **Forwarding type**: This section shows the method used for forwarding between mail flow rules, inbox rules, or SMTP forwarding (set by administrators on a mailbox). Multiple unknown inbox rules are a common indicator of a compromised user and a data leak.

- **Recipient domain**: This shows the domain to which the messages are being forwarded to; these domains can be searched using threat intelligence tools such as Threat Explorer to determine whether the domain is malicious along with helping identify the threat experienced.

- **Forwarding users**: This section shows the users with messages forwarded. A security team can help identify the impact of an incident by looking for any priority accounts in this section.

Finally, at the bottom of the report, you will see a detailed list of information on these doughnut graphs including the accounts executing the forwarding, the method used for forwarding, the final recipient of the forwarded message, the domain receiving the forwarded email, any details such as any forwarding rules used, the forwarded emails count, and the date forwarding started. As with other reports, this report can also be exported.

The Mailflow status report

From the Defender portal, the **Mailflow status report** provides a comprehensive view of email activities covering inbound and outbound communication, identifying spam and malware interceptions, and categorizing emails as *good*. Unique to this report is data on emails either allowed or rejected at the network's perimeter, also referred to as **edge protection data**. It aids in understanding the volume of emails filtered before being analyzed by **Exchange Online Protection** (**EOP**) or Defender for Microsoft 365. Key features of the Mailflow status report include the following:

- Insights on emails that are preemptively blocked before detailed scrutiny by EOP or Defender for Microsoft 365

- Distinctions between incoming and outgoing emails, successfully delivered emails, spam detections, and malware interceptions

The report can be accessed directly access via `https://security.microsoft.com/mailflowStatusReport`. Access from the Defender portal (`security.microsoft.com`) is also possible by using the left-hand menu, selecting **Reports**, then on the **Reports** page, selecting **Email & collaboration reports**, and finally selecting **Mailflow status summary** among the reports. We can see this page in the following figure.

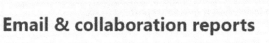

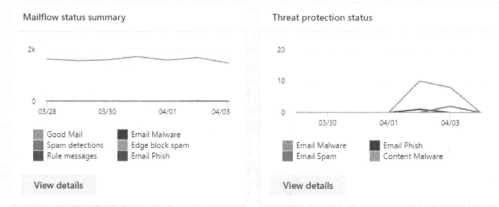

Figure 10.32 – The Email & collaboration reports page

The report is divided into three sections, which we will look at now.

The Type view

The **Type** view tab splits all messages into directions and types to provide a quick view of the level of malicious versus good email along with where it is being observed the most. Multiple columns showing different time frames are provided to help identify drastic changes. This view can provide many details about an attack that might not be immediately obvious from just looking at incidents such as an increase in internally sent phishing emails, which could indicate an active attack in progress from a compromised device or user account. We can see the **Type** view in the following figure.

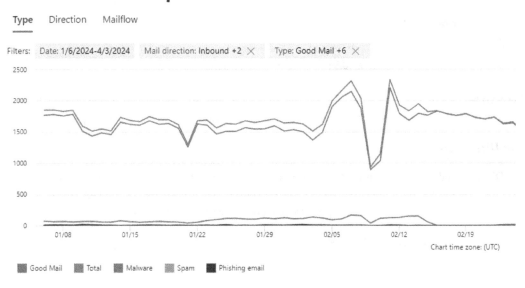

Figure 10.33 – The Type view of the Mailflow status report

The top part of this view shows us a time-based graph that can help us quickly identify spikes in malicious emails. In the bottom part, we can see detailed information about email traffic in a table divided into columns, which include the following:

- **Direction**: This indicates the direction of the message such as intra-org for internal emails, inbound and outbound.

- **Type**: This indicates the type of message seen, such as *good mail*, phishing email, spam, malware, edge protection, rule messages, and data loss prevention.

- **24 hours, 3 days, 7 days, 15 days, 30 days**: These columns show the message count for each entry at different time ranges. These columns can help us identify the need for further investigation, like in the following figure, which shows a major increase in malware-related emails in the last 3 days.

Direction	Type	24 hours	3 days	7 days	15 days	30 days
Intra-org	Good Mail	460	1531	3605	7583	14590
Inbound	Good Mail	233	726	1638	3365	6963
Outbound	Good Mail	794	2511	6041	13450	26007
Inbound	Phishing email	0	1	1	3	79
Inbound	Spam	1	1	1	1	4
Intra-org	Phishing email	0	0	0	0	1
Intra-org	Malware	5	11	11	11	11
Outbound	Malware	3	7	7	7	7

Figure 10.34 – Possible malware trend identified

As with other reports, this section can be filtered by date ranges, mail direction, and mail type. The report can also be exported if the user has the required permissions (as mentioned in the *Permissions required* section).

The Direction view

The **Direction** tab offers the same information as the **Type** tab, but the focus is on the direction of the message. The idea is that sometimes increases in certain emails could also indicate an attack trend, which the tools are wrongly identifying as false negatives. We can see this view in the following figure.

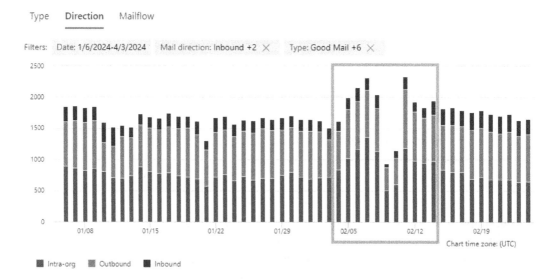

Mailflow status report

Type   **Direction**   Mailflow

Filters: Date: 1/6/2024-4/3/2024   Mail direction: Inbound +2 ✕   Type: Good Mail +6 ✕

■ Intra-org   ■ Outbound   ■ Inbound

Figure 10.35 – The Direction view in Mailflow status report

A proactive security team could look at a report like the one in the preceding picture and identify a change in the usual volume of intra-org messages over a few days, which could indicate the possible use of a system to send phishing messages internally. As with the **Type** view, this information can be filtered and exported.

The Mailflow view

The **Mailflow** tab provides a clear visual of your organization's email dynamics, courtesy of Microsoft's robust email threat protection suite. A horizontal Sankey diagram comes alive, colorfully illustrating the emails' journeys and the impact of threat protection interventions. The Sankey diagram offers a sweeping 90-day retroactive lens and paints a vivid picture with two distinctive hues representing EOP and Defender for Office 365 interventions. We can see this view in the following figure.

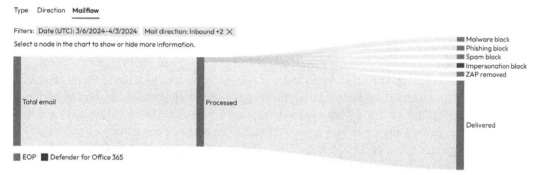

Figure 10.36 – The Mailflow view of Mailflow status report

Below the Sankey diagram, we can see a detailed breakdown of the information represented in the diagram in a table divided into columns with a row per day. Each column provides a total count for each one of these:

- **Total email**: This is the total number of emails for that day.
- **Edge filtered**: These are messages blocked before they reach any Mailflow rules. This typically is seen on inbound messages coming from blacklisted IP addresses used for spam campaigns.
- **Data loss prevention**: For organizations leveraging data loss prevention controls via Purview, these are messages on which data loss prevention controls acted. This could indicate that messages were being sent with information prohibited from distribution by the organization.
- **Rule messages**: This indicates emails quarantined due to configured Mailflow rules.
- **Anti-malware engine, Safe Attachments**: Messages filtered due to these containing malicious files.

- **DMARC, impersonation, spoof, phish**: Messages filtered due to not passing DMARC, impersonation, spoof, or phish verifications.

- **Detonation detection**: Identifies messages that contain files or URLs identified as malicious during the sandbox detonation activities.

- **Anti-spam**: Identifies spam messages detected.

- **Zero-hour auto purge (ZAP)**: Identifies messages that were removed due to zero-hour purge protection.

- **Message where no threats were identified**: These are the clean messages delivered.

The report also provides a **Show trends** button on the top of the table, which opens a side panel with time-based graphs for each one of these categories, further helping identify trends. As with other parts of this report, the data can be exported for sharing between team members.

Mail latency report

Mail latency report can also be viewed from the Defender portal in the same manner as the Mailflow status summary report, or via the direct URL: `https://security.microsoft.com/mailLatencyReport`. This report shows a comprehensive view of the timing dynamics your organization's emails are subject to during the delivery and analysis phases. It's crucial to note that several variables influence mail delivery times, making the absolute delivery time measured in seconds not always a definitive mark of operational health. This report benchmarks message delivery times to provide a contextual understanding, comparing them to the usual ebb and flow within the service. We can see this report in the following figure.

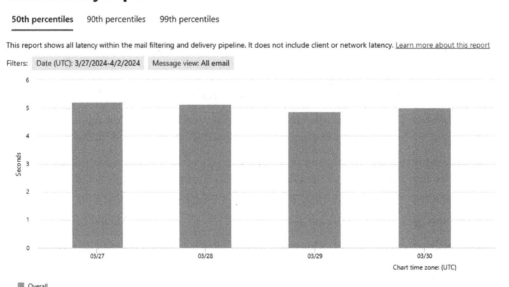

Figure 10.37 – Mail latency report

The report divides the data into three key segments via tabs:

- **50th percentiles**: The center of the message delivery timeframe distribution, acting as what you might call a standard delivery time.

- **90th percentiles**: A marker of heightened delivery latency indicating that 90% of messages did not surpass (were faster than) this delivery time.

- **99th percentiles**: A marker indicating that 99% of messages did not surpass the time indicated. The messages identified in this section might point to a possible configuration issue to verify.

All these tabs will present bar charts focused on the specific percentiles and a table shared by all tabs showing more details about the total count of messages expressed in the graph and the different times in seconds for the latency. Both the graphs and the table can be filtered by the type of latency with the options being **Overall latency**, **Inline detonation**, and **Asynchronous detonation**. A proactive security team can review these entries to validate user impact and adjust the controls as needed to prevent business disruption.

The Post-delivery activities report

Another one of the reports on the Defender portal's **Email & collaboration reports** page, the **Post-delivery activities** report, offers a critical review of actions taken by ZAP after emails reach user mailboxes. The report can be reached via the Defender portal (`security.microsoft.com`) report section, or directly via `https://security.microsoft.com/reports/ZapReport`. Proactive security teams can use this report to identify gaps in their security strategy and to include what areas users should be trained in. We can see this report in the following figure.

Post-delivery activities

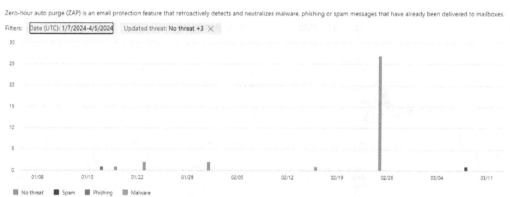

Figure 10.38 – The Post-delivery activities report

The report has a time-based bar chart that provides a quick view into the type of threat ZAP protects against including malware, spam, and phishing. At the bottom of the page, like with other reports discussed so far, there is a detailed table that helps to further understand the threats observed. The table includes the following columns:

- **Subject**: The subject line of the malicious email message. For example, "Act in the next 24 hours or risk account deactivation."

- **Received time**: The time and date the email was received.

- **Sender**: The sender's email address in the email header; be aware that malicious emails could be spoofed.

- **Recipient**: The recipient of the malicious email.

- **ZAP time**: The time at which ZAP acted on the email message. If the time between received and ZAP is too long, it might be worth it to look at delivering messages first without an attachment and having the attachment detonate in a sandbox (configurable in the safe attachments policies).

- **Original threat**: In some cases, an email could have been initially identified as a different kind of threat or not a threat at all.

We can also click on any of the items in the table for a side panel to show, presenting more details about a threat such as how it was identified as malicious, whether the email is part of a campaign, whether it passed DMARC checks, and other important details. We can see this panel in the following figure.

Figure 10.39 – Message details panel

The details panel even offers the option via the **Open email entity** button of seeing a more in-depth view of the email including a defanged (made non-malicious by making links and attachments non-clickable) copy of the email and further actions that can be taken via the **Take action** button. We can see a copy of a defanged malicious message in the following figure.

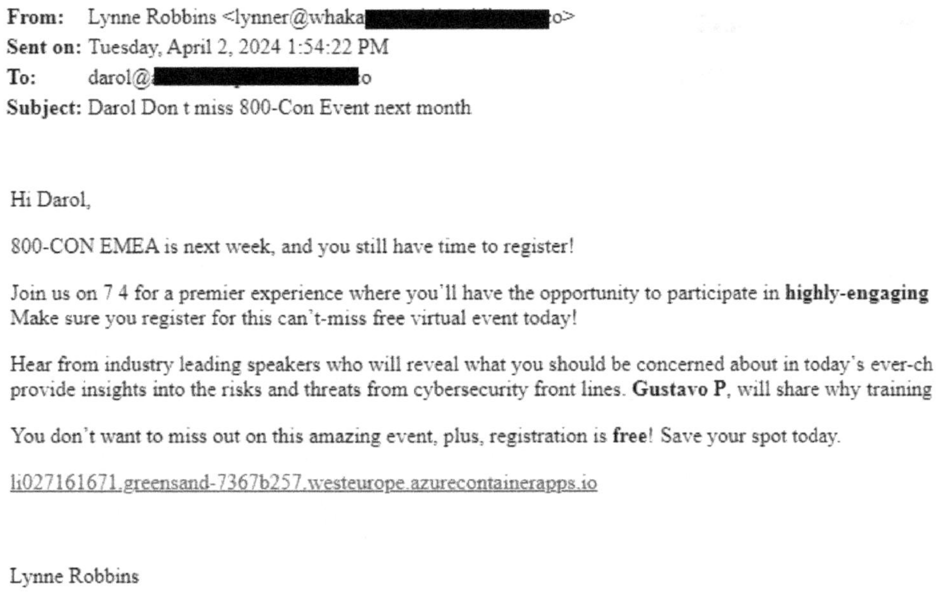

From: Lynne Robbins <lynner@whaka███████████o>
Sent on: Tuesday, April 2, 2024 1:54:22 PM
To: darol@████████████o
Subject: Darol Don t miss 800-Con Event next month

Hi Darol,

800-CON EMEA is next week, and you still have time to register!

Join us on 7 4 for a premier experience where you'll have the opportunity to participate in **highly-engaging** Make sure you register for this can't-miss free virtual event today!

Hear from industry leading speakers who will reveal what you should be concerned about in today's ever-ch provide insights into the risks and threats from cybersecurity front lines. **Gustavo P**, will share why training

You don't want to miss out on this amazing event, plus, registration is **free**! Save your spot today.

li027161671.greensand-7367b257.westeurope.azurecontainerapps.io

Lynne Robbins

Figure 10.40 – A defanged malicious message

As part of allowing security teams flexibility, the **Take action** button is provided on the details panel. Clicking on the **Take action** button on the message details panel allows for further actions, such as the following:

- **Move to mailbox folder:** In the case of wrongly identified messages, these can be returned to the recipient. Using this option is not recommended in most cases unless deep analysis has been performed on all the message artifacts and confirmed clean.

- **Submit to Microsoft for review:** In case of messages that might have been erroneously classified as malicious, this option allows for a security expert at Microsoft to further analyze the message and adjust the detection rules if the message was classified erroneously.

- **Initiate automated investigation:** This searches your environment for other malicious actions related to the message that might not have been initially detected. This is a recommended action to take on all malicious messages to ensure a proper assessment is performed.

- **Propose remediation:** A *soft delete* action for this message can be sent to an administrator for approval.

Be aware that a combination of these response actions can be performed. For example, we can send the message to Microsoft for further review while we commence an automated investigation to gather further details (and speed up our investigation if the message is identified as malicious by Microsoft). We can see the response actions panel in the following figure.

Choose response actions

Specify the actions you want to take. Only actions applicable to the selected entity are available. We've grayed out

Email message actions

☐ Move to mailbox folder

☐ Submit to Microsoft for review

☑ Initiate automated investigation

  ◉ Investigate email   ◯ Investigate recipient   ◯ Investigate sender   ◯ Contact recipients

☑ Propose remediation
  Create admin request for remediation actions ⓘ

  ◉ Create new
     This will trigger a soft delete email pending action that needs to be approved by the admin on action center

  ◯ Add to existing
     Choose an existing remediation to apply actions to this email

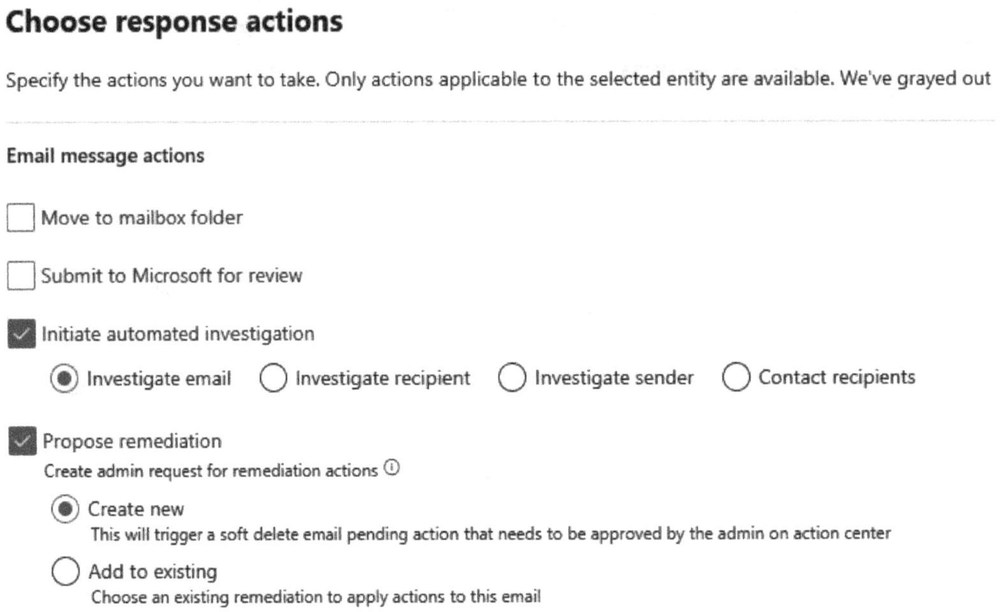

Figure 10.41 – The Choose response actions panel

As with other reports discussed previously, the **Post-delivery activities** report can be filtered to help focus investigative efforts. Export of the report is also possible if the user has the appropriate permissions.

The Spoof detections report

The **Spoof detections** report provides insights on email authentication checks on messages. This report is accessible like other reports from the Defender portal (`security.microsoft.com`) reports page alongside other email and collaboration reports or directly at `https://security.microsoft.com/reports/SpoofMailReport`. This report offers a comprehensive overview of email authentication failures, which could indicate increases in spoofing attempts or targeted attacks. We can see the report in the following figure.

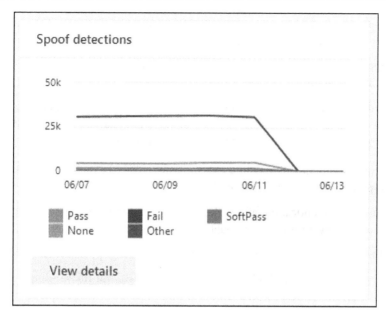

Figure 10.42 – The Spoof detections report

The report visual chart outlines the different email authentication statuses observed, including the following:

- **Pass**: Email messages that passed all the email authentication checks successfully.

- **Fail**: Email messages that failed the checks.

- **SoftPass**: Email messages that are passing some but not all the authentication checks; these might be non-spoof emails.

- **None**: Emails where no authentication checks were performed. This might be a sign of misconfiguration.

- **Other**: Messages with various, undefined results. This might be a sign of misconfiguration.

This report is another one that can provide insights into campaigns in progress or misconfiguration if an uptick in authentication failures is observed. As with other reports, a detailed table is provided at the bottom of the report to gain more details on the messages reported as spoofed to include the following:

- **Date**: The date and time the message was sent.

- **Spoofed User**: The domain the email is trying to spoof, for example, it could say contosoicecream.org to signify that the attacker was trying to make the message appear to originate from contosoicecream.org.

- **Sending Infrastructure**: The actual domain the email was sent from.

- **Spoof Type**: Indicates whether the attempt was from an external source or internal sender. An internal sender could indicate a compromise in the environment.

- **Result**: The status of spoof verification checks; a message identified as a spoof message will show **Fail** in this column.

- **Result Code**: The code used to indicate not only the reason for failing the spoof verification checks but also gives you information about the message regarding configuration settings in your environment. We can see a list of the codes in the following list:

  - `000`: Explicit authentication failure; this indicates a DMARC fail and the DMARC policy action is `p=quarantine` or `p=reject`.

  - `001`: Implicit authentication failure; this indicates that the sending domain either doesn't have an email authentication record published, or their policy has weaker measures on failure.

  - `002`: The sending domain has a record explicitly prohibiting spoofed email from that sender.

  - `010`: Explicit authentication such as error code `000`, but the sending domain is the organization's list of accepted domains. This could indicate a compromise.

  - `1xx` or `7xx` means the email authentication checks passed.

  - `2xx`: The email message soft-passed implicit authentication.

  - `3xx`: The email message wasn't checked.

  - `4xx` or `9xx` means the email message bypassed email authentication checks. This could indicate misconfiguration or compromise.

  - `6xx`: Same as code `001`, but the sending domain is in the organization's list of accepted domains.

- **SPF**: The email message either passed or failed SPF checks.

- **DKIM**: The email message either passed or failed DKIM checks.

- **DMARC**: The email message either passed or failed DMARC checks.

As with other reports discussed previously, the **Spoof detections** report can be filtered according to any of the column entries in the table. Users can also export the report if the user has the appropriate permissions.

The Submissions report

The **Submissions** report allows the security team and other administrators to see and track items reported to Microsoft for analysis due to suspicion of it being a false detection for spam, phishing, or malware. As with other reports, it can be reached via the Defender portal (`security.microsoft.com`) **Reports** section, along with other **Email & collaboration** reports. It can also be reached directly via `https://security.microsoft.com/adminSubmissionReport`. This report can show up to the last 30 days of submissions and allows for a central view of the status of the analysis. We can see this report in the following figure.

Reports > Email & collaboration reports > Submissions

Filters: Date submitted (UTC+09:00): 3/7/2024-4/5/2024

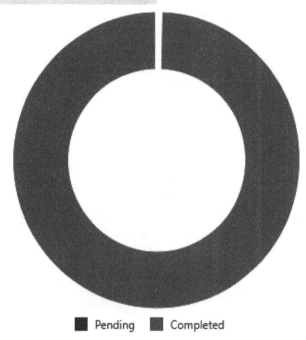

Figure 10.43 – The Submissions report

The **Submissions** report top section includes a doughnut chart that shows the status of submissions, either pending or completed. Under the chart, we can see a detailed table showing the submissions, their status, and other details such as submission name, sender, recipient, data submitted, the reason for submission, status, the result of the analysis, the reason for the block, submission ID, network message ID, message direction, sender IP, and many other related fields. It is important to note that the submission ID and network message ID are important entries to keep if further fine-tuning is needed and will be required. As with other reports, clicking on any items in the table will open a side window with more details, as seen in the following figure.

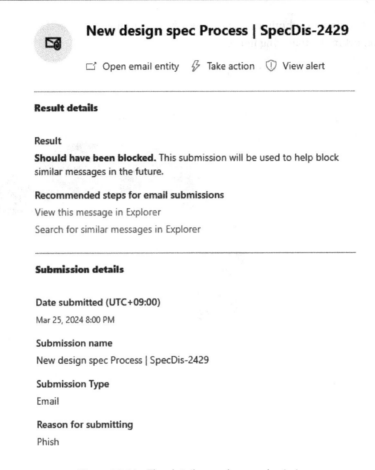

New design spec Process | SpecDis-2429

⌐┘ Open email entity   ⚡ Take action   ⓘ View alert

Result details

Result
Should have been blocked. This submission will be used to help block similar messages in the future.

Recommended steps for email submissions
View this message in Explorer
Search for similar messages in Explorer

Submission details

Date submitted (UTC+09:00)
Mar 25, 2024 8:00 PM

Submission name
New design spec Process | SpecDis-2429

Submission Type
Email

Reason for submitting
Phish

Figure 10.44 – The details panel on a submission

It is recommended to do periodic exports of the data on this report as the data only lives for 30 days and multiple parts of this report will be needed if submissions due to false detections still don't minimize enough of related false detections.

The Threat protection status report

The **Threat protection status** report is one of the most useful reports in the EOP and Defender for Office 365 set of reports. It's designed to give you a snapshot of email safety by tallying messages flagged for malicious content and providing different perspectives to help identify patterns. It is important to note that this graph focuses on messages before delivery, so ZAP-related actions will not be visible here. What sets this report apart is that it can do in-depth breakdown views of message threats according to category including phish, spam, and malware, alongside any information on messages that bypassed controls. Using a combination of the views on this report can quickly help identify issues in the policies that need fine-tuning as well as possible changes that lowered coverage

for the organization. As with other reports, it can be accessed from the Defender portal's (`security.`
`microsoft.com`) **Reports** section along with other **Email & collaboration** reports, or directly
via `https://security.microsoft.com/reports/TPSAggregateReportATP`. Let's
discuss the different views available.

Overview

The first view when opening the report is intended to give a quick high-level perspective on all the
threat protection status trends within the report. We can see this view in the following figure.

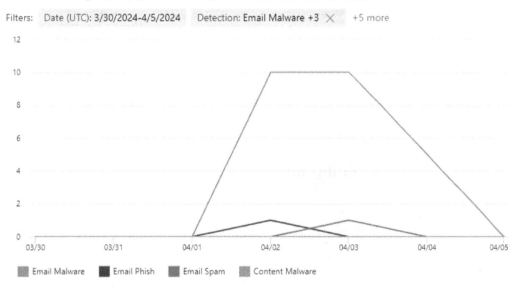

Figure 10.45 – The overview page of the Threat protection status report

A time-based graph is shown covering the increases and decreases in multiple detection types including:

- **Email Malware**: Malicious file attachments detected within emails

- **Email Phish**: Phishing attempt messages identified

- **Email Spam**: Spam emails identified

- **Content Malware**: An extra option if the safe documents option of safe attachments is enabled
  and focuses on the discovery of malware in files in SharePoint, OneDrive, and Microsoft Teams

Filtering of the data in this chart is possible by date ranges, detection type, which protection was engaged in the detection (Defender for Office 365 or EOP), tag, direction of the message, and policy type. Export of the report is not offered in this view.

Email | Phish and chart breakdown by detection technology

This part of the view focuses on phishing email messages and how these were detected. You are provided with a bar chart showing the different counts at different times for each of the detections employed by Defender for Office 365 to detect phishing. We can see this report in the following figure.

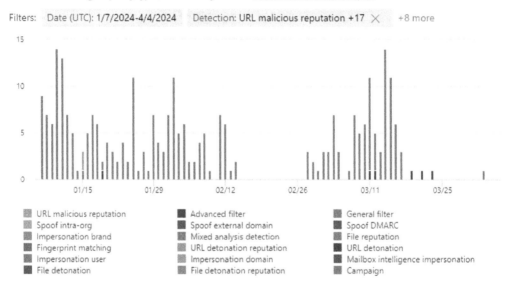

Figure 10.46 – The phishing email view

The multiple detections triggered can provide the security team with an idea of the sophistication of the phishing attacks being experienced. This allows for improved fine-tuning of anti-phishing policies and improved user education. The phishing detection technologies include the following:

- **URL detonation**: This detects malicious URLs through Safe Links analysis

- **URL detonation reputation**: This detects malicious URLs previously identified in other organizations during the sandbox detonation analysis

- **URL malicious reputation**: This spots URLs previously marked as malicious by other organizations

- **File detonation**: Messages with malicious files detected during the sandbox detonation phase
- **File detonation reputation**: Messages with malicious files detected in other organizations
- **File reputation**: This identifies malicious files flagged in other Microsoft 365 organizations
- **Impersonation brand**: Detections of sender identity mimicry of well-known organizations
- **Impersonation domain**: Detections of impersonation attempts for domains as configured in anti-phishing policies
- **Impersonation user**: Detection of impersonation attempts for users defined in anti-phishing policies
- **Mailbox intelligence impersonation**: Impersonation detections via mailbox intelligence
- **Spoof DMARC**: This notes DMARC authentication failures
- **Spoof external domain**: This identifies external domain spoofing attempts
- **Spoof intra-org**: This flags internal domain spoofing attempts
- **Fingerprint matching**: This identifies emails resembling known threats
- **Campaign**: Messages that are part of coordinated phishing campaigns
- **General filter**: This uses analyst-derived rules to spot phishing signs
- **Mixed analysis detection**: Detection triggered by a combination of multiple filters
- **Advanced filter**: Detections due to machine learning rules

As with other reports, a detailed table is included under the chart, which breaks down every phish detection including its delivery status. Clicking on any of these detections provides a side panel with details about the detection for further review, as well as the **Open email entity** button to get an example of the email and a **Take action** button for submission of false detections or recommending actions to implement. The data in this report can be exported and filtered as needed.

Email | Spam and chart breakdown by Detection Technology

Like the **Email | Phish** view, a bar chart and detailed tables are provided but focused on spam detection according to the detection technology used. We can see this view in the following figure.

Threat protection status

The Threat protection status report provides information about threats found prior to email delivery, covering relevant detection technologies, policy types, and delivery actions. Learn more about this report

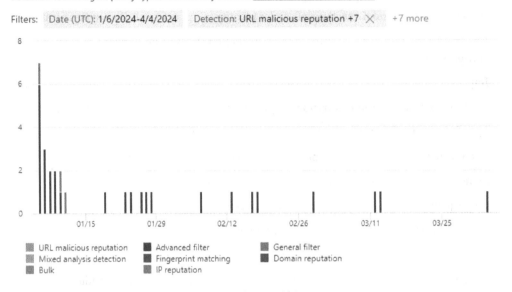

Filters: Date (UTC): 1/6/2024-4/4/2024   Detection: URL malicious reputation +7 ✕   +7 more

Figure 10.47 – The Email | Spam view

The detection technologies covered under this view include the following:

- **General filter**: These are messages that trigger the basic spam filters
- **Bulk**: This identifies emails exceeding the set **bulk complaint level (BCL)** for spam
- **Fingerprint matching**: This detects emails such as known malicious content
- **Domain reputation**: This tags messages from domains known for spam
- **IP reputation**: This marks messages from IPs with a spamming history
- **URL malicious reputation**: This identifies URLs previously recognized as malicious
- **Mixed analysis detection**: Detection due to multiple filters being triggered
- **Advanced filter**: Detection due to machine learning rules being triggered

As with the phishing view, the detailed table provides a breakdown of each spam detection including its delivery status. This data can be further filtered and exported as needed.

Email/Content | Malware and chart breakdown by Detection Technology

In the same way that other views provide visibility on the detection of other threats and the detection technology used, both the **Email | Malware** and **Content | Malware** views provide this visibility for malware detections on emails and files found on SharePoint, OneDrive, and Microsoft Teams. While on two separate views, we will discuss them together as both views are used together to tell the story of how a malicious file arrives in an environment and what happens afterward. We can see the graph at the top of the **Email | Malware** view in the following figure.

Threat protection status

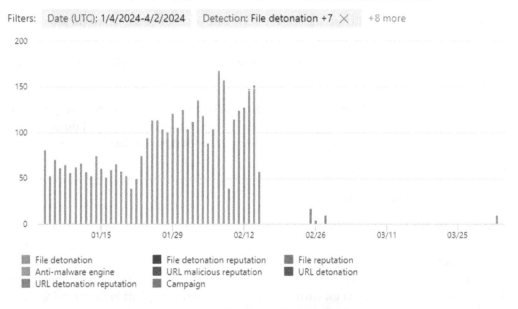

The Threat protection status report provides information about threats found prior to email delivery, covering relevant detection technologies, policy types, and delivery actions. Learn more about this report

Filters: Date (UTC): 1/4/2024-4/2/2024   Detection: File detonation +7 ✕   +8 more

Figure 10.48 – The Email | Malware view

The email malware detection technologies covered in this graph include the following:

- **Campaign**: Messages that are part of a larger campaign

- **Anti-malware engine**: Hits from traditional anti-malware scanning engines

- **File detonation reputation**: Attachments previously flagged as malicious by the Safe Attachments component of Defender for Office 365 in other Microsoft 365 organizations during their sandbox detonation phase

- **File reputation**: Files in the message are known threats

- **URL malicious reputation**: The message contains a known malicious URL

- **URL detonation**: Safe Links found a malicious URL in the message upon analysis

- **URL detonation reputation**: The URL in a message was found to be malicious by other organizations during sandbox detonation

The **Content | Malware** view also offers a bar chart with a focus on any malicious file detections in file storage locations such as SharePoint or OneDrive, as well as malicious files shared via teams. The graph provided in this view will focus on the following detection technologies:

- **Anti-malware engine**: Malicious files detected via the virus scanning engines provided to all Microsoft 365 subscriptions

- **MDO detonation**: Malicious files detected via Safe Attachments enabled for SharePoint, OneDrive, and Microsoft Teams

- **File reputation**: This pinpoints a file labeled as malicious based on intelligence gathered across various Microsoft 365 tenants

Both views provide a detailed table included under the chart, which breaks down every malware detection, and the **Email | Malware** view includes the message delivery status. Any of the entries can be clicked to view more details on the detection. The data in both views can be exported and filtered as needed.

The Top Malware Report

Top Malware Report highlights the array of malicious files intercepted by EOP controls. The report can be accessed from the Defender portal's (`security.microsoft.com`) **Reports** sections alongside the other **Email & collaboration** reports or directly via `https://security.microsoft.com/reports/TopMalware`. The data in this report is presented in a pie chart format with a breakdown of the malware strains observed in the environment. We can see this pie chart in the following figure.

Top Malware Report

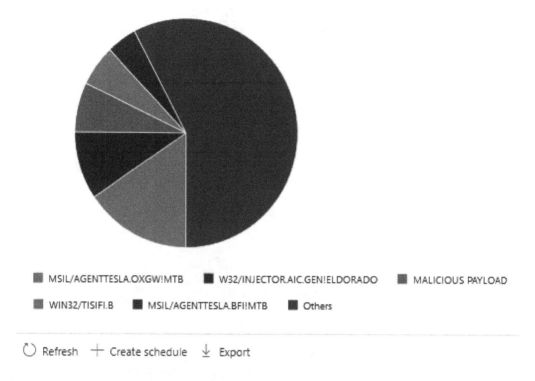

MSIL/AGENTTESLA.OXGW!MTB   W32/INJECTOR.AIC.GEN!ELDORADO   MALICIOUS PAYLOAD

WIN32/TISIFI.B   MSIL/AGENTTESLA.BFI!MTB   Others

○ Refresh   + Create schedule   ↓ Export

Figure 10.49 – Top Malware Report

This is an interesting report to review as its main purpose is to give you an insight into the types of malware strains observed, which could be beneficial during security team training or when discussing improvements to malware controls. As with other reports, the results can be filtered and exported as needed during investigations.

The Top senders and recipients report

The **Top senders and recipients** report provides insights into the most frequent targets and sources of malware, spam, and phishing attempts. Reach this report through the Defender portal's (`security.microsoft.com`) **Reports** section, along with the other **Email & collaboration** reports, or directly via `https://security.microsoft.com/reports/TopSenderRecipientsATP`. The focus of this report is the top 20 message senders within the organization and the top 20 recipients of messages flagged by EOP and Defender for Office 365 protection mechanisms. This report can help security teams identify users that might need more education and increased security controls such as being added to the priority account group due to the tendency for these accounts to be constant targets.

The report presented the data with different views to help quickly identify trends. Views provided include but are not limited to top mail senders, top mail recipients, top spam recipients, top malware recipients (as detected by EOP and Defender for Office 365), top phishing recipients (as detected by EOP and Defender for Office 365), top intra.org mail senders, top intra.org mail recipients, top intra.org spam recipients, top intra.org malware recipients (as detected by EOP and Defender for Office 365), and top intra.org phishing recipients (as detected by EOP and Defender for Office 365). We can see the **Top senders and recipients** view of the report in the following figure.

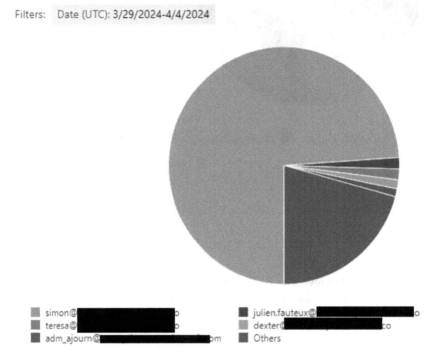

Figure 10.50 – The Top senders and recipients view for the report

The different views in this report provide a table with the item count per user. The data on these tables can be exported and filtered as needed.

The URL threat protection report

A report focused on findings from the Safe Links policy configuration, the **URL threat protection** report delivers comprehensive overviews and trend analyses related to threats identified and actions executed through Safe Links URL scanning. However, it's important to note that individual user click data won't appear if the **Track user clicks** option within the Safe Links policy is not enabled.

This report can be viewed from the Defender portal by navigating to the **Reports** section or directly via https://security.microsoft.com/URLProtectionActionReport. A periodic review of this report allows teams to identify possible misconfigurations in the Safe Links policies that require fine-tuning. An excessive number of blocked URLs could also indicate a false positive, which might require submission to Microsoft for further analysis. The security team can also change the report view to **Click by application** instead of **Protection action**, which can make apparent any attack vector being exploited by attackers. We can see this report in the following figure.

URL threat protection

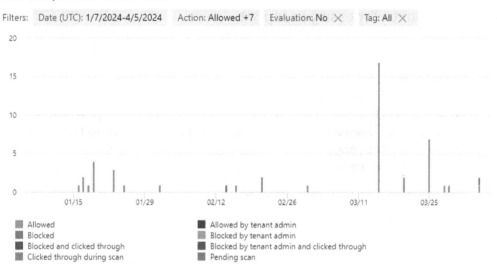

Figure 10.51 – The URL threat protection report

This report's protection actions covered include the following:

- **Allowed**: These are the instances where user clicks were permitted and indicate a non-malicious URL.

- **Allowed by tenant admin**: These clicks were specifically allowed through the Safe Links policies set by an administrator.

- **Blocked**: This denotes the malicious URLs that were successfully blocked.

- **Blocked by tenant admin**: This indicates clicks blocked due to an administrator-configured block list in the Safe Links policies.

- **Blocked and clicked through**: This refers to malicious URLs that were initially blocked but users still proceeded to the given URL due to lenient Safe Links policies.

- **Blocked by tenant admin and clicked through**: These are URLs blocked due to being added to a block list in the Safe Links policies, yet the user proceeded due to lenient Safe Links policies.

- **Clicked through during scan**: User clicks on a URL in the pending scan page that appears when a URL scan is underway. It is recommended to disallow this in the Safe Links policies.

- **Pending scan**: This represents the clicks on URLs that have not started scanning yet. It is recommended to disallow this in the Safe Links policies.

Both views provide a detailed table showing each entry along with information about the time, users, URL clicked, the action taken, and the application by which the link arrived. The data on these tables, like with other reports, can be exported and filtered as needed.

The User reported messages report

This is the final report to be discussed in this chapter and one that the security team should be reviewing daily. The **User reported messages** report provides insights into the types of emails users have flagged within their email client. Entries in this report are generated from users via either the **Report message** button in Outlook clients or on the web. Security teams should pay attention to this report often as it could indicate threats not detected by other tools, which are typically employed by sophisticated actors. This report can be accessed from the **Reports** section in the Defender portal (security.microsoft.com) or directly via https://security.microsoft.com/reports/userSubmissionReport. We can see this report in the following figure.

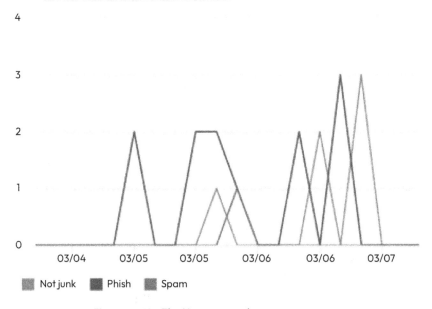

Figure 10.52 – The User reported messages report

The report offers a time-based graph that visualizes email messages that are either **Spam**, **Phish**, or **Not junk** (erroneously categorized messages). A high trend of **Not junk** might indicate a misconfiguration and need for submission of messages to Microsoft for analysis. As with other reports, a table is offered with details for each entry including the name (subject line) and type of message, the user who submitted the message, the date reported, the message sender, the reported reason (Spam, Phish, Not junk), and the results of the analysis once performed by the security team. An analyst can also filter the results during investigations and export the results as needed.

The many security reports identified in this section work as another brick in the foundation of proper security operations. From looking at many breached organizations, the lack of awareness about these reports tends to not only make their security team require more personnel to cover simpler tasks but also prevents security teams from implementing good proactive security and house cleaning. Security reports alone can provide a quantifiable representation of why the organization needs the tool and an easy way to track improvement over time.

Summary

Threat intelligence lies at the heart of protecting systems such as Defender for Office 365, and this chapter kicked off with a thorough exploration of its importance in today's cyber defense strategies. It showed how Microsoft gathers and sharpens threat data into valuable insights, utilizing a vast network of sensors, community contributions, and expert analysis. We delved into how this intelligence is woven into Defender's framework, strengthening security measures, and guiding proactive, intelligent defense approaches. Additionally, the chapter shone a light on the plethora of sophisticated threat reports from Microsoft, breaking down how each one arms security professionals with the knowledge to tackle ever-changing cyber dangers.

We took a critical look at the benefits of Microsoft's Defender TI service, underscoring its ability to elevate raw data into a strategic resource. The custom security reports for Defender for Office 365 were examined, establishing them as crucial instruments for security teams to monitor and neutralize threats within Office 365's collective environments. As firms increasingly lean on these all-encompassing, insight-led security measures, the shift to automated, efficient processes becomes increasingly apparent. The next chapter is set to continue this discussion by venturing into Microsoft's security framework's integration and API functionalities. The expectation is high as we await to probe into how these technological tools can push forward the automation of security operations, leading to quicker and more precise counteractions against the progressively intricate and swift nature of cyber threats.

References

- Microsoft. (2023a, August 18). *What is Microsoft Defender Threat Intelligence (Defender TI)?*. Microsoft Learn. https://learn.microsoft.com/en-us/defender/threat-intelligence/what-is-microsoft-defender-threat-intelligence-defender-ti

- Microsoft. (2023b, November 15). *Understand the analyst report section in threat analytics in Microsoft Defender XDR*. Microsoft Learn. `https://learn.microsoft.com/en-us/microsoft-365/security/defender/threat-analytics-analyst-reports?view=o365-worldwide`

- Microsoft. (2024, March 8). *View email security reports*. Microsoft Learn. `https://learn.microsoft.com/en-us/microsoft-365/security/office-365-security/reports-email-security?view=o365-worldwide`

- Shweta. (2023, October 12). *What is threat intelligence? Definition, types & process. Forbes Advisor*. `https://www.forbes.com/advisor/business/what-is-threat-intelligence/`

11

Integration and Artificial Intelligence

In this chapter, we delve deep into the rich set of **application programming interfaces (APIs)** offered by **Defender XDR**, providing security professionals with programmable access to their arsenal of defense mechanisms. This array of interfaces is the gateway for extracting detailed security reports, conducting proactive threat hunting, orchestrating automated response actions, and weaving the Defender XDR capabilities into bespoke security solutions tailored to organizational needs. Our exploration doesn't stop at the raw technicalities; we guide you through practical examples demonstrating how to leverage these APIs to elevate your security posture to new heights. Also, we shine a light on the game-changing **Copilot for Security**, unpacking its features and elucidating how this intelligent assistant can be harnessed to not only automate mundane tasks but also offer insightful recommendations, ensuring that security operations are both streamlined and strategically optimized. The insights gained in this chapter are key for those striving to fortify their defenses through technology-driven, integrated security frameworks.

This chapter will cover the following topics:

- The types of security APIs that are available in Defender XDR and how they work

- The automation and integration that can be accomplished with these APIs

- How Copilot for Security can improve our security operations

Let's continue our journey!

Introducing APIs

APIs are a collection of rules that allow different software to talk to each other, sharing data and functions. Think of an API as a digital messenger or middleman that helps developers build feature-rich and complex applications more efficiently and securely (*Microsoft, 2023*).

Communication between software through APIs involves a call-and-response method, such as asking a question and getting an answer. The asking part is the request sent from a user's action, such as searching for something or clicking a button, or it might even be prompted by a message from another app. We can see this in the following figure.

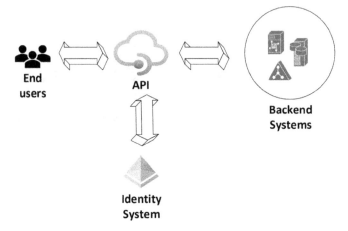

Figure 11.1 – An example of an API

When sending an API request, multiple parts must be included for the request to be understood by the recipient and for the action to be performed. At a minimum, an API request usually includes the following parts:

- **Endpoint**: This is a specific web address (URL) that stands for a particular function in the API world. For instance, if you're using a news aggregator app, the endpoint could handle the receipt and indexing of news articles.

- **Method**: This is what you want to do with the data, such as getting data (a GET request), sending new data (a POST request), or updating data (a PUT request). An endpoint could have one or more methods depending on the function of the endpoint.

- **Parameters**: These are extra bits of information you send when using methods to tell the API exactly what you want. For example, in our news aggregator app, you might tell the API that you only want to see news from a certain country and only in one language.

- **Request headers**: These include extra information that tells the API important information about either the request or the person requesting it, for example, if you're allowed to act as the API call. Typical information included is authorization details, response cookies, the format of the request and the response body, and even response caching among other information.

- **Request body**: This is the heart of the request, where you put all the specific details needed for the API to complete your request if you're sending or changing data. For example, if you are creating a new user account, you can put in the username, email address, phone number, and other information about the user.

Upon receiving the API request by an API endpoint, the request will be verified to ensure it is formatted properly and the required permissions exist. Upon validation of the API request, the action is performed, and a response is sent back to indicate the status of the request. These responses typically contain the following:

- **Status code**: These are codes that tell you whether your request worked or not, such as 200 for OK, 201 for something created, and 404 when something wasn't found.

- **Response headers**: These tell you more about the response, kind of like the request headers but for the answer you get back.

- **Response body**: This is where the data you asked for comes back to you or you get a message about what happened with your request.

The ease of implementation and flexibility of APIs make them ideal for behind-the-scenes connections and communications performed by software both on-premises and in the cloud. In a typical day, you might have experienced APIs in actions such as the following:

- **Boosting security**: APIs help with important protective workflows such as **single sign-on (SSO)** and setting automated governance policies.

- **Integration**: This is one of the most common uses of APIs—to connect systems. For example, hooking up a CRM to a marketing system to streamline your customer outreach.

- **Scaling up**: With APIs, you can build microservices, little chunks of an app that work together, letting you scale your app efficiently.

- **Adding features**: APIs let you add new functions to your apps, such as maps for delivery tracking in a food ordering app.

- **Linking IoT**: APIs are crucial for smart devices to connect to the cloud and each other, making things such as smartwatches and internet-connected fridges possible.

- **Cost-cutting**: By automating things such as emails and data sharing, APIs can help save money on operations, and they let developers reuse existing components rather than making new ones.

APIs are everywhere and most users encounter them without realizing it due to the flexibility from simple web-based communications to complex solutions. In most cases, most modern security and IT tools available offer some sort of API that can be leveraged to improve communication and visibility with the Defender security product suite via the Microsoft Graph security API.

The Microsoft Graph security API

For those working with Microsoft Defender security products, the go-to API is the Microsoft Graph security API, which provides a unified interface to integrate security solutions from Microsoft and partners. It can handle many tasks, from pulling and looking into security incidents to setting off actions based on new threats. The abilities of the Graph API are vast: it can help an organization keep an eye on and analyze threats coming from all directions, streamline alerts from various sources, automate security workflows and reporting, enable proactive risk management, and provide the tools needed for companies to respond to cyber threats effectively (*Microsoft, 2024a*). In the following section, we will cover the most used security APIs.

The advanced hunting API

The **advanced hunting API** provides an invaluable resource for cybersecurity specialists. Using KQL, this API enables the exploration of up to a month's worth of raw data. Security analysts can proactively scrutinize events, searching for signs of threats and pinpointing relevant entities. The API's capacity for unrestrained data access empowers users to hunt not only established threats but also emergent ones. This API, under the `microsoft.graph.security.runHuntingQuery` namespace, targets specific data and employs a series of operations to refine and arrange the query results. These findings are crucial for enhancing current investigations and discovering hidden threats. We can see a hunting query performed in Graph Explorer in the following figure.

Figure 11.2 – A hunting query being run in Graph Explorer

> **Note**
> It's key to note that there are resource limits: overly complex queries or heavy data pulls might trigger a `429` HTTP error. To avoid this, streamline your queries and keep an eye on the data volume and time span, targeting query executions under three minutes.

Common ways to optimize KQL queries include the following:

- **Test your query on a small data output**: Before running a complex query, use the `count` operator to assess its extent by using the `limit` or `take` operators to limit how much data is outputted. This approach can also be used when investigating the content of a table to determine the best query to write. In the following figure, we can see this query in action on the `AlertEvidence` table, which tends to have a lot of data; notice **Low**, indicating it is a low-impact query.

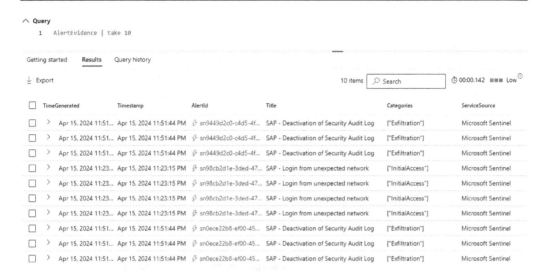

Figure 11.3 – A KQL query using the take operator

- You should opt for `has` over `contains` when pinpointing exact words. The `has` operator is preferable to `contains` here because it avoids unnecessary substring searches. We can see this in action in the following figure, where the `has` operator is used to filter the amount of data.

Query

```
1    AlertEvidence | where TimeGenerated >= ago(1d) and Categories has "Exfiltration" and DeviceName has "vhcala4hci"
```

		TimeGenerated	Timestamp	AlertId	Title	Categories
⊐	>	Apr 15, 2024 11:51...	Apr 15, 2024 11:51:44 PM	sn9449d2c0-c4d5-4f...	SAP - Deactivation of Security Audit Log	["Exfiltration"]
⊐	>	Apr 15, 2024 11:51...	Apr 15, 2024 11:51:44 PM	sn0ece22b8-ef00-45...	SAP - Deactivation of Security Audit Log	["Exfiltration"]

etting started   **Results**   Query history

Export   2 items   🔍 Search

Figure 11.4 – Using the has operator to filter data

- **Implement filters promptly**: Apply time constraints and other relevant filters promptly to streamline the data, ideally before engaging with transformation or parsing processes such as `substring()`, `replace()`, `trim()`, `toupper()`, or `parse_json()`. For instance, filter the data before utilizing the `extractjson()` parsing function in the accompanying example to manage the volume of records.

- **Search within specific fields**: Direct your search to a particular column to prevent scanning all the data fields. Refrain from using the asterisk (*) wildcard to indiscriminately examine every column.

- **Choose case-sensitive queries**: Case-sensitive search operators tend to be more precise and efficient; operators, such as `has_cs` and `contains_cs`, as well as case-specific equality operators, such as `==` as opposed to `=~`, make for more specific searches and smaller output, which can be especially effective when the output is to be used in other parts of the query. We can see an example of these case-sensitive search operators in the following figure.

Figure 11.5 – Using case-sensitive search operators

- **Prioritize parsing over extracting or regex**: Opt for the `parse` operator or parsing functions such as `parse_json()` whenever possible. Regular expressions, as used by the `matches regex` operator or the `extract()` function, should be reserved for complex scenarios.

- **Filter on table columns**: Filtering on table columns for existing data is much less resource intensive versus filtering on calculated columns. Always try to filter before performing calculations on the data when possible.

- **Use four or more characters when comparing**: Minimize the use of terms that have three or fewer characters in comparisons or filters, as these are not indexed and will require additional resources to match.

- **Project precise columns**: By selectively projecting only the necessary columns, your results will be clearer and more manageable. Doing this before executing operations such as `join` can further enhance performance. We can see the project operator in use to decrease the number of columns output in the following figure.

```
1   IdentityInfo | where TimeGenerated >= ago(1d) and AccountDisplayName has_cs "Justin"
2   | project TimeGenerated, Type, AccountName, EmailAddress, ChangeSource
```

ting started   **Results**   Query history

Export   1 item   🔍 Search   🕐 00:00.202 ▰▰▰ Low ⓘ   🗠 Chart type ∨   🖼 Customize columns

TimeGenerated	Type	AccountName	EmailAddress	ChangeSource
> Apr 16, 2024 7:47:07 AM	User	Justin Powell	Justin.Powell⬛⬛⬛⬛	AzureActiveDirectory

Figure 11.6 – Using the project operator to decrease the amount of output

- You should optimize `join` statements by having the smaller table on the left. Positioning the smaller table on the left in a `join` statement results in fewer records to link, enhancing the speed of your query.

- You should set time windows before the `join` statement. By aligning records within designated time windows before joining, you'll refine the number of records to correlate. This is especially beneficial in boosting query speed when filtering times on both sides of the `join` operation. We can see this filtering while joining tables in the following figure where identities are compared to what is found in the audit logs.

```
1   IdentityInfo
2   | where TimeGenerated >= ago(1d)
3   | join kind=inner AuditLogs on $left.AccountDisplayName == $right.Identity
4   | where TimeGenerated >= ago(1d)
```

ting started   **Results**   Query history

Export   496 items   🔍 Search   🕐 00:02.196 ▰▰▰ Low ⓘ

Figure 11.7 – Filtering by time windows when joining tables

- You should employ `join` hints. Hints guide the backend in managing the workload, especially for `join` operations that demand substantial resources.

- **Distinct values remove repetitive entries**: Use the `summarize` operator primarily to distinguish unique values in data that may repeat.

Skills in efficient KQL query creation extend beyond the advanced hunting API use case and into other tools and scenarios. If your organization uses Sentinel, efficient KQL queries will lead to improved threat-hunting efforts as complex scenarios can be more easily identified in a lower number of queries and alert rules.

The Alerts_v2 API

Microsoft and other security partners issue alerts for suspicious activities within a user's environment. These security alerts represent various attack techniques on different targets such as devices or mailboxes, often leading to many notifications from various sources. While typical security operations focus on an incident, there are times in an investigation when details on the alert might be required, such as when fine-tuning detection tools or getting more details on a specific attack among many other scenarios. During these times, the `Alerts_v2` API, located on the `microsoft.graph.security.alerts_v2` namespace, allows calls to obtain in-depth details on these alerts. We can see a glimpse of a typical reply from this API in the following figure.

```
"incidentId": 2654,
"investigationId": 401,
"assignedTo": "API-App:API Action",
"severity": "Informational",
"status": "Resolved",
"classification": null,
"determination": null,
"investigationState": "Benign",
"detectionSource": "AutomatedInvestigation",
"detectorId": "              7",
"category": "SuspiciousActivity",
"threatFamilyName": null,
"title": "Automated investigation started manually",
"description": "Renato Mendes() initiated an Automated investigation on mb-winclient.internal.
"alertCreationTime": "2024-04-13T20:06:31.6533333Z",
"firstEventTime": "2024-04-13T20:06:30Z",
"lastEventTime": "2024-04-13T20:06:30Z",
"lastUpdateTime": "2024-04-13T20:13:52.45Z",
"resolvedTime": "2024-04-13T20:13:52.2769676Z",
```

Figure 11.8 – A typical alert API response

The `Alerts_v2` API provides multiple ways for performing calls, going down in level depending on the amount of granularity required. For using this API, at a minimum, the account must have the `SecurityAlert.Read.All` permissions. This API is broken into sections for ease of use as follows:

- `Alerts_v2`: This is the level of the API and allows you to query for a list of alerts and filter this list according to the multiple properties available on these alerts such as `actorDisplayName`, `alertPolicyId`, `category`, `description`, `id`, `incidentId`, `severity`, and `status`. The freshest alerts are positioned at the top for your convenience. At this level, only the GET and POST calls are available with the POST call requiring the `SecurityAlert.ReadWrite.All` permission.

- `Alerts_v2/{alert-id}`: This allows for querying a single alert by ID number to get all the information available on this alert. It differs from the top-level call in that it provides an entire alert breakdown versus just a list. At this level, you are allowed the GET, PATCH, and DELETE calls. For the PATCH and DELETE calls, the user is required to have the `SecurityAlert.ReadWrite.All` permission.

- `alerts_v2/{alert-id}/comments`: This allows the user to append a comment to a specific alert. At this level, you are only allowed the `POST` call and it requires that the user have the `SecurityAlert.ReadWrite.All` permission.

The `Alerts_v2` API can be integrated with ticket systems to allow for easier management of tickets when an organization is already invested in an existing ticket management system. Many organizations look at implementing a service account that updates the status of the alerts in the Defender XDR portal and vice versa by using API calls.

The Alerts API (Legacy API)

This legacy API located in the `microsoft.graph.security.alerts` namespace is the old API used for combining Azure and Defender security products data into a unified solution. This API offers similar levels and calls as the `Alerts_v2` API, but it is considered deprecated, and its use is no longer recommended.

The Incidents API

The **incidents API** focuses on queries and changes to incidents. Incidents represent the totality of an attack narrative, including correlated alerts. Incident management via API might allow for some interesting automation and reporting scenarios in which the entire kill chain of an attack can be considered, which is something very difficult when just observing insights from individual alerts. This API, located in the `microsoft.graph.security.incidents` namespace, allows for effective sorting of incidents and sharing with third-party solutions. We can see this API being queried in Graph Explorer along with the permissions required in the following figure.

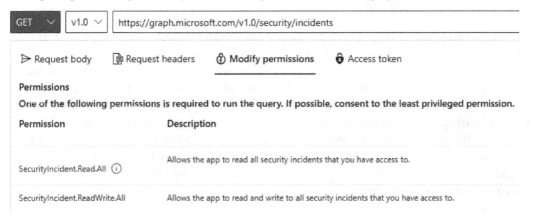

Figure 11.9 – Querying the incidents API

Like the `Alerts_v2` API, this API is divided into levels and the account using this API requires as a minimum the `SecurityIncident.Read.All` permissions. As with typical APIs, this API provides different levels of granularity as follows:

- `incidents`: This is the level of the API that allows you to query for a list of incidents and filter this list according to the multiple properties available on these incidents such as `classification`, `customTags`, `determination`, `id`, `incidentWebUrl`, `severity`, `status`, and many others. At this level, only the `GET` and `POST` calls are available with the `POST` call requiring the `SecurityAlert.ReadWrite.All` permission.

- `incidents/{incident-id}`: This allows for querying a single incident by ID number to get all the information available on it. As with the `Alerts_v2` API, it is used in combination with the top-level API when in-depth information about an incident is required versus just a list. At this level, you are allowed the `GET`, `PATCH`, and `DELETE` calls. For the `PATCH` and `DELETE` calls, the user is required to have the `SecurityIncident.ReadWrite.All` permission.

- `incidents/{incident--id}/comments`: This allows the user to append a comment to a specific incident. At this level, you are only allowed the call to be `POST` and it requires that the user have the `SecurityIncident.ReadWrite.All` permission.

When performing security operations, it is always recommended to do actions at the incident level, and the Defender portal (`https://security.microsoft.com`) already offers built-in integration for bidirectional changes to alerts and incidents through the Defender ecosystem. It is recommended to approach security this way to ensure a single source of truth for incidents and only to use this API in a read-only manner.

The attack simulation API

The **attack simulation API** is used for user training and provides multiple options depending on the type of training to be performed. While it is recommended to do attack simulation and training via the Defender portal for simplicity, some organizations like to integrate this API with other training tools to have an all-in-one training and tracking solution. This API, located in the `microsoft.graph.security.attackSimulation` namespace, allows many types of educational simulations that can be triggered via different API levels. The main aim of organizations leveraging this API is to not only bolster user defenses against social engineering attacks but also have access to valuable simulation and training data to help identify risky user behavior that needs prompt attention. When testing this API in Graph Explorer, for permission assignment, the **Permission** panel will need to be used as seen in the following figure.

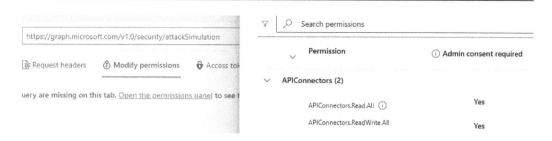

Figure 11.10 – The Permission panel when using Graph Explorer

At a minimum, the user using this API will need `AttackSimulation.Read.All` and the `APIConnectors.Read.All` permissions to leverage all the aspects of this API. This API offers the following options:

- `attackSimulation`: This section of the API provides a list of attack simulation campaigns configured for a tenant to include information such as who created the simulation, the simulation name, description, attack type, attack technique, and many other details. This level allows for `GET`, `PATCH`, and `DELETE` calls to be able to create new simulations or modify existing ones.

- `attackSimulation/endUserNotifications`: This allows for the modification of email messages sent to end users after a simulation. These notifications can include anything from positive reinforcement after the user has performed the correct action to training assignments for users failing to meet the expectations during the simulation. At this level, the focus is on listing notifications and adding new ones with allowed calls including `GET` and `POST`. Another section of the API is offered, `attackSimulation/endUserNotifications/{endUserNotification-id}`, which allows for modifying a specific notification with the `GET`, `PATCH`, and `DELETE` calls. Further discovery can be performed at the notification level with `attackSimulation/endUserNotifications/{endUserNotification-id}/details` offering `GET` and `POST` calls, and `attackSimulation/endUserNotifications/{endUserNotification-id}/details/{endUserNotificationDetail-id}` offering `GET`, `PATCH`, and `DELETE` calls.

- `attackSimulation/landingPages`: This relates to the web page the end user is presented with when they click on a payload during a simulation. This API allows for the use of out-of-the-box landing pages, also known as **global landing pages**, and custom-built landing pages, also known as **tenant landing pages**. A user is allowed to either list the landing pages available via the `GET` call or add new pages via the `POST` call. Specific tenant landing pages can be listed, modified, and deleted via `attackSimulation/landingPages/{landingPage-id}` with the `GET`, `PATCH`, and `DELETE` calls respectively.

- `attackSimulation/loginPages`: This allows for the listing and modification of login pages presented to end users during an attack simulation for the simulation of a credential-stealing attack. Pages available include *tenant login pages* and *global login pages*. A typical use case for this API is to generate and backup custom login pages for simulation involving custom applications. The API top level allows for the listing of login pages via the GET call and adding new pages via the POST call. A more granular option is provided in attackSimulation/loginPages/{loginPage-id}, which allows for the listing of all details about a specific page with the GET call, updating a specific page with the PATCH call, and removing a page with the DELETE call.

- attackSimulation/operations: This is used for querying the status of an attack simulation operation in progress. The top level allows for listing current operations underway with the GET call and adding new operations via the POST call. For a more granular check, the operation ID can be used in attackSimulation/operations/{attackSimulationOperation-id} to query information about a specific operation such as the tenant involved, when it was created, when the last action was taken, and the percentage completed, among other details. Calls available at this granular level include GET for listing details, PATCH for updating a specific operation, and DELETE to remove a specific operation.

- attackSimulation/payloads: This is used to query payloads used in attack simulations, which refers to the content of a phishing email such as the message body, any links, and any attachments. When querying this API, payloads include **global payloads**, which are built-in and unmodifiable, **tenant payloads**, which are custom payloads created by the user, and **MDO recommendations**, which include a list of payloads that simulate malicious payloads identified as capable of causing a major impact in the environment. As with other APIs, you have a top level for querying a list of payloads with the GET call, as well as adding new payloads using the POST call. Providing a payload ID to attackSimulation/payloads/{payload-id} allows for querying specific details about a payload (GET), as well as modifying it (POST), or removing it (DELETE).

- attackSimulation/simulationAutomations: This allows you to organize multiple attack simulations at the same time in an organized campaign to better mimic real-world scenarios and complex attacks. At the top level, you can list simulations underway (GET) and add a new one (POST). Providing a simulation ID in attackSimulation/simulationAutomations/{simulationAutomation-id} allows for granular querying of a simulation (GET), modification of it (PATCH), and even deleting it (DELETE).

- attackSimulation/trainings: This relates to end-user training that can be provided before an attack simulation or because of end users failing the attack simulation. The API can be used to configure multiple aspects of the attack simulation, including the language of the training and type to include but not limited to credential harvesting, link to malware, phishing, social engineering, and others. This API is very useful for global organizations that need to track training and report on it across multiple countries. As with other APIs, the top level allows for listing any training in progress (GET) as well as creating a new training (POST). Obtaining a training ID will allow the use of attackSimulation/trainings/{training-id} for more detailed queries (GET), modification (PATCH), and removal (DELETE) of existing training.

End-user training and attack simulation will be covered in detail in the next chapter. The important point to remember is that all aspects of end-user training can be controlled from both the Defender portal (`https://security.microsoft.com`), which is the easiest approach for most organizations, and programmatically via APIs.

The secure score API

Microsoft Secure Score offers a singular metric to grasp and improve your security posture within Microsoft solutions. It allows not only a comparative view with other organizations but also a quantifiable assessment of your security trend over time. By examining 90 days of security data, security improvement information is provided in the form of security and productivity-balanced actionable points. We can see the secure score of an organization as presented in the Defender portal in the following figure.

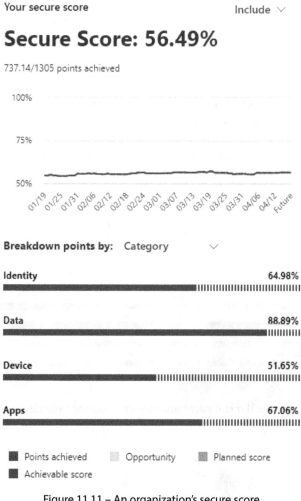

Figure 11.11 – An organization's secure score

An organization can rely on secure scores to create an improvement plan to implement in an environment and track improvements over time. This feature can be used and managed via the **secure score** and **secure score control profile** APIs and requires that the user account used for calls has at least the `SecurityEvents.Read.All` permission. We can see these APIs in the following list:

- `secureScores`: This API allows for querying (the GET call) the current secure score for the organization to include properties such as active user count, current score, maximum possible score, and enabled services, among other properties. The response provided will include a list of the controls enabled to generate the secure score presented.

- `secureScoreControlProfiles`: This API allows for querying (the GET call) all the controls used for the secure score. Control information includes properties such as threats the control mitigates, control tier, user impact, category, and many other properties. Using `secureScoreControlProfiles/{secureScoreControlProfile-id}`, a user can query a specific control (GET) and perform modifications (PATCH) such as updating its compliance status; the control status information can also be reset to a default value (DELETE).

The Threat Intelligence API

Microsoft Defender **Threat Intelligence API** provides cutting-edge intelligence to bolster your organization against cyber threats. This API enables the operationalization of intelligence found within the interface, composed of intelligence articles, IoCs, enrichment data, intel profiles, and more. We can see some of the information that can be provided by Defender TI via the Defender portal, which will also be available via the API in the following figure.

Threat analytics

Ransomware	Extortion	Phishing	Hands on keyboard	Activity group	Vulnerability	Attack campaign	Tool or technique
95	0	59	0	177	111	0	0

Latest threats

Actor profile: Silk Typhoon	0 / 0
CVE-2024-3400 command injection vulnerability	0 / 0
CVE-2024-21413 - Microsoft Outlook remote code execution vulnerability (Moniker...	0 / 0
Remote code execution exploit chain in OpenVPN	0 / 0

■ Active Alerts   ■ Resolved Alerts   ▨ No Alerts

High-impact threats ⓘ

Threat overview: On-premises credential theft	91 / 659
Tool profile: Mimikatz	74 / 356
Technique profile: Kerberos attacks	74 / 347
Technique profile: Antivirus tampering	74 / 276

■ Active Alerts   ■ Resolved Alerts   ▨ No Alerts

Figure 11.12 – Defender TI threat information as seen from the Defender security portal

The different parts of this API behave similarly at the top level, allowing the generation of a list (GET) and the creation of new entries (POST). Supplying an ID allows for other actions such as listing more properties in detail (GET) or updating an entry (PATCH). This API requires the user to have as a minimum the APIConnectors.Read.All and ThreatIntelligence.Read.All permissions along with a license to Defender TI. The parts of this API and what can be queried on these include the following:

- articles: This relates to threat intelligence articles that provide insights on threat actors and the tools, attacks, and vulnerabilities these use.

- articleIndicators: This relates to the indicators of compromise identified in articles. A user can query more details about the indicator to include its properties and relationship to a TI article.

- hosts: This relates to indicator artifacts identified by Defender TI, in particular a malicious hostname or IP address identified on the internet.

- hostComponents: Related to indicators artifacts, this is a web component such as a web page or server identified as hosting malicious content or being used by a bad actor.

- hostCookies: Related to cookies, this is the data used on clients during web communications that has been identified as an artifact left on victims of malicious sites.

- hostPairs: This provides information on relationships between different malicious hosts, in particular, a parent-child relationship usually observed when malicious sites perform redirects.

- hostPorts: This provides responses related to ports that have been identified as having been used by malicious hosts for malicious services.

- hostSslCertificates: This provides information on SSL certificates identified in TI articles and their relationship to a malicious host.

- hostTrackers: This provides information on unique identifiers found in malicious hosts' web page code. As many bad actors just copy malicious sites with minimal modifications, this can be a common indicator to find. Actors also might deploy tracker IDs to keep track of campaigns in progress, which this part of the API can provide information on.

- intelProfiles: This provides information on threat actor profiles, which can be leveraged on a security operations workflow to improve detection and further customize their controls.

- intelligenceProfileIndicators: This relates to indicator artifacts identified by Defender TI identified in threat actor intel profiles.

- passiveDnsRecords: This provides DNS records for previously identified threat actor infrastructure to include the location and time the record was captured. This information can be very useful during the creation of an incident timeline or even when threat hunting.

- `sslCertificates`: This provides information on SSL certificates identified previously by Defender TI via web crawlers for comparison to detect malicious fake SSL certificates.

- `Subdomains`: This provides information on subdomains previously identified by Defender TI via web crawlers for comparison to detect malicious fake subdomains.

- `Vulnerabilities`: This provides context and in-depth information on vulnerabilities of interest. Context information might include dark web chatter, key observations, related articles, how these vulnerabilities are used in attacks by threat actors, and much more.

- `whoisHistoryRecords`: This contains WHOIS information about a malicious host previously identified by Defender TI.

As seen in the amount of information offered by the Threat Intelligence API and other security-related APIs mentioned in this section, organizations have a lot of flexibility to create interesting solutions that leverage this data such as enrichment of other security tooling data, automation, and even customized training for the security team. One important thing to note is that organizations should aim at not reinventing the wheel with these options and instead look at leveraging the product already provided by Microsoft first as their security maturity grows. The following sections will touch on ways to test and integrate these APIs along with the kind of overhead expected on average.

API exploration and integration

Microsoft provides a range of tools for organizations to harness the power of APIs, and among these is **Graph Explorer**. This web-based application acts as a hands-on platform for learning and experimenting with the Microsoft Graph APIs, allowing users to prototype and test applications and integrations with ease. You can dive into Graph Explorer by accessing it via `https://developer.microsoft.com/en-us/graph/graph-explorer`, which will open Graph Explorer and automatically load the environment you are already logged into. For switching tenants, just add the tenant parameter to the URL like this (replace the entry after `tenant=` with your desired tenant): `https://developer.microsoft.com/en-us/graph/graph-explorer?tenant=mytenant.onmicrosoft.com`. Upon opening Graph Explorer, you'll notice the interface is cleverly organized into three main sections:

- **The queries and resources panel**: There, you will find three tabs for navigating the Microsoft Graph API. The **Sample queries** tab contains many ready-to-use examples that you can modify as needed to kick-start your testing. Clicking on any of these will populate the **Run query** section with the premade query. We can see the **Sample queries** tab in the following figure.

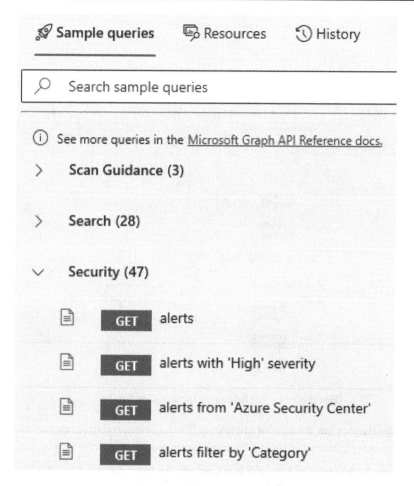

Figure 11.13 – The Sample queries tab in Graph Explorer

The **Resources** tab lists the available APIs, including the sub-levels in these APIs and operations available such as GET, POST, PATCH, and DELETE. Clicking on any of these entries will populate the **Run query** section with the initial parts of a query to be used on that API. We can see this tab in the following figure.

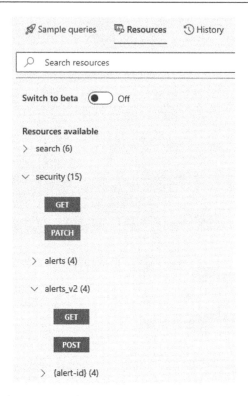

Figure 11.14 – The Resources tab in Graph Explorer

Finally, the **History** tab shows a list of previously run queries and the response code obtained. Clicking on any of these will populate the results section of the page with the old results obtained.

- **Run query**: This section is where you select, edit, and execute your queries. It includes an array of options such as the HTTP method selection (GET, POST, PUT, PATCH, DELETE), the version of the API to use, the URL input field for API endpoints, and the ability to append additional headers or change permissions. We can see the **Run query** section with a prefilled query in the following figure.

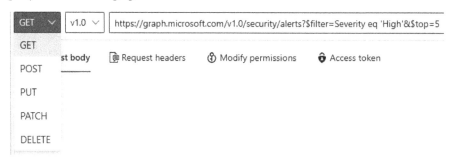

Figure 11.15 – The Run query section of Graph Explorer

This section also allows for further modification of the request such as modifying the request body, adding values to the request headers, modifying permissions (might require tenant admin approval depending on the tenant configuration), and a section to copy the access token, which can be translated on jwt.ms to see the embedded claims.

- **The results section**: Following an API call, this is where outcomes are displayed, showcasing the response codes and data. We can see an example of a response in the following figure.

Figure 11.16 – The Response section in Graph Explorer

This section is not just about the results but also other features such as the **Code snippets** tab, which translates your API interactions into code blocks in various programming languages, such as C#, Java, and Python, among others. We can see an example of the contents of the **Code snippets** tab in the following figure.

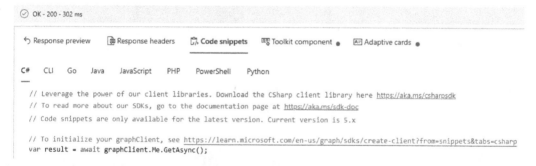

Figure 11.17 – The Code snippets tab in the results section

The results section provides a quick way to verify whether an approach via the API is possible and how it should be written. Being very familiar with this section will save your team countless hours of rework when developing new integrations.

While Graph Explorer serves as a solid starting point, it's just one piece of the puzzle. Microsoft Graph Toolkit complements it by offering ready-made web components for building applications with Microsoft Graph. We can see an example of these web components in the **Response** section's **Toolkit component** tab, as shown in the following figure.

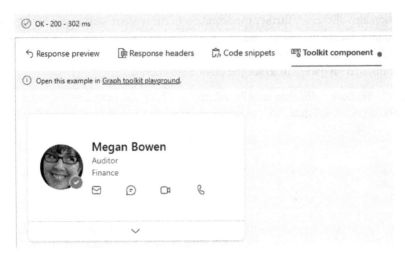

Figure 11.18 – The Toolkit component tab in the Response section

These components are not only easy to use but they also work with any web framework, providing prewritten code for elements such as agendas, people and file information cards, authentication, and many more elements requiring minimal modification. We can also see examples and perform testing by using Microsoft Graph Toolkit Playground located at mgt.dev. We can see the **Playground** page in the following figure with an example of one of these elements.

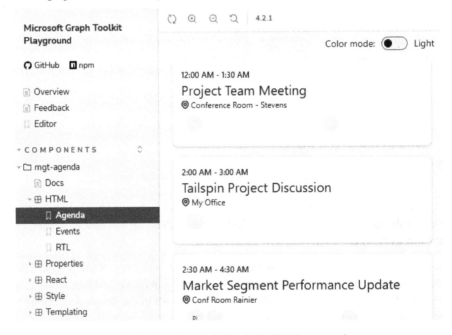

Figure 11.19 – The Microsoft Graph Toolkit Playground page

These elements are easy to use as they offer prebuilt integrations that simplify authentication and Graph interactions and support many common environments such as SharePoint. Whether these components are used in unison with the already integrated providers (organizations that expose data via API) or standalone, they streamline access to Microsoft Graph. You can explore these components and their documentation by visiting the Microsoft Graph Toolkit's GitHub repository at `https://github.com/microsoftgraph/microsoft-graph-toolkit`. The repository provides comprehensive guides to help you get from the start to a fully functional prototype swiftly.

For those who wish to create proprietary connectors, Microsoft provides the `ms-graph` connectors **software development kit** (**SDK**), available on GitHub at `https://github.com/microsoftgraph/msgraph-connectors-sdk`. This repository includes information on configuring components such as the graph connector agent, which allows for testing the connector, configuration information for developers that prefer to use a Visual Studio Code development environment or third party alternative, and even code samples to kickstart development.

When searching for native connectors from Microsoft's partners, it's advised to directly contact the providers for the latest versions and usage guidance. Some common native connectors include the following:

- **PowerShell module and PowerBI connector**: This module and connector let users craft custom visualizations and reports and can take automated remedial actions directly through PowerShell scripts.

- **Azure Logic Apps / Microsoft Flow**: While these are part of Microsoft's offerings, they are not included with the Defender suite by default. They can be utilized to automate responses and streamline data analysis in security-related workflows.

- **ServiceNow**: For those managing alerts through ServiceNow, this integration enriches the tracked data with insights from Microsoft Defender.

- **Splunk Enterprise and Splunk Cloud**: This connector funnels alerts from Microsoft's Defender into Splunk, bolstering its SIEM capabilities.

- **QRadar**: Like Splunk, QRadar gains enhanced SIEM functionality by incorporating alerts from Microsoft Defender.

Always remember that API integration should only be performed when a gap has been identified that cannot be resolved via the Defender portal (`https://security.microsoft.com`) and has a real business impact. Due to the overhead required to maintain custom solutions and integrations, opting for reengineering or overengineering a solution not only takes valuable resources away from security work but also might negatively impact the security posture due to decreased visibility and a possible increased attack surface. As our security matures, a greater focus will go toward decreasing unnecessary work and that is where **artificial intelligence** (**AI**) shines, which will be discussed in the next section.

AI to the rescue

In the digital world, the pace and sophistication of cyber threats are escalating quickly, posing challenges for organizations to keep up. Security teams are inundated with warnings, resulting in alert fatigue. Even with automation and the use of APIs, some threats still slip through the net. Though threat hunting can aid in spotting these elusive threats, it's a time-intensive process that security experts need to master. Additionally, the scarcity of skilled cybersecurity professionals has led companies such as Microsoft to harness AI to strengthen their cyber defense.

Copilot for Security

Copilot for Security is a state-of-the-art AI tool built in collaboration with OpenAI. This tool assimilates the immense data gathered from Microsoft's security signals and threat intelligence with intel from various Microsoft products within a given environment. This comprehensive information enables a **large language model** (**LLM**) to swiftly analyze data and identify overlooked patterns that either match known attack vectors or deviate from standard behaviors. The distinguishing feature here is that the model and its inquiries are custom-tuned to the signals from Microsoft products within the environment. Continuous refinement improves detection accuracy, and integration with Defender XDR means no additional deployment is necessary to access its features (*Microsoft, 2024b*).

Copilot for Security performs many tasks in the background that used to take security teams many man-hours to complete, which in turn frees security personnel to perform more impactful security work. Some of the automated tasks include the following:

- Converting complex datasets to actionable advice and insights, providing clear guidance during incident probes

- Producing easy-to-comprehend reports and presentations that analysts can use to communicate security incidents within the organization

- Responsively addressing inquiries about incidents or vulnerabilities in a clear format that could include natural language or visuals

Copilot for Security amplifies the efficiency of a unified security operations platform by merging capabilities from Microsoft Defender XDR, the entire Defender suite, and Microsoft Sentinel. While integration already exists, Copilot for Security goes a step further by using this data in the same manner as a security analyst would to enhance the scrutiny of current hunting processes and identify areas of improvement. This integration and enhancement go a step further by using data from other Microsoft products in use in the environment to give a clearer impact assessment of security events. We can see an example of Copilot for Security performing a risky user analysis in the following figure.

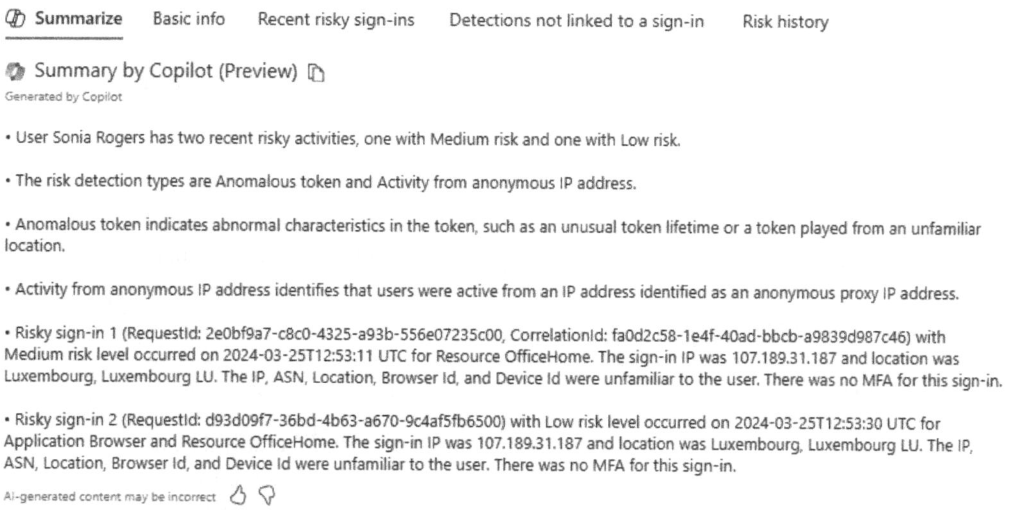

Figure 11.20 – A Copilot for Security summary about a risky user

Some examples of this integration include the following:

- **Microsoft Entra**: Copilot for Security can scrutinize identity-related risks and ease the troubleshooting of identity issues, offering instant summaries and remediation guidance in understandable terms.

- **Microsoft Defender for EASM**: This tool allows security teams to quickly understand their external attack surface with Copilot's help, which means coping with vulnerabilities becomes more efficient, benefiting priority asset protection and remediation prioritization.

- **Microsoft Intune**: With Copilot, insight across security data, including full device context, is clearer. It assists in formulating policies and provides AI-driven recommendations for rapid response to security mishaps.

- **Microsoft Purview**: This system streamlines the examination of diverse data types in various operational aspects such as data loss prevention and insider risk management. This speeds up responses and empowers analysts to tackle intricate tasks using AI-powered support.

- **Microsoft Defender for Cloud**: With Copilot, identifying critical cloud resource threats becomes more straightforward. It helps administrators enhance threat response by creating recommendation summaries and remediation scripts in their preferred language.

Thanks to the Microsoft Graph API integration capabilities, Copilot's scope can go beyond just Microsoft products and into third-party products via plugins, allowing for more comprehensive protection and enhanced recommendations. For security teams with advanced operational needs, custom API and KQL-based plugins are supported, enabling the querying of data using natural language prompts.

Deploying Copilot for Security

To set up Copilot for Security with Defender XDR, the process is straightforward with components deployed within the Azure environment. Using Azure resources for the solution means that securing it also follows the usual security measures and monitoring as typically performed on other Azure resources. The user deploying the solution will require either Global Administrator or Security Administrator privileges in the tenant, as well as the Azure owner or contributor permissions for the resource group where cloud resources are deployed. Deployment involves a two-part process with the first one involving the provisioning of the Azure cloud resource and the second one being the configuration of the solution.

Provisioning capacity

This part involves the deployment of the **security compute units** (**SCUs**) that will be used by Copilot for Security to perform queries. Keep in mind the number of SCUs is tailored to query complexity, typically 10 simultaneous queries can be performed per SCU. This step can be performed from the Copilot for Security portal (`https://securitycopilot.microsoft.com`) by clicking on **Get started** and following the setup wizard. We can see this wizard in the following figure.

Figure 11.21 – The Copilot for Security setup wizard

In this wizard, choose an **Azure Subscription** option, a **Resource group** option, provide a unique name for the **Capacity name** option, select the location where you'll evaluate prompts (make sure to comply with your organization's data restrictions), and decide on the number of SCUs. Accept the terms and conditions and proceed, which will trigger the deployment of Azure cloud resources. The same wizard and actions can be performed from the Azure portal, the only difference is that you need to search for `Microsoft Copilot for Security compute capacities` and then click on **Create Microsoft Copilot for Security compute capacity**, which will start a similar wizard.

Configuring the solution

Once capacity provisioning is complete, it is time to configure the solution. Follow these steps to complete the configuration wizard.

1. Navigate to the Copilot for Security portal (`https://securitycopilot.microsoft.com`), select the capacity previously created, and click **Continue** to start the configuration wizard. The wizard will indicate the location where Copilot for Security data will be stored and if it aligns, select **Continue**. We can see an example of this location notice in the following figure.

Figure 11.22 – Copilot for Security data location notification page

> **Note**
> Verify that the data storage location complies with any geographical restrictions your organization may need to adhere to. If in doubt, confirm this requirement before moving forward. Failing to verify this could result in your organization violating legal requirements, which can lead to costly fines and other costs for migration and clean-up of the data.

2. You will be given options for sharing data with Microsoft to improve the Copilot AI model. Choose your data-sharing preferences according to your organizational policies and proceed. We can see this page in the following figure.

Figure 11.23 – Configuration wizard data sharing options page

3. Review the **Copilot for Security** role assignment. By default, the **Contributor** role is assigned to all users, and the **Owner** role for those with **Global Administrator** and **Security Administrator** roles. The **Contributor** role is the role required to use Copilot for Security, while the **Owner** role is intended for those who manage Copilot's configuration and permissions. For now, we can proceed with the default configuration by clicking on **Continue**. You can change these role assignments later as necessary. You can see the role assignment page in the following figure.

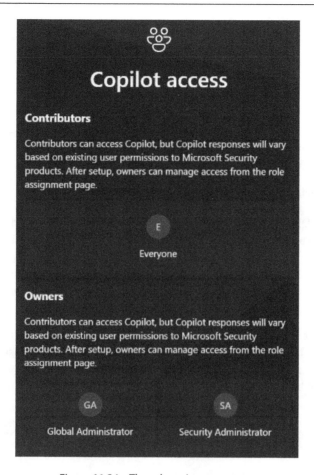

Figure 11.24 – The role assignment page

4. A screen showing the configuration selected will appear to confirm the configuration is completed. Confirm your setup is correct and save the provided link for the **Copilot for Security** access. If everything aligns with what was configured, click **Finish**.

Now, Copilot for Security should be operational, but it's worth taking an extra step to fine-tune roles. While Copilot for Security has specific roles for accessing the platform, the data queried and viewable by a user will still depend on the user's permissions to the data. These restrictions are due to Copilot employing on-behalf-of authentication to gather security data via active Microsoft plugins. As such, it is important to think about the entire set of permissions users will need and configure these via groups or other central management approaches to ensure ease of management and prevent privilege creep. Also, note that roles used for accessing the Copilot for Security platform are independent of those within Entra or Azure IAM; they apply solely to the Copilot platform. For broader product access, you'd still use Entra roles, and Azure IAM roles remain essential for any changes to security capacity, such as SCU adjustments.

Using Copilot for Security

The features of Copilot for Security integrate seamlessly not only as a standalone platform but also within other Microsoft security solutions. Analysts typically begin their investigative process through the embedded functionalities accessible via other Microsoft products. Should an investigation require deeper analysis, analysts can employ natural language queries in the comprehensive standalone Copilot interface, found at the Copilot for Security portal (`https://securitycopilot.microsoft.com`). In the subsequent sections, we will delve into the particulars of using Copilot for Security alongside Defender XDR.

Summarizing incidents

Microsoft Defender XDR enhances its incident handling with Copilot for Security, streamlining the process for incident response teams. In the digital defense arena, time is critical, and comprehending the complexities of a cyberattack swiftly is vital for safeguarding an organization. The integration of Defender XDR with Copilot for Security's AI-driven analyses allows teams to swiftly make sense of chaotic situations involving multiple alerts and impacted assets, thus pinpointing the scale and severity of the threat.

Each incident, encompassing up to a hundred alerts, is combined into a cohesive summary. This summary encapsulates the attack narrative, making clear that the beginning stages, the antagonists, the affected assets, and any IoCs provided the data permits. Accessible with ease, these summaries are auto-generated and housed in the **Copilot** pane, visible within the **Incident summary** card to the right. We can see an example summary in the following figure.

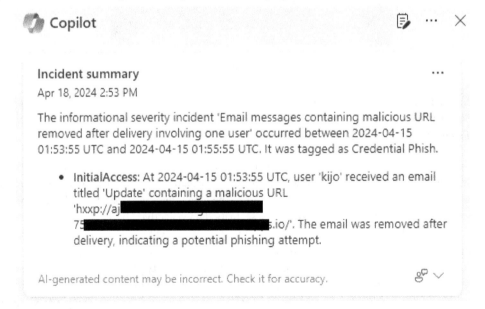

Figure 11.25 – Incident summary generated by Copilot for Security

For detailed exploration, an analyst might proceed with further probing using natural language in Copilot for Security's standalone interface. By navigating through the *more actions ellipsis* (**…**) atop the incident summary card, one can opt for **Open in Copilot for Security**, which will open the Copilot for Security portal in a separate tab, ready for deeper analytical prompts and additional tools.

Analyzing scripts and files

The integration of AI-driven investigative tools within the Microsoft Defender portal, courtesy of Copilot for Security, empowers security teams to expedite the scrutiny of dubious or potentially dangerous scripts and command line inputs. Sophisticated cyber threats, such as ransomware, tend to slip past detection barriers using complex scripts and PowerShell commands. These are frequently obfuscated, making detection tougher and more intricate. For security operations teams, it's critical to rapidly dissect these scripts to grasp their intent and implement measures to halt further spread within a network. The script analysis features empower security teams, allowing them to inspect scripts internally rather than relying on outside resources, creating a streamlined analysis process. The Microsoft Defender XDR plugin extends this analytical prowess even further when used in the Copilot for Security standalone interface. We can see the script analysis tool in the following figure.

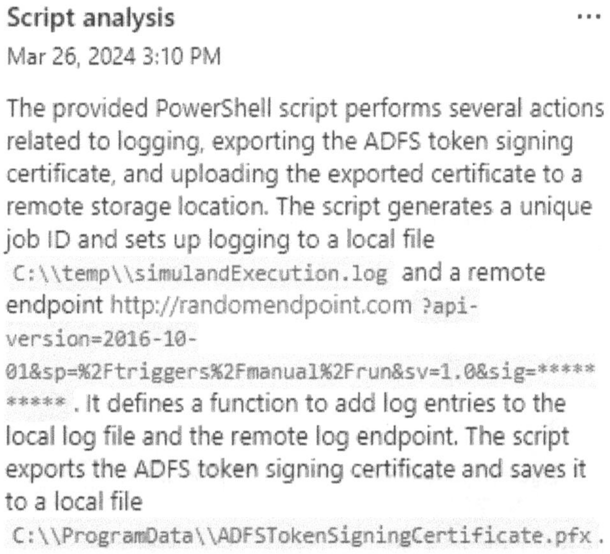

Script analysis   **…**

Mar 26, 2024 3:10 PM

The provided PowerShell script performs several actions related to logging, exporting the ADFS token signing certificate, and uploading the exported certificate to a remote storage location. The script generates a unique job ID and sets up logging to a local file `C:\\temp\\simulandExecution.log` and a remote endpoint http://randomendpoint.com `?api-version=2016-10-01&sp=%2Ftriggers%2Fmanual%2Frun&sv=1.0&sig=********** `. It defines a function to add log entries to the local log file and the remote log endpoint. The script exports the ADFS token signing certificate and saves it to a local file `C:\\ProgramData\\ADFSTokenSigningCertificate.pfx`.

Figure 11.26 – Script analysis performed by Copilot for Security

To engage the script analysis tool, navigate to the **Attack story** tab, and in the **Alerts** panel, choose a suspicious event with a script. Once the **Event** panel has opened under the incident graph, click **Analyze** to prompt Copilot to begin the analysis. Results are presented in the **Copilot** pane. For a deeper examination, you can expand the script by clicking **Show code** or retract it with **Hide code**. The *more actions ellipsis* (**…**) presents options such as copying, regenerating results, or delving into the analysis in the Copilot for Security standalone environment.

Likewise, Microsoft Copilot for Security enhances file assessment through AI-aided analysis capabilities. It is especially useful for security analysts, new and seasoned, who are tracking and resolving cyber incidents, where rapid file evaluation is necessary. Complicated attacks may employ files that masquerade as benign or system files, evading initial detection measures. Copilot cuts through this camouflage by delivering swift, accurate evaluations. This proves invaluable for less experienced analysts who are still familiarizing themselves with various analysis tools. The file intelligence report furnished by Defender's Copilot feature includes a comprehensive overview of the suspicious file, insights into its content, and a conclusive assessment. This functionality, accessible with a Copilot for Security license, enhances efficiency and reduces turnaround times for investigations. We can see this analysis in the following figure.

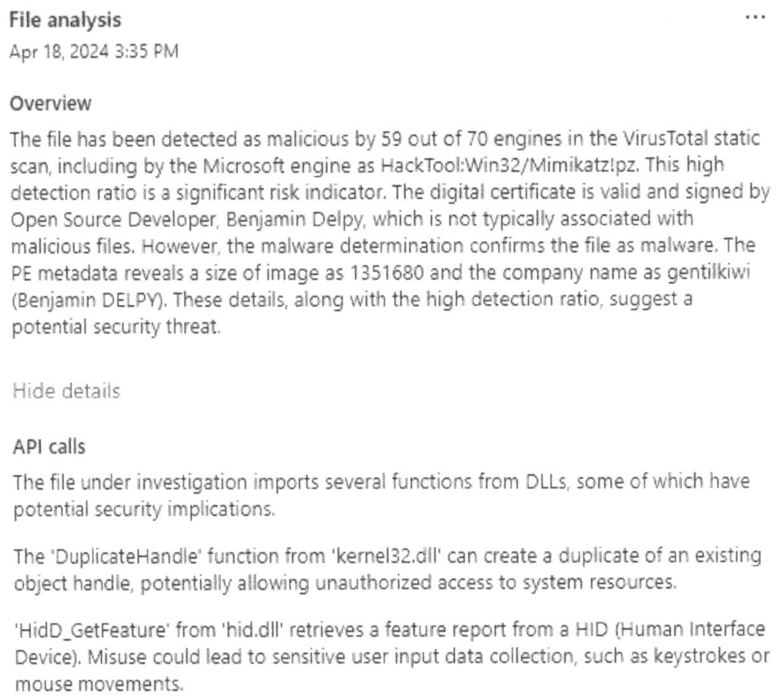

File analysis   ...

Apr 18, 2024 3:35 PM

Overview

The file has been detected as malicious by 59 out of 70 engines in the VirusTotal static scan, including by the Microsoft engine as HackTool:Win32/Mimikatz!pz. This high detection ratio is a significant risk indicator. The digital certificate is valid and signed by Open Source Developer, Benjamin Delpy, which is not typically associated with malicious files. However, the malware determination confirms the file as malware. The PE metadata reveals a size of image as 1351680 and the company name as gentilkiwi (Benjamin DELPY). These details, along with the high detection ratio, suggest a potential security threat.

Hide details

API calls

The file under investigation imports several functions from DLLs, some of which have potential security implications.

The 'DuplicateHandle' function from 'kernel32.dll' can create a duplicate of an existing object handle, potentially allowing unauthorized access to system resources.

'HidD_GetFeature' from 'hid.dll' retrieves a feature report from a HID (Human Interface Device). Misuse could lead to sensitive user input data collection, such as keystrokes or mouse movements.

Figure 11.27 – Copilot for Security | File analysis

Copilot's generated file analysis provides a broad view and critical assessment, flagging potential threats with detection names, drawing emphasis on key file attributes such as certificates and signatories, and summarizing related contents. To utilize this feature, click on the file in question to open its profile. Copilot will automatically initiate analysis, showcasing an overview by default in the **Copilot** pane. You have the **Show details** option for extensive information or the **Hide details** option to simplify the view. You can also select a file from the **Attack story** tab to investigate and click **Analyze** to start the process, which will lead to results being displayed in the **Copilot** pane. Results can be copied, regenerated, or viewed in-depth in the Copilot for Security portal through the *more actions ellipsis* (**...**). It's essential to review the Copilot in Defender's results thoroughly and provide feedback using the designated icon at the base of the **Copilot** pane. This will help improve the accuracy of the model.

Device summary

Copilot for Security accelerates the assessment of devices with advanced AI-based investigation features. These features free the security teams from the tedious task of combing through vast device data and identifying signs of dubious activities or entities that could signify a breach. Such teams must distill copious amounts of intricate data to understand, triage, and link the device's status and activities with potential security threats swiftly. The device summary tool offered by Copilot for Security presents a concise view of a device's security health, listing potentially vulnerable software, and highlighting any abnormal behavior detected. We can see an example device summary in the following figure.

Device summary   ...
Apr 18, 2024 4:01 PM

The device named "mb████████████████████████co" has a medium exposure level. There is no information available about the device's owner, type, or compliance state.

- The device's Tamper Protection is off, which could allow malicious apps to change important Microsoft Defender for Endpoint settings.
- Real-Time Protection (RTP) is also off, disabling important antivirus functionalities.
- Potentially Unwanted Application (PUA) Protection is in non-blocking mode, leaving the device vulnerable to unwanted applications with potentially malicious behavior.
- The device's controlled folder access is off, making key system folders vulnerable to changes made by malicious and suspicious apps, including file-encrypting ransomware malware.
- Microsoft Defender Firewall notifications are disabled for the Private profile, which could allow an attacker to more easily exploit a weakness in a network service.
- Merging of local Microsoft Defender Firewall connection rules with group policy firewall rules for the Public profile is disabled.
- Attack Surface Reduction (ASR) is off, making the device more vulnerable to common attack techniques used in cyber attacks and malicious software.

Vulnerable software

- The device has a critical vulnerability due to the installed version of PowerShell (7.4.1.0).
- The installed version of .NET (6.0.0.0) also presents a critical vulnerability.
- The installed version of Visual Studio Code (1.78.2.0) presents a high-level vulnerability.

Figure 11.28 – Device summary provided by Copilot for Security

The device summary crafted by Copilot features crucial details about a device. It includes the functionality status of paramount Defender XDR protections such as attack surface reduction rules and tamper-proof safeguards, accounts of significant user activities such as abnormal login patterns, an inventory of vulnerable software on the device, the condition of various security parameters such as firewall settings, which factor into the device's overall risk profile, and further important insights such as the device's last active moment. Additionally, insights provided by Microsoft Intune, such as the primary user, device group membership, or discovered applications, are also integrated into this summary.

To use the device summary function, one simply opens an incident and selects the **Assets** tab, which will display all impacted assets. Select **Devices** from the list of assets and select the device to investigate from the given list. This will open a side panel with details on the device; select **Open device page** to be taken to the device page when Copilot will automatically encapsulate the pertinent device information and display it within the **Copilot** pane. One can also select a device on the incident graph (inside the **Attack story** tab) to open the device details panel and have the option to open the device page. As with the previous method, the device page will provide a device summary. This summary can be accessed within the **Copilot** pane, and analysts have the option to copy to clipboard, regenerate, or delve deeper by launching the Copilot for Security portal via the *more actions ellipsis* (**...**) present on the device summary card. A careful review of these summaries ensures that security teams remain well informed and proactive in maintaining the integrity of the systems they safeguard.

Advanced hunting

Copilot for Security introduces an elegant solution for advanced hunting: the query assistant feature. This tool is particularly beneficial for threat hunters and security analysts new to KQL as users can simply enter their inquiries conversationally, such as by instructing, `Get all alerts involving user admin123`. In response, Copilot for Security artfully crafts a corresponding KQL query utilizing the advanced hunting data schema, effectively streamlining the query creation process. This feature allows threat hunters and security analysts to channel their energy and time more efficiently into the pivotal tasks of threat detection and analysis. Any user with access to Copilot for Security can leverage this query assistant within the advanced hunting feature to get quickly started on threat hunting. We can see this feature in action in the following figure.

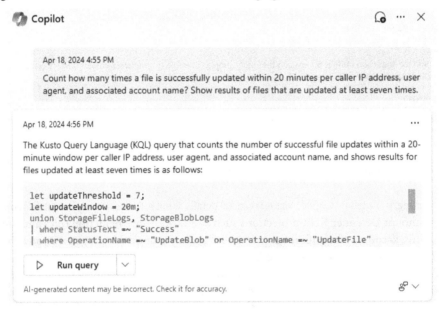

Figure 11.29 – The query assistant features in Copilot for Security

For the generation of KQL queries optimized for threat-hunting activities, one would begin by navigating to the advanced hunting section via the Defender portal (`https://security.microsoft. com`) left-hand menu, under the **Investigation & response** section and **Hunting** subsection. When the **Advanced hunting** page opens, the Copilot for Security sidebar will emerge on the right and provide sample queries that the user can ask. A query bar is also provided, where you can express any threat-hunting query; then, with a click on the arrow icon or by pressing *Enter*, Copilot springs into action, translating your natural language instruction into a well-structured KQL query.

Once the query is generated, just select the **Run query** button to execute it. From the same **Run query** button, a drop-down menu is provided with the **Add to editor** option, which places your query within the query editor and allows for further customization and execution. For security teams looking at increasing their efficiency, this feature allows for building a library of ready-to-use queries for the most common scenarios that can be imported as alert rules to Microsoft Sentinel or used for validation of findings.

Guided responses

Guided responses are another feature that makes your security teams more efficient during incident response efforts. This remarkable feature facilitates swift incident resolution through guided responses that consider not only data related to your environment and detections performed by the tools but also best practices guidance from Microsoft security experts. With AI and machine learning at the helm, Copilot in Defender assesses incidents within its context and draws on learnings from past inquiries to recommend relevant response actions. We can see an example of guided responses in the following figure.

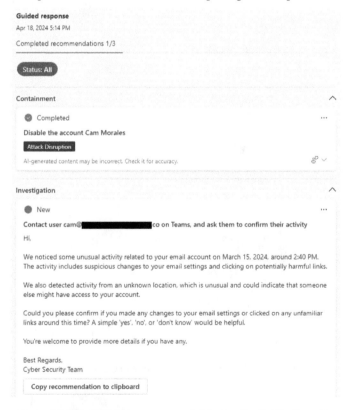

Figure 11.30 – Guided responses provided by Copilot for Security

From novice to experienced incident responders, initiating an appropriate incident response might be a challenging and time-consuming effort, but guided responses empower teams of all experience levels to apply response actions decisively and expediently. The recommendations focus on different areas of the investigation process:

- **Triage**: This is used to train the model and lower false detections. The analyst must classify the finding as either informational, a true positive, or a false positive.

- **Containment**: These are actions to help prevent further impact on the environment by containing the incident.

- **Investigation**: This pertains to advice on additional exploratory steps for a thorough assessment.

- **Remediation**: This pertains to specific response actions tailored to the entities involved in an incident.

Each card detailing these actions provides critical insights, such as the targeted entity and the rationale behind a recommended response. The cards are also marked to highlight whether an action stems from automated investigations such as attack disruption. Users can sort through the guided response cards by status, selecting a specific status to view via the **Status** option. We can see the **Status** option in the following figure.

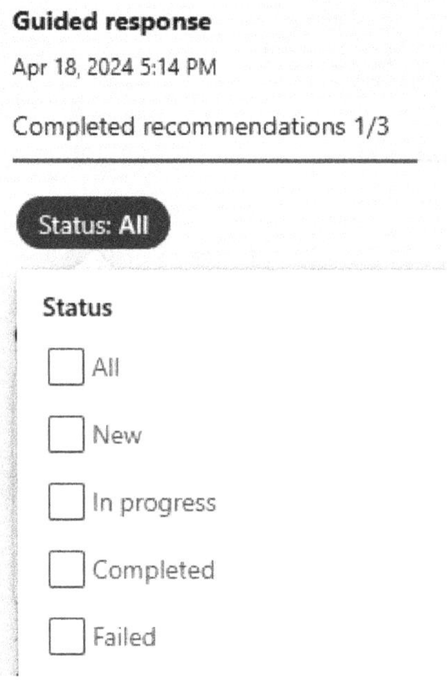

Figure 11.31 – The Status options available in Guided response

To deploy guided responses, navigate to the incident page; Copilot dynamically generates relevant suggestions. The guided response cards are visible in the **Copilot** pane on the right of the incident page. Before implementing any recommendations, carefully review each card. The actions available for each recommendation can be found under the *more actions ellipsis (...)* on each card. We can see an example of actions available for a remediation recommendation in the following figure.

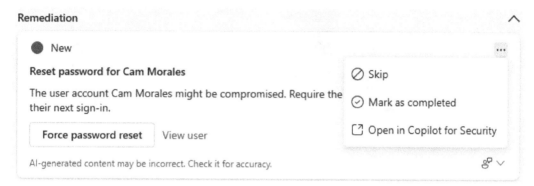

Figure 11.32 – Available actions for a remediation recommendation

The Copilot in Defender's interface not only proposes actions but also enriches analysts' understanding of the incident by providing additional insights. Particularly for remediation responses, teams can access more data, for instance, by selecting **View similar incidents** or **View similar emails**, which become available when parallel incidents within the organization have been detected. Selecting it presents a list of similar incidents identified by Defender's machine learning algorithms. This list aids in incident classification and serves as a resource for reviewing actions taken in related cases.

Incident reports

Copilot for Security enhances efficiency by streamlining the process of crafting incident reports for security operations teams. Through the AI-driven capacities of Copilot for Security, these teams can swiftly generate detailed incident reports with a mere click. Crafting an inclusive report detailing key incident insights is a pivotal tool for security teams and their management but often demands significant time for compilation. Amassing data, systematizing it, and distilling it into a cogent summary typically involves considerable attention. Copilot in Defender eradicates this tedium, allowing for the instant creation of comprehensive incident reports within the portal itself. We can see a sample of parts of an incident report generated by Copilot for Security in the following figure.

Incident report

Apr 18, 2024 5:57 PM

Incident title

BEC financial fraud attack was launched from a compromised account (attack disruption)

Incident details

Analysts	Mimik Emails - A▮▮▮▮▮e
	adm_ajourn@▮▮▮▮▮on...
	renatal▮▮▮▮▮...
	Defender Experts
	ha▮▮▮▮▮om#...
	Reema ▮▮▮
	Hannah▮▮▮
Time created	03/15/2024 14:26:31
First log	03/15/2024 14:43:50
Last log	04/11/2024 18:53:31
Time closed	-

Incident summary

The high severity incident 'BEC financial fraud attack was launched from a compromised account' occurred between 2024-03-15 14:26:31 UTC and 2024-03-15 14:40:38 UTC. The incident was tagged as Attack Disruption, BEC Fraud, and Credential Phish, and triggered an automatic Attack Disruption action.

- **DefenseEvasion:** The incident began at 2024-03-15 14:26:31 UTC when user 'cam' set a suspicious inbox rule on their own inbox, possibly indicating a compromised account. The rule was set to move messages to a folder named RSS Feeds. This activity was associated with the IP addresses 192.42.116.187 and 13.92.62.207 and involved the cloud apps Microsoft Exchange Online and Microsoft 365. At the same time, user 'cam' accessed the Tor IP address 192.42.116.187.

Figure 11.33 – Incident report generated by Copilot for Security

An incident summary delivers a snapshot of an event assembled by Copilot fusing data from various sources such as Defender for Office 365, Defender for Endpoint, Microsoft Sentinel, and Defender XDR into a thorough narrative. Reports include steps taken by analysts, automated processes, response team notes, and pertinent commentary. Irrespective of the security solutions in deployment, Copilot for Security integrates all pertinent incident data into the report it crafts. This automated report construction happens based on the interplay of automated and manual actions and insights recorded by analysts throughout the incident's life cycle. Reports generated by Copilot in Defender encompass the following:

- **Time-stamped key actions**: From incident initiation to resolution

- **First and last log details**: Discerning between analyst-driven and automated logs

- **List of analysts involved**: A list of all analysts who performed actions related to the incident either on the incident itself or on any asset including comments

- **Incident categorizations**: Categorization includes the analysts' rationale for classification

- **A list of actions taken**: These include investigative and remedial actions related to the incident and its assets

- **List of follow-up actions**: Based on recommendations, any environmental data analyzed by Copilot, and best practices, a list of recommendations and unresolved issues is provided for follow-up by the security team.

The report includes typical remedial measures such as isolating devices, user deactivation, and safeguarded email deletions, indexed in **Action Center**. Although Microsoft Sentinel playbooks are addressed in the reports, commands initiated during live response and reactions imparted from public APIs or bespoke detections are not encapsulated.

To initiate a report with Copilot in Defender, one must access an incident page and on the **Copilot for Security** pane, select the **Generate Incident Report** option. Upon a report's creation, a detailed card is presented in the **Copilot** pane. It is important to note that the depth of the report relies on the availability of data via enabled plugins in Copilot for Security, proper user permissions, and any data available in Microsoft Defender XDR. Finally, these reports can be exported as a PDF file, which might be of great value as a training tool for security teams as well as a low-cost approach to keeping records of important incidents.

Standalone experience and prompts

While the Copilot embedded experience available in many Microsoft products, such as Defender XDR, Entra, and Intune, is very useful and convenient, users can do more in-depth investigations by leveraging the standalone experience. This experience, available at the Copilot for Security portal (`https://securitycopilot.microsoft.com`), allows users access to a powerful resource for conducting in-depth and adaptable security examinations via natural language prompts. These prompts act as the main ingredient for Copilot for Security to conjure up responses aiding users with their security tasks. This experience is available immediately after finishing the Copilot for Security setup and all it requires is to submit a prompt at the prompt bar in the portal. Be aware that if, after the configuration of Copilot, Copilot via the Defender portal was used, you will instead be greeted by any of the prompts and sessions that were generated in the portal, in which case you will need to start a new session to get the prompt bar. We can see the portal and the prompt bar in the following figure.

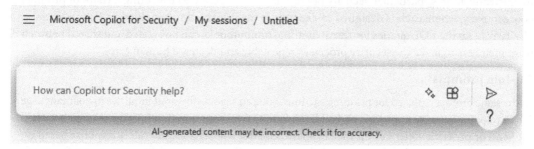

Figure 11.34 – The Copilot for Security prompt bar

If we look at the prompt bar, you will see the option to either create your own custom prompt or use a prebuilt prompt from the promptbooks available by clicking on the sparkle icon (*2 stars icon*) found within the prompt bar. We can see what happens when you click on the sparkle button in the following figure.

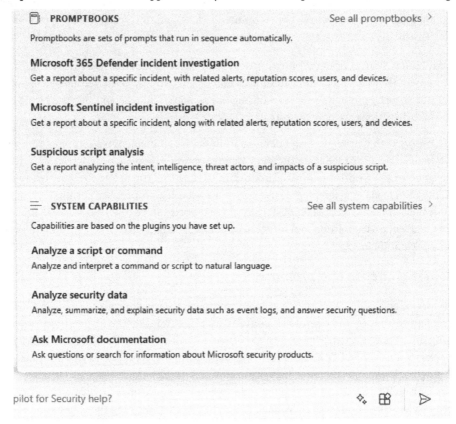

Figure 11.35 – The menu that appears when clicking the sparkle button

Promptbooks serve as curated collections of prompts, carefully fashioned and bundled together to efficiently address particular security operations. They operate by executing a chain of prompts consecutively, where each is designed to expand on the information provided by its predecessor. Similarly to saving KQL queries for threat hunting, promptbooks can be created and shared between team members to make the security processes repeatable and decrease human error.

Custom prompts

In the same prompt bar used for prior interactions, you can seamlessly input inquiries in your language for Copilot for Security to process and address. Simply type your question into the prompt bar, then hit **Send** or press the *Enter* key on the keyboard. Copilot will start to generate a response based on your configured process log preferences. While you wait, the sequence of actions being performed in the background will be displayed. We can see this in the following figure.

how many users have received malicious phishing emails in the last 90 days?

> ✅ **Chose Natural language to KQL for Microsoft Defender XDR** 7 seconds

> ↷ **Processing your request ...** 9 seconds

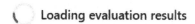

Loading evaluation results

Figure 11.36 – Background processes listed during the prompt evaluation

This process list is essentially a window into the procedure, showing each step in real time, the sources being consulted, and the timeline for the response formulation. You have the flexibility at any point to cancel, adjust, or remove your prompt. When Copilot for Security completes its response, it will be output for your review. At this stage, you can verify its accuracy and determine whether further modification is required to answer a specific security inquiry. We can see a typical reply in the following figure.

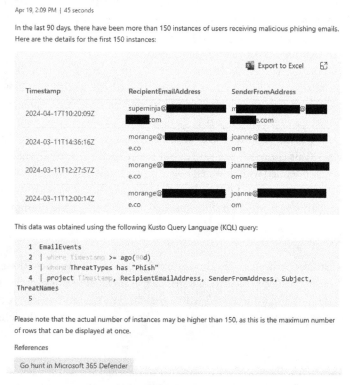

Figure 11.37 – A Copilot for Security reply

Should further queries arise, the dialogue with Copilot for Security can smoothly continue with additional prompts for progressive responses. In case the initial response doesn't fully meet your expectations, consider rephrasing your question, adding more context or illustrative details, or specifying a preferred format of response, and then resubmit it. To facilitate impactful responses from Copilot for Security, structure your prompts with these key elements:

- **Goal**: Clearly define what specific security information you require. For example, you can ask how many users received a phishing email from a specific user.

- **Context**: Clarify the purpose for which you need this information and how you'll utilize it. For example, you can indicate that the information will be used for incident triage and provide an incident number.

- **Expectations**: Describe how you want the response formatted and who it's intended for. For example, you could say that it should be formatted as a table and output as a CSV file.

- **Source**: Mention specific data sets, records, or tools that Copilot for Security should reference. For example, you can indicate time ranges.

Aim to be explicit and precise to make the prompt effective while ensuring conciseness. You may begin with a straightforward prompt and progress to more elaborate requests as you become more comfortable with the system. For example, you could start by asking, `how many users received a phishing email in the last 90 days?` Or, you could improve it by asking, `how many priority user accounts received a phishing email from internal users in the last 90 days with a focus on Midnight Blizzard-identified indicators of compromise as per their intel profile? Format the output in a table format with columns for the recipient, sender, email subject, attachments, and date and time. The output will be used to determine the impact of activities observed on incident id 23456-987-123456.` This approach will lead to not only concise results but they will also be in a format that can be used as intended with minimal noise. Prompts often require iteration for refinement or alternative perspectives to align more closely with your needs. Just as with all language model systems, Copilot for Security may yield variably nuanced results to identical prompts.

Prebuilt promptbooks

Copilot for Security features prebuilt promptbooks, akin to security playbooks. These collections of prompts aid in accomplishing tailored security actions, such as incident response or investigation tasks. Each promptbook is designed to be activated with a certain type of information, such as a snippet of code or the name of a threat actor. We can see one of these promptbooks in the following figure.

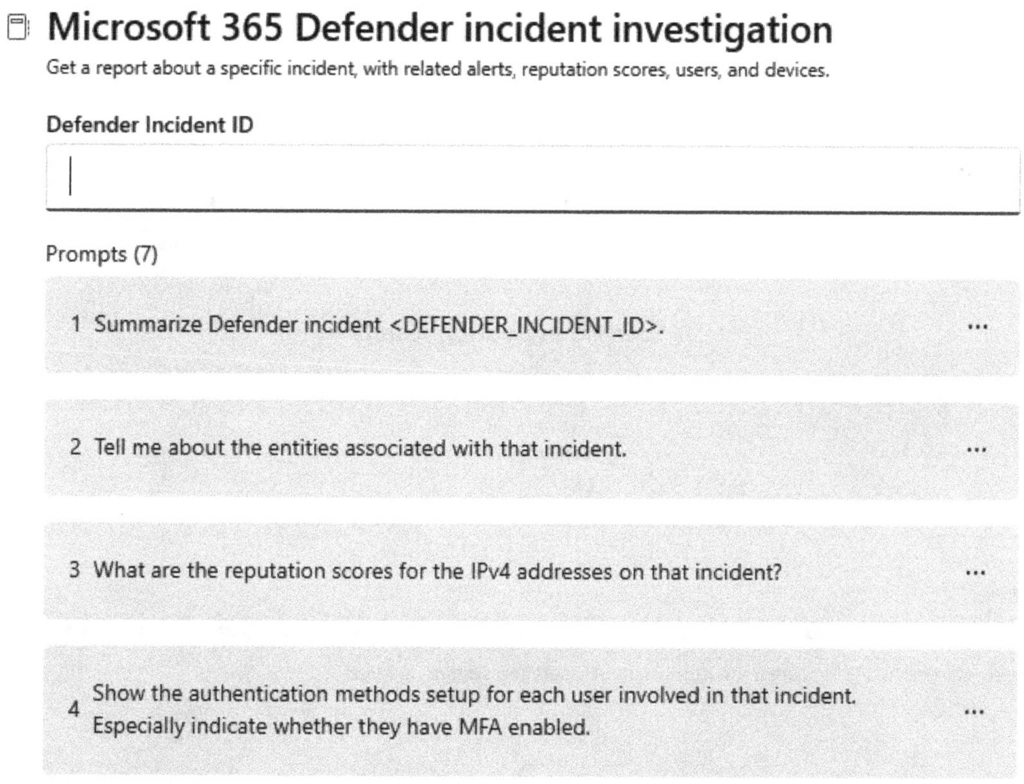

🗂 Microsoft 365 Defender incident investigation

Get a report about a specific incident, with related alerts, reputation scores, users, and devices.

Defender Incident ID

Prompts (7)

> 1 Summarize Defender incident <DEFENDER_INCIDENT_ID>.   ...

> 2 Tell me about the entities associated with that incident.   ...

> 3 What are the reputation scores for the IPv4 addresses on that incident?   ...

> 4 Show the authentication methods setup for each user involved in that incident. Especially indicate whether they have MFA enabled.   ...

Figure 11.38 – A Copilot for Security promptbook

A commonly used prompt is the Microsoft 365 Defender incident investigation promptbook. In this promptbook, you can input a specific defender incident ID and a sequence of prompts will be executive to compile an executive report meant for a non-technical audience. As when running any other promptbook, the prompts will be executed in a sequence, taking the output from the previous prompt to perform analysis. We can see an example of this execution in the following figure.

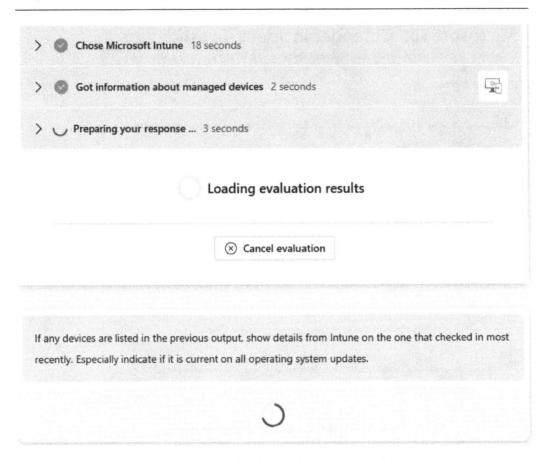

Figure 11.39 – A promptbook executing a sequence of prompts

The final output of this promptbook will be a report for non-technical audiences containing an overview of the investigation process and important points to know about how the security incident occurred. An analyst can output this report into a Word file or copy it to share with executives. We can see an example of the final output in the following figure.

- The incident triggered a total of 42 alerts. Out of these, 38 were chosen for a detailed analysis.

- Two devices were implicated in the incident. These are 'avoriaz-win10v', a Windows 10 machine hosted on Azure, and 'AVORIAZ-DC', a domain controller. Both devices are part of the 'Avoriaz' group.

- A user named 'jeff', who is part of the 'AVORIAZ' domain, was involved in the incident. The Azure AD User ID for 'jeff' is '29a26039-3456-4b02-82c7-0e9ed1859e7d'.

- Several processes were detected on the 'avoriaz-win10v' device, all associated with the user 'jeff'. These processes were part of the malicious activities that took place during the incident.

- Two files, 'mimikatz.exe' and 'Get-KRBTicket.ps1', were involved in the incident. The 'mimikatz.exe' file is associated with a known credential theft tool called Mimikatz.

The entities involved in the incident were part of a multi-staged attack. This attack included stages of credential theft, lateral movement, and defense evasion. This incident is a reminder of the importance of maintaining strong security measures and vigilance in our digital environment.

Figure 11.40 – Incident report promptbook output

Another valuable resource is the threat actor profile and the vulnerability assessment promptbooks. The threat actor profile promptbook compiles an executive abstract on a given threat actor, drawing from existing threat intelligence that includes their TTPs and indicators, with remediation suggestions. The vulnerability impact assessment promptbook serves an essential role in risk assessment and threat hunting by analyzing CVE numbers or documented vulnerabilities. It determines whether the vulnerability in question is known publicly, whether it has been exploited, and any associated threat actor campaigns. Additionally, it furnishes recommendations for mitigating the threat. Other promptbooks are available with an expectation that the quantity and quality of these promptbooks will continue to increase as time goes on.

Creating your own promptbooks

While prebuilt prompts and promptbooks are a good starting point, as your team gains familiarity with Copilot for Security and LLMs, custom promptbooks will prove beneficial during investigations. Copilot for Security provides a promptbook builder to allow teams the flexibility of creating prompts that streamline their investigation workflows and help tailor investigations. Creating a custom promptbook is a simple process that involves opening a session that already contains the prompts you'd like to incorporate. Select your desired prompts or choose all of them by ticking the top checkbox, which activates the **Create promptbook** option. We can see the **Create promptbook** option in the following figure.

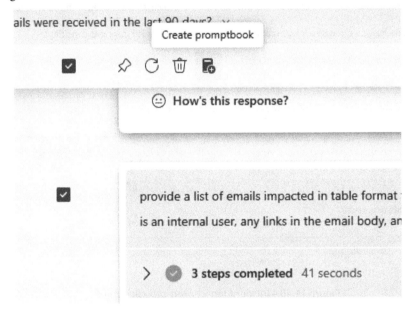

Figure 11.41 – The Create promptbook option

Click on it to start the **Create a promptbook** wizard and provide a name, tags, and a descriptive overview for your promptbook to clarify its purpose. The prompts will be listed in the lower part of the wizard with a pencil icon available for fine-tuning the prompts and an ellipsis (…) icon available to either change the order of the prompt or remove it from the promptbook. We can see the wizard and one of the prompts with the ellipsis icon clicked in the following figure.

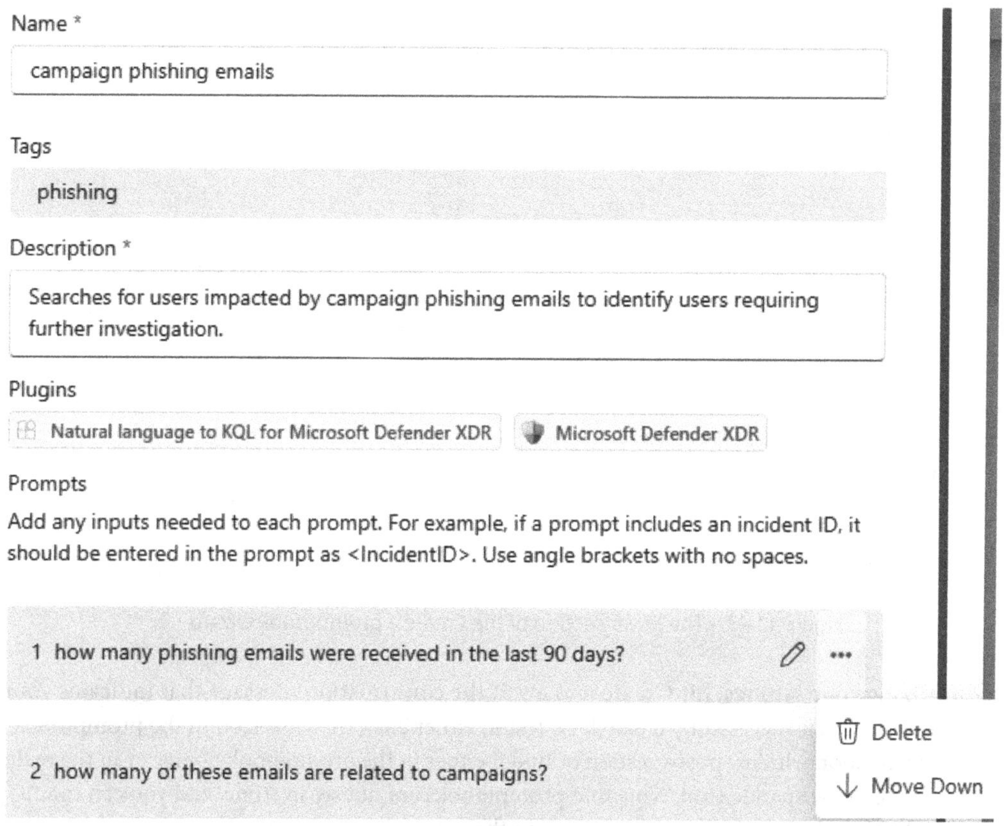

Name *

campaign phishing emails

Tags

phishing

Description *

Searches for users impacted by campaign phishing emails to identify users requiring
further investigation.

Plugins

Natural language to KQL for Microsoft Defender XDR   Microsoft Defender XDR

Prompts

Add any inputs needed to each prompt. For example, if a prompt includes an incident ID, it
should be entered in the prompt as <IncidentID>. Use angle brackets with no spaces.

1 how many phishing emails were received in the last 90 days?

🗑 Delete

2 how many of these emails are related to campaigns?   ↓ Move Down

Figure 11.42. The Create a promptbook wizard

For reusable promptbooks in various situations, include input parameters within your prompts. An example could be replacing specific threat actor names with an input parameter such as <ThreatActorName>. Review under **Inputs you'll need** to ensure your parameters are correctly formatted. If the input does not show, it means an incorrectly written input parameter, such as one with spaces or missing angle brackets. Next, decide who has access to this promptbook, either just yourself or everyone in your organization. We can see this section of the wizard in the following figure.

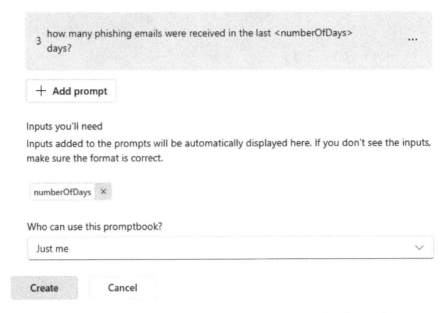

Figure 11.43 – The lower section of the Create a promptbook wizard

After finalizing your settings, hit **Create** and await the confirmation message that indicates your promptbook has been successfully created. To test it, either click the *view* icon in the promptbook-created confirmation window post-creation or find it either in the promptbook library or in the main prompt bar using the sparkle icon. Note that promptbooks are not set in stone, and you can modify, duplicate, share, and delete them at any time from the promptbook library.

Creating a promptbook is a skill developed over time as your team understands better how LLMs work and what is possible with the data available. For novice users, the following are some good recommendations for creating effective promptbooks:

- **Aim for precise, effective prompts**: Setting clear objectives will guide you in formulating pointed questions.

- **Balance is key**: A multitude of prompts might lead to higher SCU consumption; conversely, overly complex prompts can also be demanding on resources. Assess whether each prompt is essential and if any might be combined or simplified.

- **Ordering is very important**: While reorganizing prompts might seem trivial, it can affect both the output and the SCU usage. Remember that promptbooks progress linearly, utilizing the outcome of one prompt to inform the next.

As always, remember that the dynamic nature of artificially intelligent LLMs will cause users to receive different outputs from the same prompts. As such, being precise in our prompts not only in what needs to be analyzed, but the output expected can help decrease this variable and provide for similar results among team members.

Using custom data

Incorporating custom data into Copilot for Security is effortless via the Azure AI search plugin or a simple file upload. This capability allows Copilot to tap directly into your organization's data through resources such as wikis, policies, templates for procedures, and even KQL scripts. By blending this collective intelligence into Microsoft Copilot, it gains the ability to understand and interact with your knowledge base, leading to more precise and operationally relevant assistance.

To begin linking your knowledge base via the Azure AI search approach, proceed to the Copilot for Security portal (`https://securitycopilot.microsoft.com`) and start a new session. In the prompt bar, select the square icon labeled **Sources**. We can see this icon in the following figure.

Figure 11.44 – The Sources icon in the prompt bar

The **Manage sources** panel will appear. Under the **Plugins** section, ensure that **All** is selected to make all plugins visible. Select the **Setup** button next to **Azure AI Search** to open the configuration panel. We can see the configuration panel in the following figure.

← **Azure AI Search (Preview) settings**

Manage this plugin. Learn more

> ### Settings ⓘ
>
> **Name of Azure AI Search service** ⓘ *
>
> Enter Instance
>
> **Name of index** ⓘ *
>
> Enter Index
>
> **Name of vector field in index** ⓘ *
>
> Enter Vector

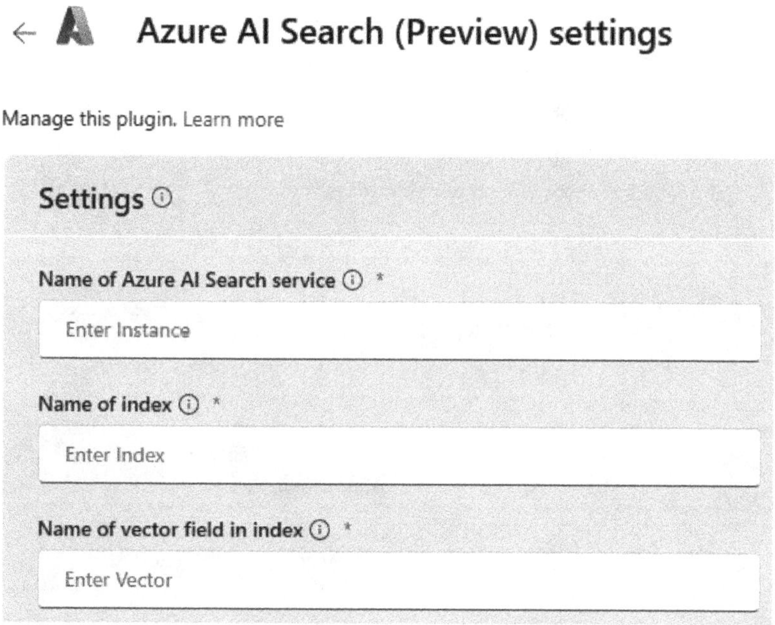

Figure 11.45 – The Azure AI Search plugin configuration panel

Enter the required information to include a name for the service, index, vector, and other information as requested. There is an information button next to each field explaining what information is required. After entering all the necessary information into the fields, click save and close the **Settings** window. The plugin will show as configured in the **Manage sources** panel and Copilot is now ready to sift through the business intelligence you've put in. It's crucial to include the *Azure AI Search* term when you ask Copilot to fetch info. When new employees are learning the ropes, they have a tool at their disposal. They can simply inquire within their workflow about company protocols, streamlining their integration into the team.

The second approach, file upload, is also simple to use. When using this feature, it is important to remember that while Copilot for Security is flexible in the file formats that can be used, one should stick to commonly used formats. At the time of this writing, Copilot supports Word (`.doc`, `.docx`), Excel (`.xls`, `.xlsx`), PowerPoint (`.ppt`, `.pptx`), PDF files, text (`.txt`, `.md`, `.log`), HTML (`.html`, `.htm`), `.csv`, `.xml`, OpenDocument files (`.odt`, `.ods`, `.odp`), `.json`, `.yaml`, and some other text-based files. No support is provided currently for images, video, audio, and executable files. File sizes are also restricted to 3 MB per file and a total of 20 MB when uploading multiple files at a time. To start uploading files, go to the Copilot for Security portal (`https://securitycopilot.microsoft.com`) and start a new session. On the prompt bar, click on the sources icon to open the **Manage sources** panel. On the left-hand side, select **Files** to open the **Files** section and the **Uploads** dialog. We can see this section in the following figure.

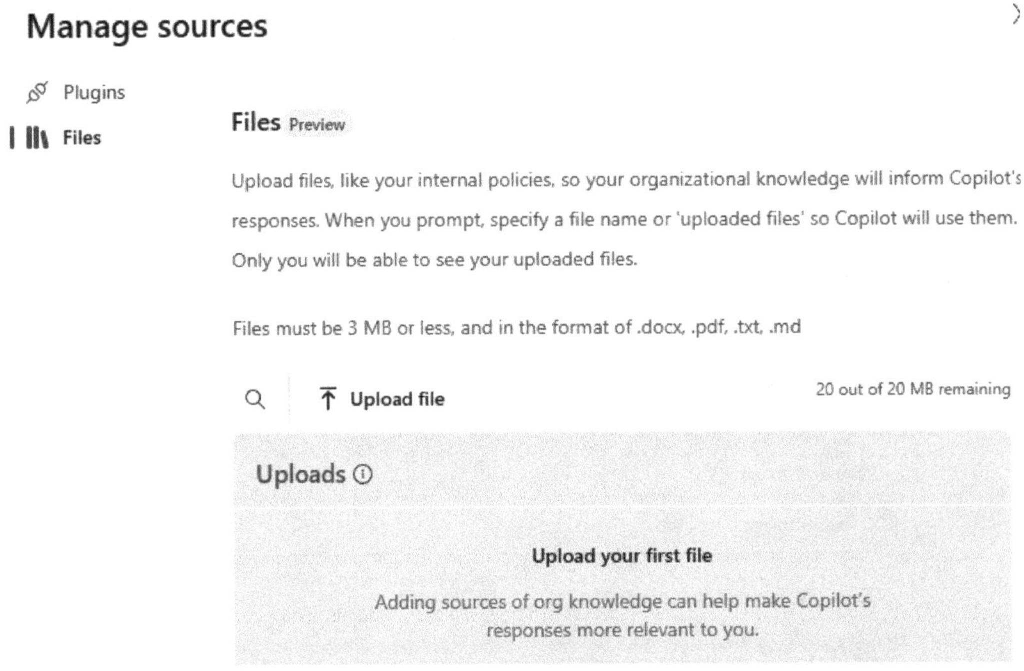

Manage sources ⟩

🖋 Plugins

| Ⅲᛁ Files

Files Preview

Upload files, like your internal policies, so your organizational knowledge will inform Copilot's responses. When you prompt, specify a file name or 'uploaded files' so Copilot will use them. Only you will be able to see your uploaded files.

Files must be 3 MB or less, and in the format of .docx, .pdf, .txt, .md

🔍   ↑ **Upload file**   20 out of 20 MB remaining

Uploads ⓘ

Upload your first file

Adding sources of org knowledge can help make Copilot's responses more relevant to you.

Figure 11.46 – The Files | Uploads section

Click on **Upload file**, navigate to the desired file, and wait for it to pop up in the **Uploads** list. If any errors are shown, try to reupload the files. To apply a file to your present session, flip its toggle to the right-hand side of the file listed and close the **Manage sources** panel. When prompting Copilot, call out **Uploaded files** for it to list the uploaded files enabled for use in prompts. Continue your prompt and ensure to specify the file name to use to direct its focus to a particular document and perform analysis using the data inside it.

The addition of Copilot for Security to the security tool arsenal is sure to have an impact on the speed of detection and execution of security operations for years to come. While teams should not blindly trust AI tools at this stage in time, their capability to analyze large amounts of data and identify patterns is something that all security teams need, but until recently only the high-paying incident response teams could afford it. The level of capabilities for these tools will keep on increasing as time goes on, therefore it is highly recommended that all security teams spend the time and resources to understand the logic of how these tools work, the connections they can have to organization data, how this data is protected, and what areas the tool shines versus what areas just augments security work.

Summary

In this chapter, you learned how to utilize the powerful APIs offered by Microsoft Defender XDR. Discussion included the steps to merge security operations across various platforms, highlighting the APIs' role in secure data sharing and elevating an organization's rapid response to threats. The guidance included vital procedures for backing up and restoring security data to ensure resilience post-incident. This chapter went on to delve into integrating Microsoft Defender XDR with both third-party and in-house solutions, shedding light on the processes this integration entails. The Copilot for Security features were covered, including its capability to streamline investigations through both its embedded and standalone experiences, allowing even novice security analysts to play a key role in their organization's security efforts. Looking ahead, the subsequent chapter will pivot to discuss training for the end user. It will address various strategies and tools offered by the platform to enlighten users. This education allows end users to recognize and respond to security incidents aptly, making them part of an active, front-line defense.

References

- Microsoft. (2023, March 16). *Microsoft Graph overview - Microsoft Graph*. Microsoft Learn. `https://learn.microsoft.com/en-us/graph/overview`

- Microsoft. (2024a, March 8). *Microsoft Graph security API overview - Microsoft Graph*. Microsoft Learn. `https://learn.microsoft.com/en-us/graph/security-concept-overview`

- Microsoft. (2024b, April 1). *What is Microsoft Copilot for Security?* Microsoft Learn. `https://learn.microsoft.com/en-us/copilot/security/microsoft-security-copilot`

12

User Awareness and Education

In this chapter, we will explore the transformative potential of attack simulation training, a cutting-edge approach that immerses employees in realistic scenarios, reflecting actual cyber threats. By exposing end users to simulated phishing emails, social engineering attempts, and other common attack vectors, organizations can provide hands-on experience that enhances threat recognition and response capabilities. We will look at the key components of effective attack simulation training, including available tools and platforms, techniques to create custom simulations, and strategies to automate and optimize the training process.

This chapter will cover the following topics:

- How attack simulations help support end user training requirements
- How to customize simulations and payloads
- How to automate simulations and training
- How to understand the simulation findings

Let's continue our journey!

Why we need to train users

Ensuring that employees are well-educated and trained is crucial for companies to lower the risks that come with human mistakes and bad habits in cyber defense. These weak spots are often the most vulnerable parts of a defense plan, highlighting the need for a strong strategy to tackle them. The first step is making sure workers know the basics, as if employees aren't aware of cyber threats and how they work, they won't be able to spot or avoid them. People who haven't been trained properly might not know how to react when they encounter malware or something suspicious. Giving access to important systems and networks to those who lack experience could cause serious problems, especially when they work remotely. For example, just one ransomware attack, launched by clicking a harmful link, can cause serious trouble and shut down a business for a long time.

To protect themselves, companies need to make teaching their employees about cybersecurity a top priority. This includes training that helps them identify phishing emails, recognize strange behavior, and respond correctly to cyber threats. Creating a culture of cybersecurity and teaching the best practices in information technology can be straightforward and doesn't have to break the bank. It's all about encouraging employees to be aware of security, which helps reduce mistakes and bad habits. For most organizations, even assigning 5% of available resources to educate their workforces can greatly improve their proactive defenses and visibility against cyberattacks.

It is common for business leaders to view employee security violations as either deliberate or unintentional, leading them to design security policies based on this assumption. However, there is an important middle ground that often goes unnoticed, the space between ignorance and malicious intent. It is crucial for organizations to recognize and address this middle ground by adapting their training programs and policies accordingly. Instead of solely focusing on malicious attacks, security policies should acknowledge that many breaches occur as employees try to strike a balance between security and productivity. This calls for educating employees and managers about non-malicious violations and providing clear guidance on how to handle situations where security practices conflict with work requirements.

Organizations should actively involve employees in the development and user testing of security policies. It is essential to equip teams with the tools to effectively adhere to these policies. Too often, IT departments create protocols in isolation, with a limited understanding of how these rules might disrupt workflows or contribute to increased stress. Considering the shift to remote work especially, which has transformed how many people operate, IT leaders should actively engage employees who will be affected by the creation, evaluation, and implementation of additional security measures. By embracing employee involvement, organizations can cultivate a stronger security culture that not only protects against threats but also aligns with employees' needs and workflows.

The prevailing mindset often undervalues security compared to productivity. Under normal circumstances, this might not pose a significant issue, as employees typically can allocate sufficient resources to both aspects. However, amid the multitude of challenges that arose from the COVID pandemic, maintaining productivity has become increasingly strenuous. Security often takes a backseat to essential tasks that directly influence performance evaluations, promotions, and bonuses.

Addressing this imbalance requires managers to acknowledge the intrinsic connection between job design and cybersecurity. It is crucial to recognize that adherence to cybersecurity protocols can contribute to employees' workloads. Therefore, it is essential to integrate cybersecurity considerations into workload determinations, considering and incentivizing compliance with security policies on par with other performance metrics. By intertwining job design with cybersecurity practices effectively, organizations can prioritize security without compromising productivity, fostering a culture that values both aspects as complementary components of a successful and secure work environment.

Altruism emerges as a significant factor that influences security initiatives within organizations. While most managers view employee altruism positively, recognizing it as a commendable trait in fostering a supportive work environment, it also introduces risks. Often, policy violations stem from employees' well-intentioned efforts to assist their colleagues. The COVID pandemic exacerbated the daily challenges faced by individuals, creating additional opportunities for employees to inadvertently compromise organizational security while attempting to aid their peers. Cyber attackers capitalize on this dynamic, leveraging social engineering tactics to exploit employees' willingness to bend rules under the guise of offering help.

To tackle this vulnerability, managers need to set up security policies designed to block these kinds of attacks. It's important for them to find a balance, making sure these rules don't disrupt employees' work too much. They should also clearly explain why these policies are necessary. This reduces any inconvenience caused by the new rules and helps employees understand them better, leading to higher follow-through and a stronger security stance overall. This broad strategy defends against vulnerabilities linked to good intentions and builds a workplace where being alert and following security rules are part of the culture.

Be it a fully remote environment or everyone working at the office, each employee could potentially open a door to threats. For an organization to be secure, its leaders must deeply understand what makes someone ignore rules and give hackers a way in. Often, the story of an upset employee trying to hurt their company gets a lot of attention, but often, it's employees who don't realize what they're doing that end up accidentally causing breaches.

To reduce the rising threat of cyberattacks and the varied dangers of a workforce that's more mobile and has more autonomy than ever, leaders need to launch efforts to tackle the root causes of not knowing about security risks at work. Part of this effort should include creating training programs that fit smoothly into employees' daily routines. By confronting these basic issues directly, companies can strengthen their protection against digital threats and foster an environment where everyone is more aware of and serious about cybersecurity.

Next, we will see what tools are available to train our end users.

Introducing attack simulation training

Organizations equipped with Microsoft Defender for Office 365 Plan 2 (either through add-on licenses or as part of packages such as Microsoft 365 E5) can use Attack simulation training within the Microsoft Defender portal (`security.microsoft.com`). This feature allows you to execute realistic but safe attack scenarios within your environment, which are crucial for pinpointing and aiding vulnerable individuals before a genuine cyber incident occurs. An attack simulation training exercise involves deploying seemingly authentic yet harmless phishing emails to users in the hopes of determining if the users would fall for this phishing attempt (*Microsoft, 2024a*). When configuring these simulations, the security team will have to configure the following aspects:

- Who will receive the training via simulated phishing emails and when? A controlled approach ensures that results can be better sorted and training results are much more useful.

- What specific training will be provided to end users? This should include what happens when there is interaction and no interaction with the simulation. Training should adapt to a user's behavior and schedule for maximum effectiveness.

- The format of the simulated phishing email, which should align with what end users typically see in an environment. Emails could either have links or attachments, look like official company correspondence, or include topics that exploit other aspects of human nature, such as urgent requests, account issues, or even winning a prize.

- The social engineering technique that will be used in the attack simulation. Things such as threat intelligence, news, and other sources can be used to influence the most common attacks experienced by your environment. This decision will define the choice of payload.

Attack simulation training offers a variety of social engineering methods to simulate threats. Aside from the **how-to guide** type, which is more educational, these methods are aligned with the MITRE ATT&CK® framework and involve different payloads, tailored to each technique. The following techniques are available when configuring the simulation:

- **Credential harvest**: This method involves an attacker emailing a malicious link to a user. Clicking this link redirects the user to a website that mimics a reputable site and asks them to input their username and password. This is a very commonly used approach, and many attack toolkits are available to generate these malicious websites effortlessly. A credential harvest payload during a configuration can be seen in *Figure 12.1*; the variable strings with a dollar sign character ($) are replaced with the actual values during execution.

Payload Description

This payload looks like it comes from Office 365, asking the user to confirm the continued use of their current password.

From name
System notification

Email subject
Your office 365 password expires today

Figure 12.1 – A credential harvest payload example

- **A malware attachment**: A common scenario seen in the wild for many decades involves an email containing a malicious attachment. Opening the attachment will trigger arbitrary code, such as a macro, which can enable an attacker to install more malicious software or gain a stronger foothold. We can see a typical malware attachment simulation payload in the following figure, where an attachment is included in the email message.

You have received a document from a Xerox Scanner.

It was scanned and sent to you using a Xerox WorkCentre on Office 365 Portal.

Number of Images: 1
Attachment File Type: PDF

Device Name: WorkCentre 5819

For more information on Xerox products and solutions, please visit Xerox Service Center.

Payload Description

This payload looks like it comes from a hotel employee sending along a scanned document from a Xerox scanner.

From name	**From email**
Noreply	noreplyadmin99@techidal.es
Email subject	**Source**
Scanned from a Xerox	Global

Figure 12.2 – A malware attachment simulation payload example

- **A link in an attachment**: A variation of the credential harvesting technique, here an attacker encloses a URL within an email attachment rather than in the email body itself. Opening the attachment and clicking the link leads a user to a website that requests their login details, usually masquerading as a trustworthy site to gain their trust. We can see an example of this attack while configuring the payload in the following figure; be aware that the string value with the dollar sign ($) is replaced by a valid entry during execution.

Figure 12.3 – An example of a link in an attachment payload

- **A link to malware**: This approach sees an attacker send a link to a file hosted on a popular file-sharing platform, such as OneDrive, SharePoint, an S3 bucket, or Dropbox. When a user clicks the link, the downloaded file runs arbitrary code that can assist the attacker in installing malicious software or securing their presence on a device. We can see an example of this simulation payload in the following figure.

Figure 12.4 – An example of a link to a malware payload

- **Drive-by URL**: Here, an attacker emails a link that directs the recipient to a site, attempting to execute code in the background that will collect information about a user or deploy harmful software. This attack, also referred to as a watering hole attack, uses a compromised reputable site or a convincing clone, making the link appear safe for the user. We can see an example of the payload for this simulation in the following figure.

Good Day

Please kindly received 50% deposit for goods ordered.
Do advice accordingly .

DOWNLOAD POP ONLINE

Disclaimer

The information contained in this communication from the sender is confidential. It is intended solely for use by the recipient and others authorized to receive it. If you are not the recipient, you are hereby notified that any disclosure, copying, distribution or taking action in relation of the contents of this information is strictly prohibited and may be unlawful.

This email has been scanned for viruses and malware, and may have been automatically archived by **Mimecast Ltd**, an innovator in Software as a Service (SaaS) for business. Providing a **safer** and **more useful** place for your human generated data. Specializing in; Security, archiving and compliance. To find out more Click Here.

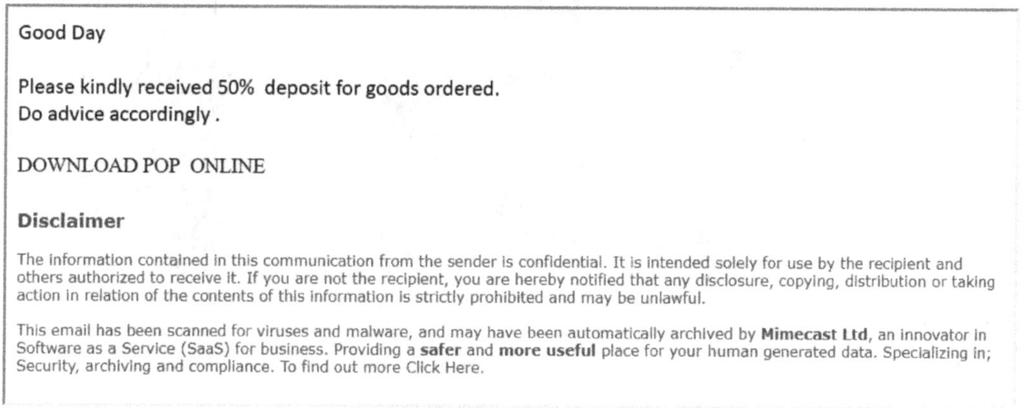

Figure 12.5 – An example of a drive-by URL payload

- **OAuth consent grant**: Here, an attacker creates a rogue Azure application, aimed at accessing data, that dispatches an email with a malicious link. Clicking the link triggers a request for data access permissions (e.g., access to a user's inbox) via the application's consent grant mechanism. The attacker's aim is to trick the end user into providing the rogue application with high privileges in an environment, or just enough privileges to not raise suspicion but allow for other attacks or even enumeration to occur. We can see an example of this payload in the following figure.

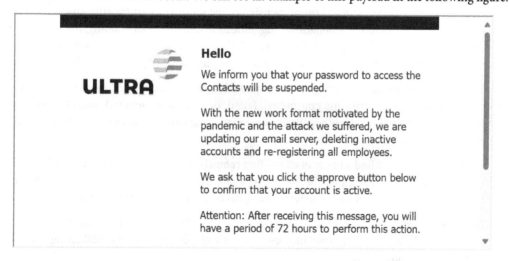

ULTRA

Hello

We inform you that your password to access the Contacts will be suspended.

With the new work format motivated by the pandemic and the attack we suffered, we are updating our email server, deleting inactive accounts and re-registering all employees.

We ask that you click the approve button below to confirm that your account is active.

Attention: After receiving this message, you will have a period of 72 hours to perform this action.

Figure 12.6 – An example of an Oauth consent grant payload

- **A how-to guide**: This is an instructional guide that offers directions for users on actions such as reporting phishing attempts. We can see an example in the following figure.

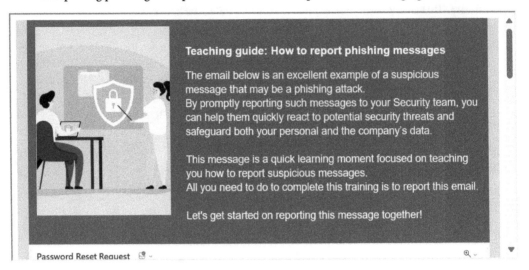

Teaching guide: How to report phishing messages

The email below is an excellent example of a suspicious message that may be a phishing attack.
By promptly reporting such messages to your Security team, you can help them quickly react to potential security threats and safeguard both your personal and the company's data.

This message is a quick learning moment focused on teaching you how to report suspicious messages.
All you need to do to complete this training is to report this email.

Let's get started on reporting this message together!

Password Reset Request

Figure 12.7 – An example of a how-to guide payload

From the **Attack Simulation Training** page, you'll find a variety of tabs designed to help streamline and monitor your training initiatives. We can reach this page from the Defender portal (`security. microsoft.com`), and then, on the left-hand menu under **Email & collaboration**, navigate to **Attack simulation training** to open the page, or directly to it via `https://security.microsoft. com/attacksimulator`. Here's what each tab offers:

- **Overview**: This tab presents a comprehensive snapshot of your attack simulation training program. It includes recent simulations, suggestions to enhance your training, the number of users who have gone through at least one simulation, the count of users trained versus those yet to be trained, identified repeat targets, and highlights of those most vulnerable to threats.

- **Simulations**: This segment catalogs all your simulation exercises, including those being drafted, scheduled, currently running, completed, failed, or canceled. Detailed insights provided here can aid you in effectively organizing and managing simulations, along with launching new simulations.

- **Training**: Here, you'll find a roster of all training campaigns in various stages – draft, scheduled, active, completed, or failed. Like the **Simulations** tab, you can also launch new campaigns from this tab. The list in this tab can help reveal any gaps in training and organize training activities.

- **Reports**: This section gathers and displays all key metrics from ongoing and concluded attack simulations. Data is broken down into user-friendly graphs and reports, facilitating analysis and sharing among teams.

- **Automations**: Details on automated workflows to run simulations are found in this tab, streamlining the process. New automations can also be configured from this tab.

- **Content Library**: This functions as a repository for all materials used during simulations, such as notification messages, payloads, landing pages, fake login pages, and training modules, all organized in a left-hand menu.

- **Settings**: Here, you can configure the repeat offenders threshold and training threshold. These thresholds are used by functions such as training campaigns to assign user training.

For most training efforts, the **Attack Simulation Training** page will be the centralized location where efforts will be tracked and launched. Each tab is engineered to provide a central, thorough, and efficient pathway to answer any executive questions about training and provide a rationale for changes to the training approach. Now that we understand the basics of the **Attack Simulation Training** interface, we need to learn how to use it for training.

Using simulations and payloads

The key parts of any simulation training involve how to configure and launch a simulation, as well as the payload used in it. A security team will typically have a training schedule that has been approved across an organization and will ensure a training cadence that does not impact security operations. This training should also include any extra basic training that users might need to take, as well as additional training if they fail an attack simulation. For extra impact, many organizations also include prizes for end users who react well to the simulations.

Understanding payloads

A **payload** refers to either a link or an attachment in a message used in an attack simulation. Out of the box, a comprehensive built-in catalog of payloads is available, which represents various social engineering tactics. For those looking to tailor this experience more closely to their organization's needs, there's an option to craft custom payloads. To browse the available payloads, visit the Microsoft Defender portal (`security.microsoft.com`), and on the left-hand menu, navigate to **Email & collaboration**, choose **Attack simulation training**, click on the **Content library** tab, and then click **Payloads**. Alternatively, for direct access, go to `https://security.microsoft.com/attacksimulator?viewid=contentlibrary`. Within the **Content Library** tab, you'll find three sections:

- **Global payloads**: You can find the out-of-the-box built-in payloads here, which are unmodifiable. This is what most organizations use to get started.

- **Tenant payloads**: If your organization has the time and resources, you can create custom payloads, which will be listed here.

- **MDO recommendations**: By using information about your environment and previous user behaviors, Defender for Office 365 identifies and lists in this tab the built-in payloads that would be highly impactful to your organization when used by adversaries. This list is updated monthly.

Each section provides a list of payloads available, including details (depending on the section) such as the payload name, predicted compromise rate (%), type, technique, and language. It is recommended that organizations stick to the built-in payloads because of ease of use and proven effectiveness. For organizations pursuing the use of custom payloads, it is important to note that various trademarks, logos, symbols, badges, and other identifiers are safeguarded extensively by local, state, and national regulations. Misusing these symbols may lead to severe consequences, including criminal penalties. When using third-party trademarks, there's always a level of risk involved. This is especially true when creating materials such as payloads in cybersecurity training exercises. Opting to use your organization's trademarks and logos is safer, provided your organization has specifically allowed their use.

Creating a custom payload

To craft custom payloads in the Microsoft Defender portal (`security.microsoft.com`), navigate to **Email & collaboration | Attack simulation training | Content library | Payloads | Tenant payloads**, or for direct access, use `https://security.microsoft.com/attacksimulator?viewid=contentlibrary&source=tenant`. We can see this tab in the following figure.

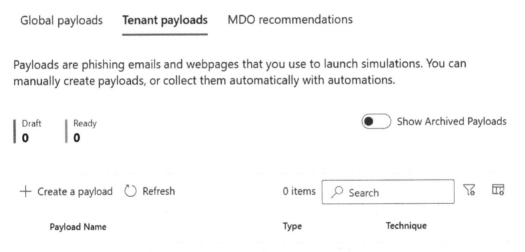

Figure 12.8 – The Tenant payloads tab

This page will be empty on any new Defender for Office 365 deployments. To start the process, on the **Tenant payloads** tab, click on **Create a payload** to launch the new payload setup wizard:

1. **Select type**: Choose either **Email** or **Teams,**and click **Next**. Note that, at the time of writing, the **Teams** option is grayed out for many tenants, as it is in a private preview. We can see this page in the following figure.

Select type

Select a payload type to create.

Email
Create a phish email payload

~~Teams~~
~~Create a phish teams message payload~~

Figure 12.9 – Selecting the type of payload

2. **Select technique**: These options are identical to those in the simulation wizard, which includes techniques such as **Credential Harvest**, **Malware Attachment**, and **Drive-by URL**. Once a selection has been made, select **Next**. We can then see the following screen.

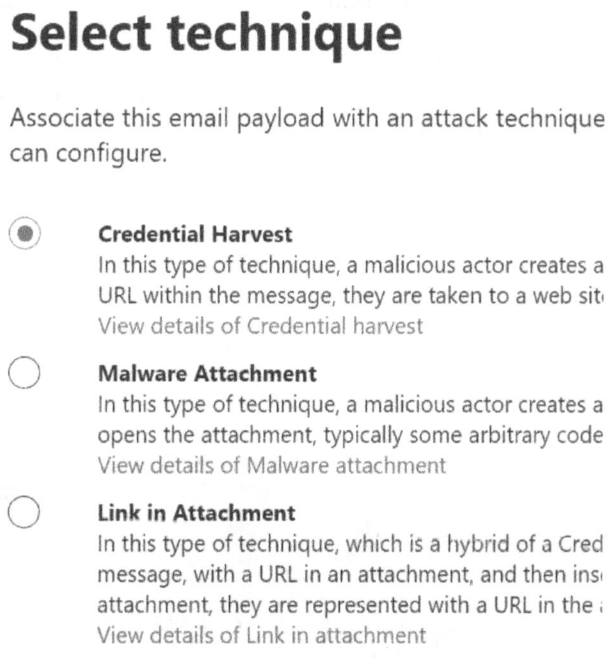

Select technique

Associate this email payload with an attack technique
can configure.

◉ **Credential Harvest**
In this type of technique, a malicious actor creates a
URL within the message, they are taken to a web sit
View details of Credential harvest

○ **Malware Attachment**
In this type of technique, a malicious actor creates a
opens the attachment, typically some arbitrary code
View details of Malware attachment

○ **Link in Attachment**
In this type of technique, which is a hybrid of a Cred
message, with a URL in an attachment, and then ins
attachment, they are represented with a URL in the ;
View details of Link in attachment

Figure 12.10 – Selecting the technique

3. **Payload Name**: Enter a unique name for the payload and, optionally, a description. Select **Next** to proceed. We can see an example in the following figure.

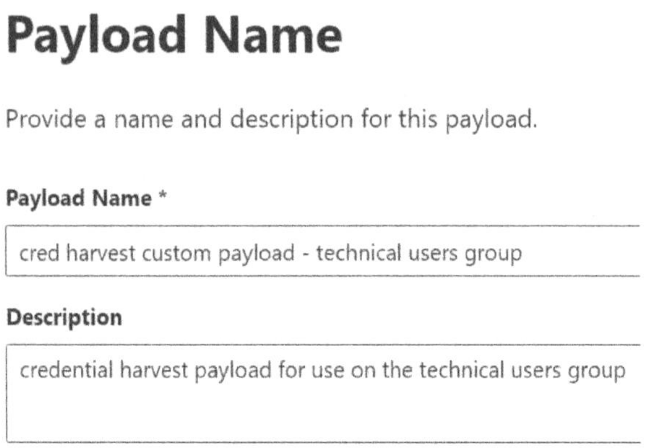

Payload Name

Provide a name and description for this payload.

Payload Name *

cred harvest custom payload - technical users group

Description

credential harvest payload for use on the technical users group

Figure 12.11 – An example of naming a custom payload

4. **Configure the payload**: The specifics here depend on the previously chosen technique (e.g., links versus attachments). The typical fields for all the techniques are as follows:

    • **Sender details**: Customize the sender's info, such as the name, email, and email subject. You can opt to tag emails as external and use internal addresses for realistic simulations. We can see this section in the following figure.

Sender details

From name *

Enter Sender's Name

☐ Use first name as display name

From email *

Enter Sender's Email

Email subject *

Enter Email Subject

☐ Add External tag to email

Figure 12.12 – The Sender details section

    • **Tags** (except for links in attachment payloads): Add relevant tags and select the theme, brand, and industry that best matches your payload. Define whether the event is current or controversial, and then choose the payload's language. We can see these options in the following figure.

Add Tag(s)

Theme

Theme

Brand

Brand

Industry

Industry

Current event

Current event

Controversial

Controversial

Figure 12.13 – The Tags section

- **Email message:** Import text messages or use the text editor provided to craft your message. You can also insert dynamic tags or links as needed. We can see this section in the following figure.

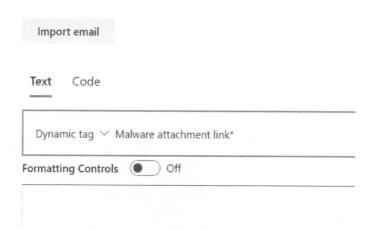

Email message

Import an email or paste the code below to use it. Be sure to insert the

Import email

Text   Code

Dynamic tag ∨ Malware attachment link*

Formatting Controls ⦿ Off

Figure 12.14 – The Email message section

For **Credential Harvest**, **Link in Attachment**, **Drive-by URL**, and **Oauth Consent Grant**, the following field is included:

- **Phishing URL**: The phishing URL can be selected from a list here. Be aware that in the email message text area, you can change the link's appearance to add realism to the attack, including dynamic tags. We can see the option to change the appearance of the phishing URL in the following figure.

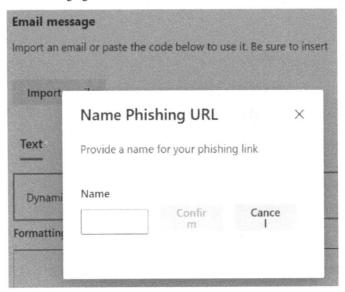

Figure 12.15 – Changing the appearance of a phishing URL

For **Malware Attachment**, **Link in Attachment**, and **Link to Malware** the following field is available:

- **Attachment Details**: Here, you can add a name for the attachment and provide a type. We can see the **Attachment Details** fields in the following figure.

Attachment Details

Name your attachment *

Select an attachment type *

Select an attachment type ∨

Figure 12.16 – The Attachment details section

After detailing your payload, proceed to the **Predicted Compromise Rate** section and click on **Predict Compromise Rate** to estimate its effectiveness. You can either wait for the results or proceed to the next section of the wizard by selecting **Next**. We can see the predicted compromise rate in the following figure.

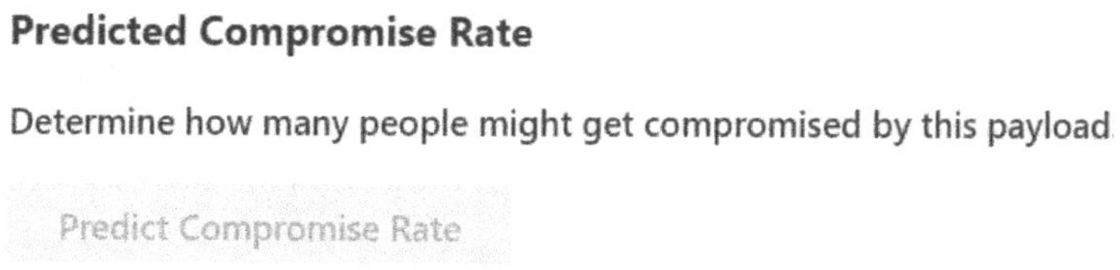

Predicted Compromise Rate

Determine how many people might get compromised by this payload

Predict Compromise Rate

Figure 12.17 – Predicted Compromise Rate

5. **Add indicators**: This step is crucial for techniques such as **credential harvest**, as indicators serve as hints to recognize phishing attempts performed by your payload, and they further help the security team not confuse the simulation with a real attack when looking at logs and investigating the simulated attack. You can select the indicator to use, where they appear, and customize descriptions, as shown in the following figure.

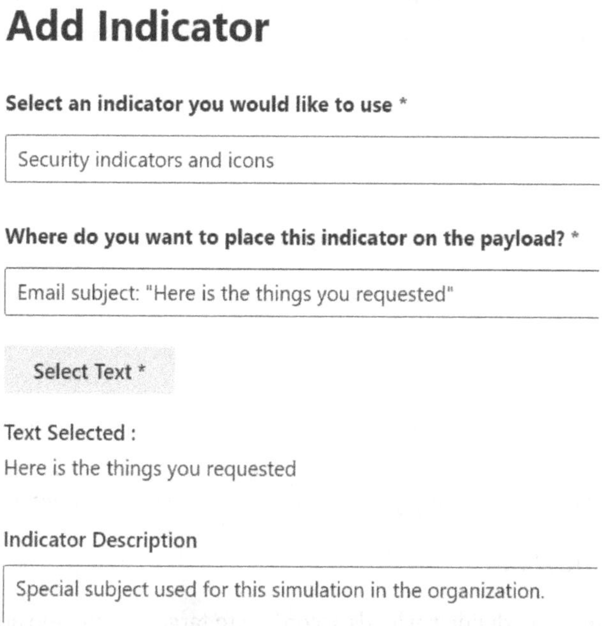

Add Indicator

Select an indicator you would like to use *

Security indicators and icons

Where do you want to place this indicator on the payload? *

Email subject: "Here is the things you requested"

Select Text *

Text Selected :
Here is the things you requested

Indicator Description

Special subject used for this simulation in the organization.

Figure 12.18 – Adding indicators to a custom payload

Once the indicator has been added, it will be listed on the page, and you can either add more indicators or select **Next** to proceed.

6. **Review**: Inspect the payload's configurations. Send a test email to yourself or preview the payload with indicators. Edit the payload if needed, and once completed, click on **Submit**.

7. The payload will be created, and you will be presented with a confirmation page where you can select **Done** to finish the wizard. Your new payload, marked as **Ready**, will now be listed under **Tenant payloads**.

Due to the number of attacks possible, an organization should understand the importance of custom payloads and when to use them. As previously mentioned, most organizations should start with the built-in payloads (global payloads) and see how well their organization reacts to these simulations. Once the team has enough experience and understands the gaps in the end user security posture, custom payloads can be visited.

Modifying custom payloads

Either due to errors discovered, or just to improve on payloads, there will be a time when payloads will need to be modified. Changing, copying, archiving, and restoring custom (tenant) payloads within the Microsoft Defender portal is a straightforward process. To edit an existing payload, on the **Tenant payloads** tab, click the checkbox beside the payload's name, and then click the **Edit payload** option that appears. Alternatively, click on the payload name to bring up the details flyout, and then select **Edit payload** at the bottom. A third way is to click on the three-dots icon on the last column of the table and, from the menu that appears, choose **Edit payload**, as shown in the following figure.

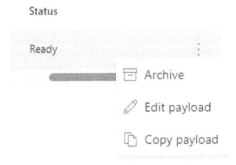

Figure 12.19 – The menu that appears when clicking the three-dots icon

The payload wizard will launch, allowing you to adjust the settings and values as required.

Copying a custom payload

Many organizations attempt to divide payloads according to target groups and add slight differences to make a payload more effective. Instead of creating many payloads from scratch, copying a custom payload can save a lot of time, and it is a very simple process. From either the **Tenant payloads** or **Global payloads** tabs, click the checkbox next to the desired payload's name, and then select the **Copy payload** action that appears. In the same manner as when modifying payloads, you can also click on the three-dots icon on the last column and, on the menu, select **Copy payload**. If doing this from the **Global payloads** tab, be aware that this will be the only option, as we cannot modify global payloads, as shown in the following figure.

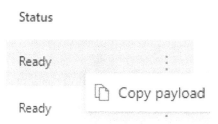

Figure 12.20 – The three-dots icon menu in the Global payloads tab

The creation wizard will open, auto-filled with the chosen payload's specifics, ready for modification.

Archiving and restoring custom payloads

From time to time, it is good to clean up the list of custom payloads to make management simpler. During this process, your team might prefer to keep the custom payload somewhere to either make a copy or review its configuration. Instead of deleting it, a custom payload can be archived; while on the **Tenant payloads** tab, look for the payload to archive, click on the three-dots icon, and select **Archive** from the menu. This changes the payload's status to **Archive**, making it invisible on the **Tenant payloads** list unless **Show Archived Payloads** is turned on, as shown in the following figure.

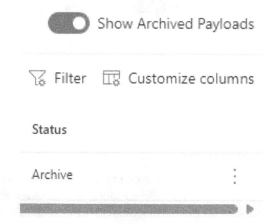

Figure 12.21 – Turning on the Show Archived Payloads switch

Bringing an archived payload back into circulation is also a very simple process. With the **Show Archived Payloads** switch turned on, click on the three-dots icon beside the payload, and from the menu, select **Restore**, as shown in the following figure.

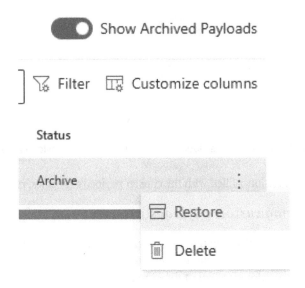

Figure 12.22 – Restoring an archived payload

Post-restoration, the payload becomes visible and it's status becomes **Draft**. To use the payload again, we need to make its status switch back to **Ready** by editing the payload, reviewing any changes to the settings if necessary, and clicking **Submit** to complete the payload creation wizard. Through these mechanisms, maintaining an organized, efficient payload catalog in the Microsoft Defender portal remains manageable and streamlined, allowing for the easy adaptation and reuse of cybersecurity training materials.

Creating a simulation

Setting up and starting simulations is a straightforward task, accomplished through the Defender portal (`security.microsoft.com`) or by using APIs. When you create a simulation, it's crucial to remember that many prevalent attacks are pre-configured and ready to use as payloads. When creating a simulation, the process is simple to follow, especially if you have attempted to create a custom payload before. To start a simulated phishing campaign, follow these steps:

1. Visit the Microsoft Defender portal (`security.microsoft.com`) and head over to **Email & Collaboration** | **Attack simulation training** | the **Simulations** tab, or go directly via `https://security.microsoft.com/attacksimulator?viewid=simulations`.

2. On the **Simulations** tab, click on **Launch a simulation**, which opens a new simulation wizard. We can see the **Launch a simulation** option in the following figure.

Attack simulation training

Overview   **Simulations**   Training   Reports

A list of all your simulations and their status.

Draft	Scheduled	In progress	Completed
0	**0**	**0**	**2**

+ Launch a simulation

Figure 12.23 – The Launch a simulation option.

3. On the **Select technique** page, choose from a variety of social engineering techniques for your simulation, including **Credential Harvest**, **Malware Attachment**, **Link in Attachment**, **Link to Malware**, **Drive-by URL**, **OAuth Consent Grant**, and **How-to Guide**. For a deeper understanding of each technique, the **View details of** link reveals more information about the selected technique and the steps it entails. Select a technique and click on **Next** to proceed. We can see this page in the following figure.

Select technique

Select the social engineering technique you want to
selection, you will be able to use certain types of

◉ **Credential Harvest**
In this type of technique, a malicious actor
are taken to a web site, the website often...

View details of Credential harvest

○ **Malware Attachment**
In this type of technique, a malicious actor
typically some arbitrary code such as a mac

View details of Malware attachment

Figure 12.24 – The Select technique page

4. On the **Name simulation** page, provide a unique name for your simulation and, optionally, a brief description, and then click on **Next** to proceed. We can see this page in the following figure.

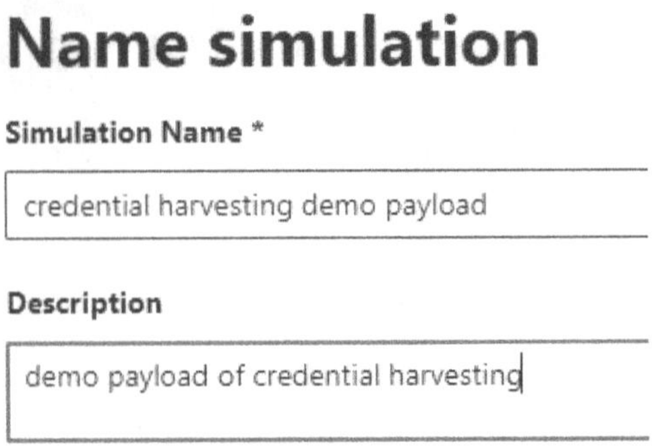

Name simulation

Simulation Name *

credential harvesting demo payload

Description

demo payload of credential harvesting

Figure 12.25 – The Name simulation page

5.   Navigate to the **Select payload and login page** section. Here, you can select what payload to use for the simulation (only one can be selected); the available payloads include the following:

- **Global payloads**: Includes built-in payloads crafted by Microsoft, detailing information such as the name, language, and an estimate of the compromise rate based on Microsoft 365's historical data.

- **Tenant payloads**: Here, you will find any custom payloads specific to your organization, along with an option to create new ones by selecting **Create a payload**.

Most organizations are recommended to start with global payloads and progress to using tenant payloads once a need has been identified. We can see a list of the payloads available in the following figure.

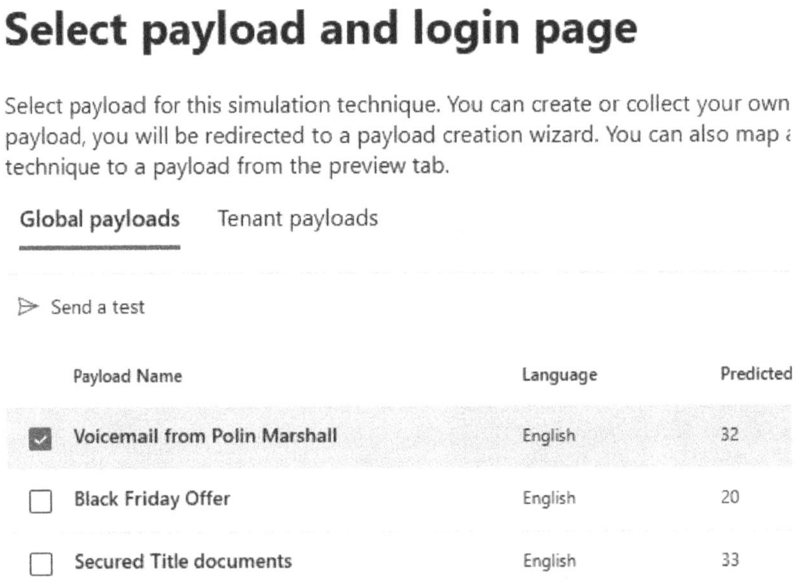

Select payload and login page

Select payload for this simulation technique. You can create or collect your own payload, you will be redirected to a payload creation wizard. You can also map a technique to a payload from the preview tab.

Global payloads   Tenant payloads

▷ Send a test

Payload Name	Language	Predicted
☑ **Voicemail from Polin Marshall**	English	32
☐ Black Friday Offer	English	20
☐ Secured Title documents	English	33

Figure 12.26 – A list of the global payloads available

For specific payload searches, input a part of the payload's name in the search box on the top right of the list and press *Enter* on your keyboard. Payloads can also be filtered using various attributes, such as **Source**, **Complexity**, **Language**, **Theme**, **Brand**, **Industry**, **Current event**, and **Controversial**. Upon choosing a payload, the **Send a test** option becomes available, as shown in the previous figure, letting you send a test email of the payload to yourself for review.

We can also review the details of the payload by selecting it, which will open a flyout panel that can house up to four tabs, depending on the payload. The tabs you might encounter are as follows:

- **Overview**: This gives you a comprehensive look at what the payload entails, detailing the message content, the sender's email, and the subject line, among other vital information. You'll find this tab when working with techniques such as **malware attachment**, **link to malware**, **drive-by URL**, **OAuth consent grant**, and **how-to guide**. We can see an example in the following figure.

Invoice from Beyond Diet

Social Engineering• Malware Attachment

Delivery Platform : Email

Overview   Attachment   Simulations launched

Thank you for your vigilance!

The email "[External]Invoice from Beyond Diet" that you reported is safe.

If you don't want this email, we recommend using the unsubscribe link in Outlook by going to Junk -> Block Sender

If you feel we categorized this message in error, please call the service des

For more information on legitimate internal messages and to see malicious

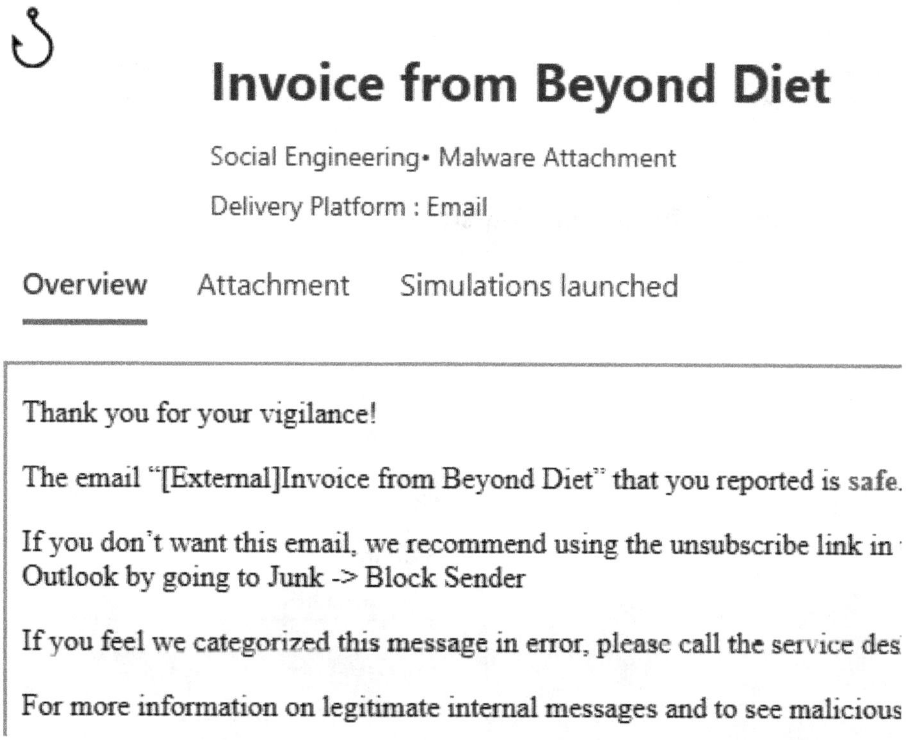

Figure 12.27 – The Overview tab of a payload

- **Attachment**: You'll see a list of attachments designated for the recipient. This tab appears for techniques such as malware attachment, link to malware, and OAuth consent grant. We can see this tab in the following figure.

Figure 12.28 – The Attachment tab of a payload

- **Login page**: This is only visible with the credential harvest and link in attachment techniques. This tab offers tools to view and change the login page being used. By selecting **Change login page**, a new panel appears, showcasing available login pages, including their names, languages, creators, last modification dates, and previews under the **actions** column. Here, you also have the option to click **Create new** to initiate the login page creation process. We can see the **Login page** tab in the following figure.

Figure 12.29 – The Login page tab

- **Simulations launched**: This tab catalogs details about previously started simulations that used similar techniques as a reference. This tab is accessible across all techniques, and we can see it in the following figure.

Payload   Login page   **Simulations launched**

Simulation Name	Click rate	Compromised rate (%)	Action
test	--	--	View details

Figure 12.30 – The Simulations launched tab

After you've picked a payload, click the **Next** button to move onto the **Target users** page. If you're setting up a simulation with the OAuth consent grant technique, you'll first see a prompt to set up the OAuth payload. This involves naming the OAuth payload app, adding an optional logo, and defining the app's permission scope. Once the OAuth payload is configured to your liking, click **Next** to advance to the **Target users** page.

6. On the **Target users** page, you can decide who will partake in the simulation. You have the option to involve all users in your organization or narrow it down to include only specific users and groups. If you go for the latter, you add these individuals or groups using the **Add users** or **import** buttons. The **import** button allows you to upload a list of users for inclusion. If adding users manually via the **Add users** button, you can locate users to add by searching their name, group, tags, or even location. After making your selections, click **Next**.

7. On the **Exclude users** page, you can specify any users that you'd like to leave out of the simulation. Once you've made your exclusions, hit **Next** to continue.

8. On the **Assign training** page, you are presented with the opportunity to assign specific training tailored to a simulation. Assigning training is highly recommended, as it significantly enhances employee resilience against similar future threats. We can see the **Assign training** page in the following figure.

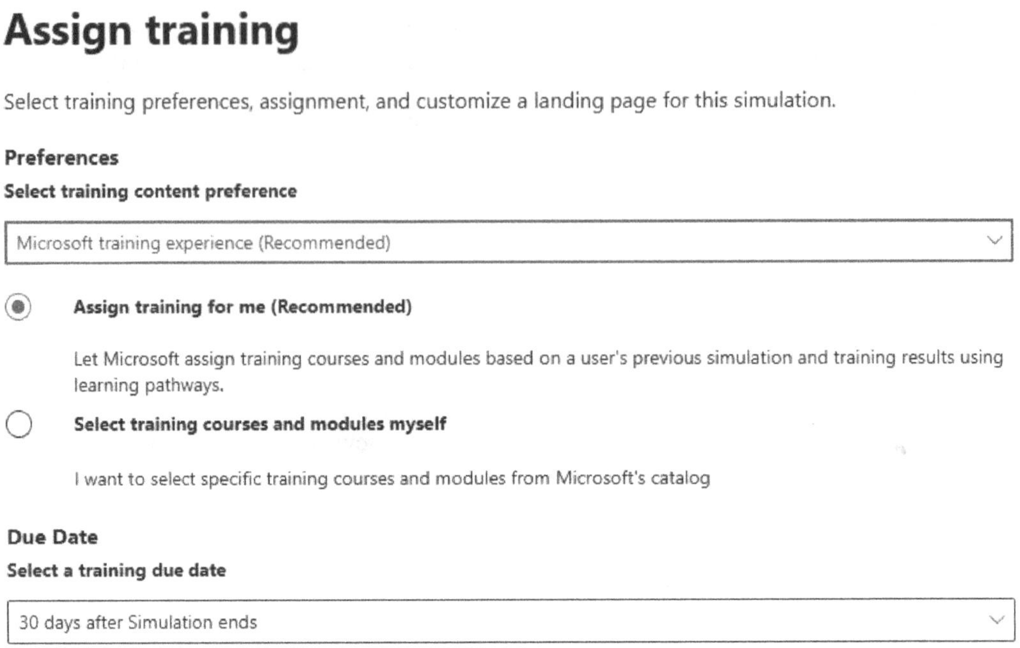

Assign training

Select training preferences, assignment, and customize a landing page for this simulation.

Preferences

Select training content preference

Microsoft training experience (Recommended)	∨

◉ **Assign training for me (Recommended)**

Let Microsoft assign training courses and modules based on a user's previous simulation and training results using learning pathways.

○ **Select training courses and modules myself**

I want to select specific training courses and modules from Microsoft's catalog

Due Date

Select a training due date

30 days after Simulation ends	∨

Figure 12.31 – The Assign training page

Three main options are available for training assignments:

- **Microsoft training experience (Recommended)**: This option is chosen by default and offers either a recommended training plan, curated by Microsoft (the **Assign training for me (Recommended)** choice), or the option to pick specific built-in modules and courses (the **Select training courses and modules myself** choice). You can set the deadline to complete the training to either 7, 15, or 30 days (the latter being the default setting) following the end of the simulation.

- **Redirect to a custom URL**: Opting for this will guide the end user to a unique training program developed in-house. To use this option, you'll need to input the training URL, its name, an optional description, and the length of the training. Like **Microsoft training experience**, you can set the training's completion deadline to 7, 15, or 30 days after the simulation concludes.

- **No training**: Choosing this option means no training will be provided for the user, which is not recommended.

Once you have set your training assignments on the **Assign training** page, click **Next** to proceed.

9. On the **Select Phish landing page** section, you can decide the web destination that users will encounter if they engage with a payload during the simulation. You can either use one of the pre-existing landing pages (global landing pages) or input a custom URL for a unique page. When opting for a library landing page, you can add a logo and change the default language, making it look more authentic. You can also build a custom landing page, also known as a **tenant landing page**, if required. You can also click on any landing page to open the details flyout panel, which provides a preview of the page and details on it. After selecting a landing page, click **Next** to proceed. We can see the preview of a phishing landing page in the following figure.

Microsoft Landing Page Template 1

Preview   Details

Select language *

English

${DisplayName}, you were just **phished** by your security team.

It's okay! You're human. Let's learn from this.

Close

PHISHING MESSAGE!!

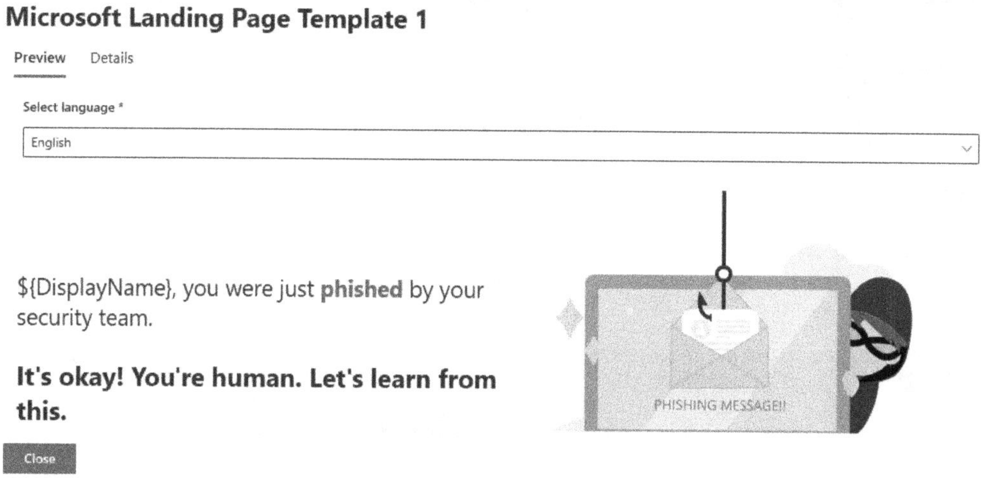

Figure 12.32 – A phishing landing page example preview

10. On the **Select end user notification** page, you can configure the types of alerts that users will receive during the simulation. These notifications are multifaceted; they might include positive feedback for reporting suspicious messages, an alert about assigned training, or reminders about upcoming training deadlines. The choices include opting out of delivering notifications, selecting Microsoft's default notification, and tailoring customized end user notifications. Should you opt for Microsoft's default notification, you're presented with additional customization options, such as the language selection, scheduling the delivery of the notification, and a chance to preview what users will see. Once you've made your choice, clicking the **Next** button will navigate you forward. We can see this selection in the following figure.

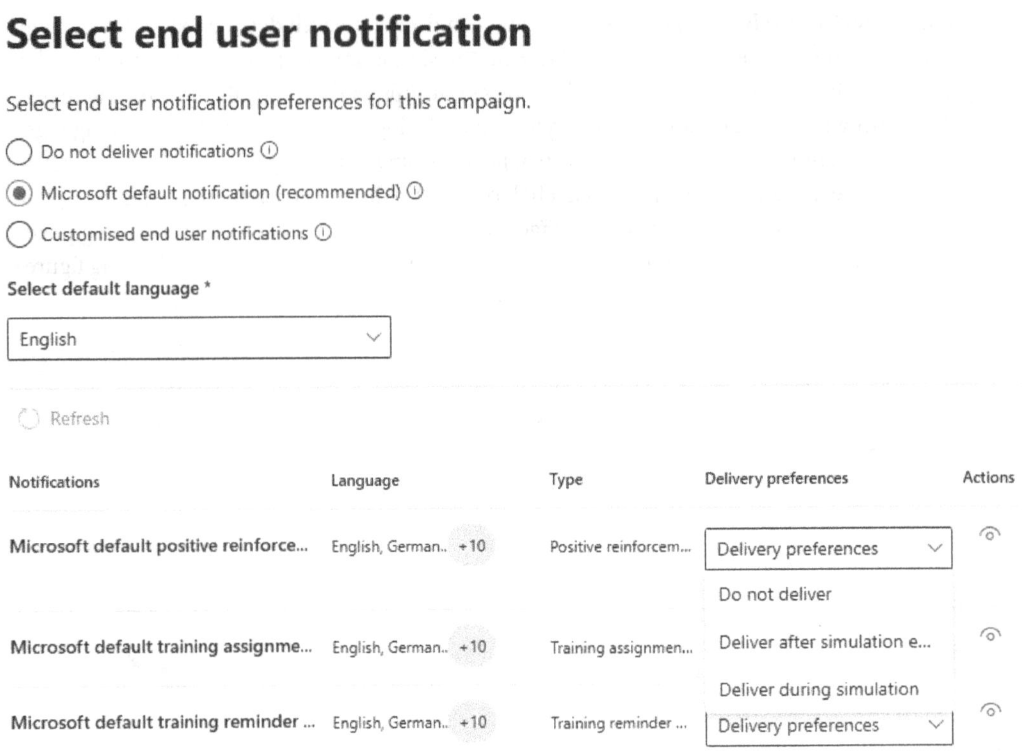

Select end user notification

Select end user notification preferences for this campaign.

○ Do not deliver notifications ⓘ

◉ Microsoft default notification (recommended) ⓘ

○ Customised end user notifications ⓘ

Select default language *

English   ⌄

⟳ Refresh

Notifications	Language	Type	Delivery preferences	Actions
Microsoft default positive reinforce…	English, German.. +10	Positive reinforcem…	Delivery preferences ⌄	⟨⟩
			Do not deliver	
Microsoft default training assignme…	English, German.. +10	Training assignmen…	Deliver after simulation e…	⟨⟩
			Deliver during simulation	
Microsoft default training reminder …	English, German.. +10	Training reminder …	Delivery preferences ⌄	⟨⟩

Figure 12.33 – Microsoft's default end user notifications options

Should you choose the **Customized end user notifications** option, you will be brought to pages dedicated to training assignment notifications, training reminder notifications, and positive reinforcement notifications. Here, you can either pick from a selection of pre-designed notifications or click on **Create new** to craft a custom notification, tailored to your needs. After making your selection, click **Next** to continue with the setup process.

11. The **Launch details** page allows you to set the beginning and end of your simulation. Remember that after the simulation concludes, the collection of data relating to it will cease. You can kick off the simulation immediately or schedule a specific start date and time. Additionally, you can determine the duration of the simulation, ranging from 2 to 30 days. For a touch of authenticity, there's an **Enable region aware time zone delivery** option, which ensures that the simulated attack messages reach users within their local working hours. We can see this page in the following figure.

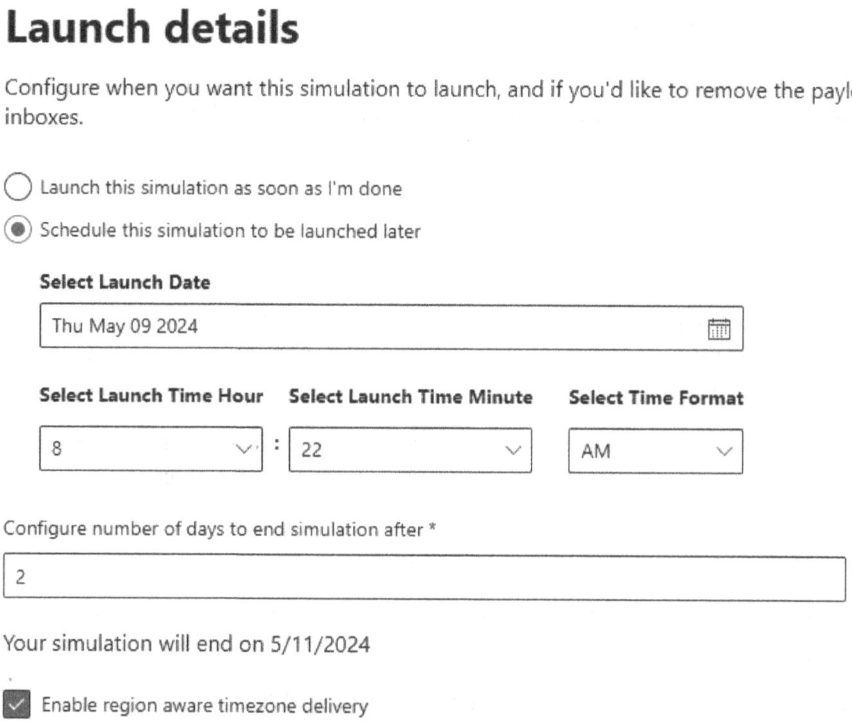

Launch details

Configure when you want this simulation to launch, and if you'd like to remove the payloads from user inboxes.

○ Launch this simulation as soon as I'm done

◉ Schedule this simulation to be launched later

Select Launch Date

| Thu May 09 2024 | 🗓 |

Select Launch Time Hour	**Select Launch Time Minute**	**Select Time Format**
8 ⌄	:	22 ⌄	AM ⌄

Configure number of days to end simulation after *

| 2 |

Your simulation will end on 5/11/2024

☑ Enable region aware timezone delivery

Figure 12.34 – The Launch details page

Once you've finalized your choices on the **Launch details** page, click **Next** to proceed.

12. On the **Review simulation** page, it's crucial to go over the details of your planned simulation meticulously, as this is your opportunity to ensure that everything is set up exactly as intended. If all entries are accurate, click **Submit** to complete the simulation configuration. Following your submission, a confirmation screen will appear, indicating that your simulation is now fully configured and scheduled for initiation. To complete the process, simply click **Done**.

Once the simulation is launched, its information will be visible under the **Overview** and **Simulations** tabs on the **Attack Simulation Training** page. It's important to note that the campaign will continue to be marked as in progress, even after all targeted users have interacted with the email. The status will only update to reflect the campaign's completion after reaching the designated end time set on the **Launch details** page.

Next, let's see how we can automate our training efforts to minimize human error.

Automating the training

Now that we understand how attack simulations and payloads work, it is important to understand their automation aspect, which makes for a more realistic scenario. Automation is a good way to expose users to more unexpected attacks by launching multiple payloads, and it opens the possibility of identifying gaps in training (*Microsoft, 2024c*).

Simulation automation

Simulation automation allows you to automate a group of simulations together and add randomization of the payloads used. The process of creating simulation automations is the same as creating a single simulation, with the following differences:

- A simulation automation is created from the **Attack simulation training** page, **Automations** tab, **Simulation automations** section, or via `https://security.microsoft.com/attacksimulator?viewid=automations`

- You can select multiple techniques to use in the simulation

- You can select multiple payloads or select **Randomize** to let Defender for Office 365 select the appropriate payloads to use

When using simulation automations, ensure that the times, duration, techniques, and payloads are discussed and planned with all the members of the security team. During these simulations, any reports to the help desk, even if part of the simulation, will be treated as a typical user report. Members of the security team might receive notifications of emails being reported, which might lead to confusion unless the process has been previously discussed and team members have been instructed on how to treat these reports, along with what indicators to search for to identify them.

Payload automation

Payload automation allows you to create customized payload workflows that harvest real-world attacks, according to conditions experienced by your organization, and generate custom payloads. Many conditions can be specified, including targeting a specific number of users in an organization, specifying what technique to use, specifying sender details, and other conditions. This customization allows you to create replicas of attacks seen in the field, using similar messages and payloads, which allows you to randomize payload usage while staying within defined constraints. To view any payload automations you've set up, navigate to the Microsoft Defender portal (`security.microsoft.com`), and then **Email & Collaboration** | **Attack Simulation Training** | the **Automations** tab, and then click on **Payload Automations**, or go directly via `https://security.microsoft.com/`

`attacksimulator?viewid=automations`. Payload automations can also be created from the **Attack simulation training** page, **Automations** tab by clicking on **Create automation** to launch the payload automation wizard and then following these steps:

1. **Name the workflow**: Enter a unique automation name and, optionally, a detailed description, and then click on **Next** to proceed.

2. **Configure the run conditions**: Decide on the specific scenarios of a phishing attack under which the automation should run. Click on **Add condition** and choose one of the set conditions, such as the number of users targeted, phishing techniques (such as credential harvest or malware attachment), specific sender details, or targeted user and group recipients, as shown in the following figure.

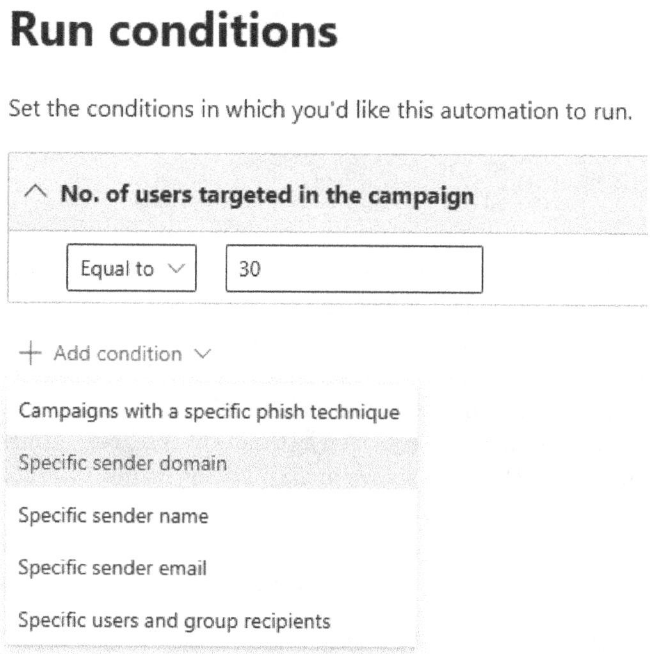

Run conditions

Set the conditions in which you'd like this automation to run.

∧ **No. of users targeted in the campaign**

| Equal to ∨ | 30 |

+ Add condition ∨

Campaigns with a specific phish technique

Specific sender domain

Specific sender name

Specific sender email

Specific users and group recipients

Figure 12.35 – Adding some run conditions to payload automations

Only one instance of each condition can be added, and if you use multiple conditions, they're combined using **AND** logic. To add another condition, select **Add condition** again, and if you need to remove a condition, you can do so by selecting the removal option next to it. Once all conditions have been entered, click on **Next** to proceed.

3. **Review and submit**: On the **Review automation** page, review your settings and edit any section as needed. Once everything looks good, click on **Submit**. Following submission, you'll land on the **New automation created** page, where options to activate the automation or proceed to the **Simulations** page are provided. Click **Done** when finished.

4. Now, back on the **Payload automations** page under the **Automations** tab, your newly created payload automation will be listed and shown as **Ready** under the **Status** column. If you did not turn on the automation during the setup wizard, you will be presented with an option to do so, as shown in the following figure. This will activate the payload.

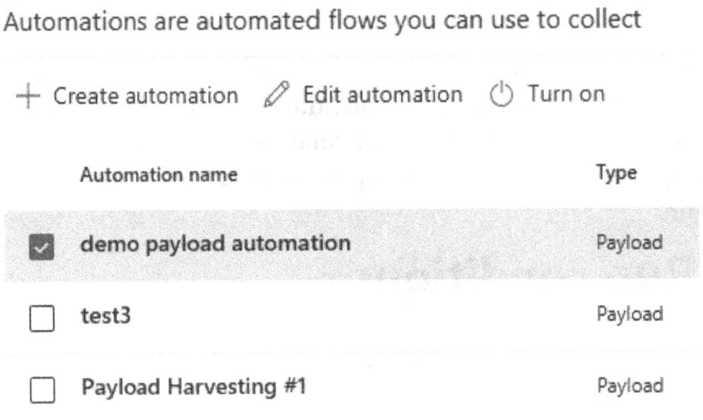

Automations are automated flows you can use to collect

+ Create automation   ✎ Edit automation   ⏻ Turn on

Automation name	Type
☑ **demo payload automation**	Payload
☐ test3	Payload
☐ Payload Harvesting #1	Payload

Figure 12.36 – The payload automation page

Remember that while setting up payload automations, specifying clear conditions ensures the effectiveness in capturing real-world attacks that might occur in your organization and how well the generated payloads mimic real-life scenarios. Any activated payload can be quickly deactivated if problems are encountered by ticking the checkbox next to its name, clicking the **Turn off** action that appears, and confirming your choice in the ensuing dialog. In the same manner, payload automations can be edited or deleted, by selecting the automation by ticking its checkbox, turning it off, and either clicking on **Edit automation** or **Delete**, as shown in the following figure.

Automations are automated flows you can use to collect payloads to launch simulations.

+ Create automation   ✎ Edit automation   ⏻ Turn on   🗑 Delete

Automation name	Type	Items collected	Last modified
☑ demo payload automation	Payload	0	5/9/2024, 2:55:51 PM
☐ test3	Payload	0	4/26/2024, 1:33:50 PM
☐ Payload Harvesting #1	Payload	0	3/31/2021, 5:50:33 PM

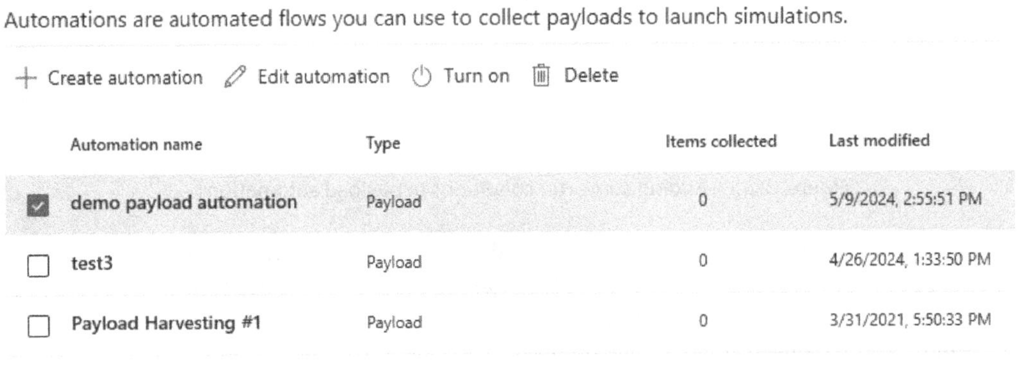

Figure 12.37 – The options to edit or delete payload automations

Clicking on any payload automation marked as **Ready** opens a flyout panel that provides general details on the payload automation and allows you to edit it. A **Run history** tab is provided, with details on previous runs of the payload automation workflow.

Next, we will see how we can further leverage automation to create customized approaches to end user training, via training campaigns.

Training campaigns

Training campaigns offer a swift and straightforward method to deliver security education to users and work as a more organized approach, instead of sending random simulated attacks in the hope of identifying the proper training sessions required. These campaigns allow you to handpick and assign one or more of over 70 available training modules directly to users, which ensures campaigns can meet diverse educational needs. For example, you can create a training campaign for all newly hired employees that covers foundational topics to learn during the first week of employment. To explore the training campaigns at your disposal, navigate to the Microsoft Defender portal (`security.microsoft.com`), and on the left-hand menu under **Email & Collaboration**, go to **Attack Simulation Training**. Once the **Attack Simulation** page opens, select the **Training** tab. You can also go directly via `https://security.microsoft.com/attacksimulator?viewid=trainingcampaign`. On this page, you'll find detailed listings for each training campaign you've started. This includes crucial details such as the campaign name, a brief description, the total duration in minutes, the completion date, the tally of users who have successfully completed the training, and the number of training modules involved. We can see this page in the following figure.

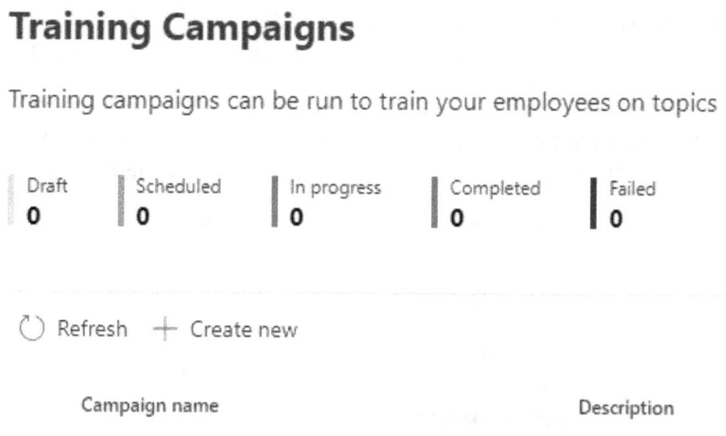

Training Campaigns

Training campaigns can be run to train your employees on topics

Draft	Scheduled	In progress	Completed	Failed
0	0	0	0	0

○ Refresh   + Create new

Campaign name	Description

Figure 12.38 – The Training Campaigns page

To create a new campaign, while on the **Training Campaigns** page, click on **Create new** to initiate the training campaign wizard. Be aware that this wizard will be very similar to other configuration wizards seen before in this chapter, so similar steps will be covered briefly. The first step involves naming your campaign and, if you wish, providing a description. After completing this, click **Next** to move on to the **Target users** page, and select the users that will be targeted by the campaign. You will have the option to either include all users within your organization or add specific users manually, or via a fixed list (the **Import** option). Once completed, click **Next** to proceed to the **Exclude users** page, select any users to exclude, and then click on **Next**. You will now be on the **Select training modules** page, as shown in the following figure.

Select training modules

◉ Training catalog

◯ Redirect to a custom URL

+ Add trainings

Training name   Source

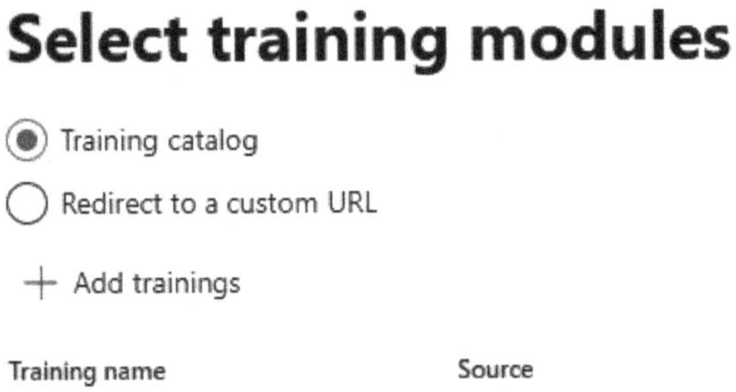

Figure 12.39 – The Select training modules page

On the **Select training modules** page, you will be able to select what training the targeted users will need to complete. You will be able to select from the following options:

- **Training catalog**: Click **Add trainings**, and in the popup, choose one or several training modules by marking the adjacent checkboxes, before clicking **Add**. These modules align with those found under the training modules in the **Content library** tab at https://security.microsoft.com/attacksimulator?viewid=contentlibrary. Note that the training module selection is sometimes slow to load and could take one to two minutes for the list to populate.

- **Redirect to a custom URL**: After selecting **Add trainings**, a panel will appear to input all the details about the custom training URL, including the URL itself, the name, an optional description, and the duration of the training in minutes. We can see this panel in the following figure.

Custom Training URL

Custom Training URL *

Custom training URL

Custom Training Name *

Custom training name

Custom training description

Custom training description

Custom training duration (in minutes)

0

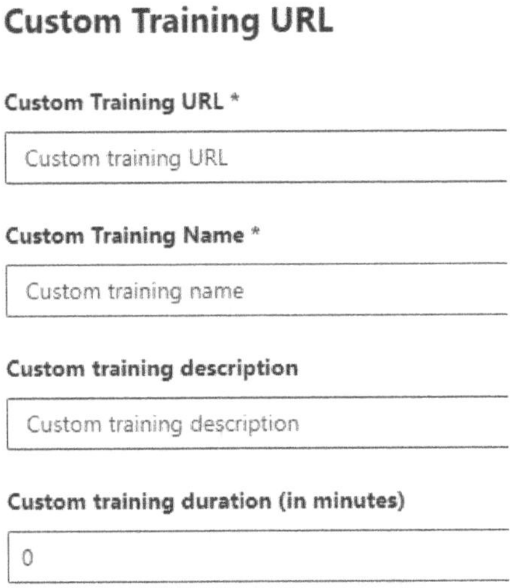

Figure 12.40 – The Custom Training URL configuration panel

Once completed, select **Add**, and the custom URL information will be reflected on the **Select courses** page.

After selecting the training modules to use, click **Next** to proceed to the **Select end user notification** page. As with other wizards, you are offered either the Microsoft default notification option, which provides built-in notifications for assignments and reminders, or customized end user notifications, which allow you to assign pre-existing notifications or create new ones for training reminders. Once you have made your selection, click **Next** to proceed. If you chose customized end user notifications, you will be presented with the pages to modify existing notifications or create new ones. Once you have finished creating the end user notifications as required and clicked on **Next**, you will be taken to the **Schedule** page, as shown in the following figure.

Schedule

Select the launch/end date & time for your training campaign

◉ Launch this training campaign as soon as I'm done

○ Schedule this training campaign to be launched later

☑ Send training with an end date ⓘ	Set the campaign end date	Launch Hours	Launch Minutes	AM/PM
	Select Date 🗓	hh ⌄	mm ⌄	Select Time Format ⌄

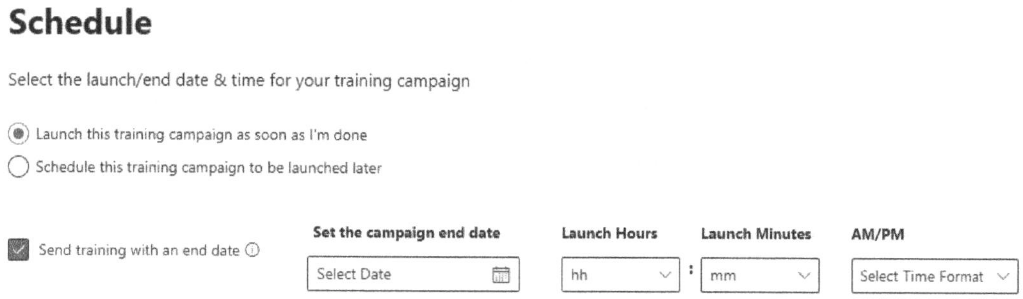

Figure 12.41 – The Schedule page

On the **Schedule** page, you will be presented with options for when to start the campaign. Options to start the campaign include right after completion of the wizard or at a specified date and time. Optionally, you can set a time by which users will need to complete the training. With selections made, click **Next** to be taken to the **Review** page. Here, a **Send a test** button enables you to test the campaign on yourself before final submission, and editing options are available throughout for any adjustments needed.

Once everything is reviewed, click on the **Submit** button to return to the **Training campaign** page, where your newly created campaign is now visible. Its status, **In Progress** or **Scheduled**, reflects your choices at the scheduling stage. Once the scheduled time appears, targeted users will receive an email like the one in the following figure, which will include a calendar file (an ICS extension) for the end user to add to their calendar.

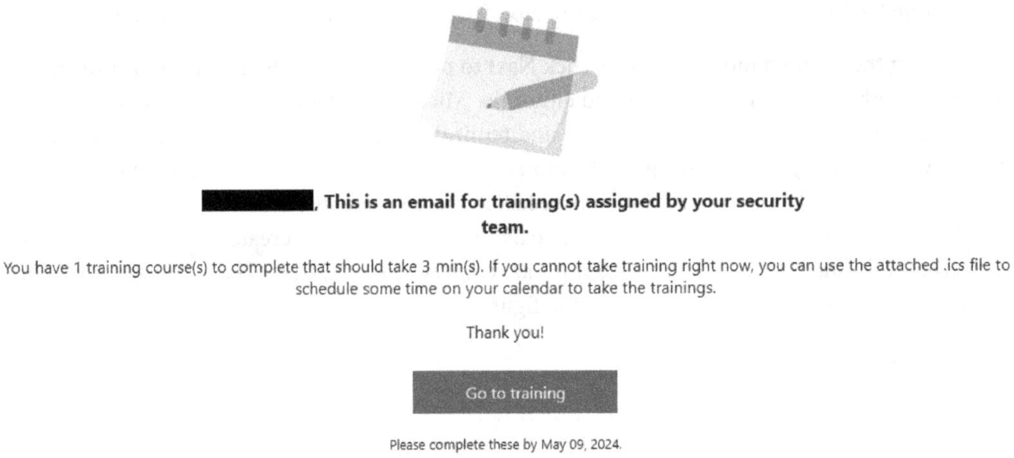

, This is an email for training(s) assigned by your security team.

You have 1 training course(s) to complete that should take 3 min(s). If you cannot take training right now, you can use the attached .ics file to schedule some time on your calendar to take the trainings.

Thank you!

Go to training

Please complete these by May 09, 2024.

Figure 12.42 – An end user campaign email

The email will include a blue button with the words **Go to training**, clicking on this button will open an URL link which will take the user to the training assignments page in the **Defender** portal (security.microsoft.com), where all the scheduled training modules will appear. The user will be able to click on any of the training modules in the list and training will start, as shown in the following figure.

When handling printed QR codes, you will want to observe the following best practices:

✓ **Inspect:** Visually inspect physical QR codes and look for signs of manipulation.

✓ **Verify:** Be cautious with QR codes and SMS links, especially from strangers or in public places. Verify the source.

✓ **Preview:** Use a trusted QR code scanner app that provides security features such as an option to preview link before navigation.

Figure 12.43 – The training module

During the campaign, as well as when it is completed, the security team will be able to click on it and review its details, including the completion status and details on the training being provided. The information provided can serve as a good approach to ensure that users stay up to date on required security refresher courses, and it can also be used to automate training. Launching end user training is not enough; we need to understand the data generated by this training to understand how well our end users understand the organization's security expectations.

Next, we will look at how to read the reports generated and use the insights provided to improve our efforts.

Understanding reports and insights

Microsoft delivers valuable insights and reports based on simulation outcomes and subsequent training activities. This essential information keeps you up to date on the progress of your users' threat readiness, offering guidance on the next steps to enhance their preparedness for potential future cyberattacks. The insights are presented succinctly within the reports, aiding in the quick identification of ways to refine an organization's security strategies (*Microsoft, 2024b*). Through the Microsoft Defender portal (`security.microsoft.com`), go to the **Attack Simulation Training** page, where these insights will be visible from both the **Overview** and **Report** tabs. Direct URL access is also possible – for the **Overview** tab, via `https://security.microsoft.com/attacksimulator?viewid=overview`, and for the **Report** tab, via `https://security.microsoft.com/attacksimulationreport`. Insights are provided via cards and organized into categories such as **simulation coverage**, **training completion**, **repeat offenders**, and **behavior impact on compromise rate** on both tabs, and the **Overview** tab offers extra information under the **Recent simulations** and **Training completion** cards, as shown in the following figure.

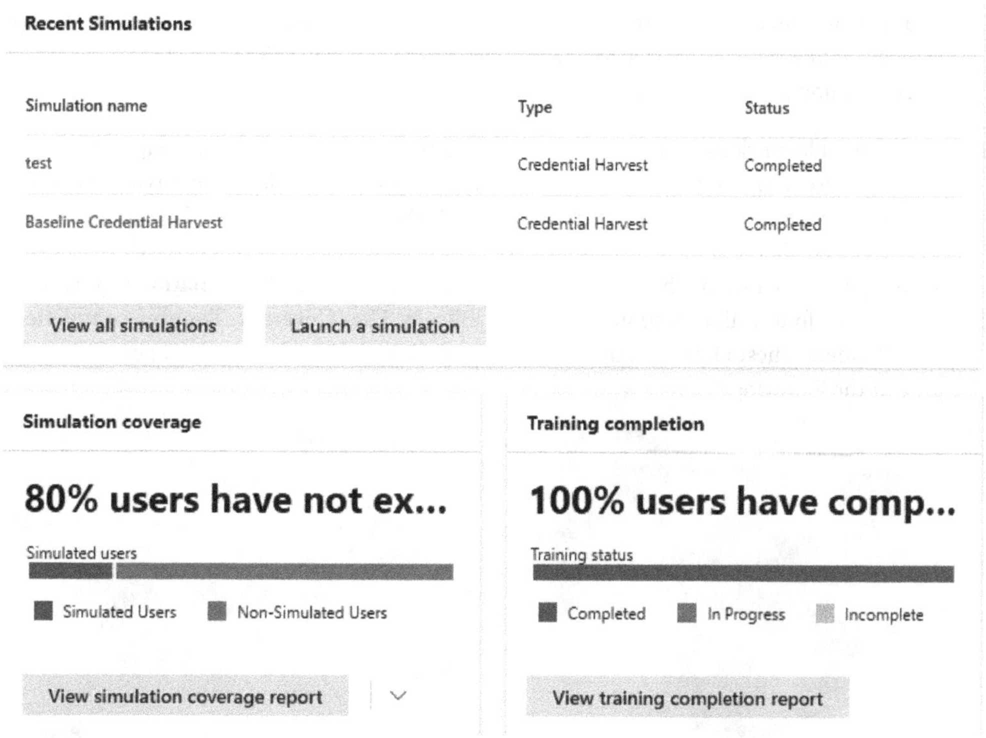

Recent Simulations

Simulation name	Type	Status
test	Credential Harvest	Completed
Baseline Credential Harvest	Credential Harvest	Completed

View all simulations   Launch a simulation

Simulation coverage

80% users have not ex...

Simulated users

■ Simulated Users   ■ Non-Simulated Users

View simulation coverage report   ⌄

Training completion

100% users have comp...

Training status

■ Completed   ■ In Progress   ▨ Incomplete

View training completion report

Figure 12.44 – Some of the insight cards in the Overview tab

The insights cards categories cover multiple important aspects of training. The contents of each card include the following:

- The **Recent Simulations** card: Displays the latest three simulations conducted within your organization, offering a detailed view when selected.

- The **Recommendations** card: Proposes various simulation types to consider, with a **Launch now** feature that initiates the new simulation wizard, selecting the advised simulation type by default.

- The **Simulation coverage** card: Illustrates the proportion of users who have received a simulation versus those who haven't, with details available upon hovering. Opting to view the simulation coverage report or launch a simulation for non-engaged users directs you to further details or the simulation wizard, respectively.

- The **Training completion** card: Categorizes users who underwent training post-simulation into **Completed**, **In Progress**, and **Incomplete**. Selecting to view the training completion report directs you to more comprehensive data.

- The **Repeat offenders** card: Details users who failed consecutive simulations, defining a repeat offender typically as a user compromised in two successive attempts. Viewing this card provides insights into users who need support during their training to decrease the risk of a compromise.

- The **Behavior impact on compromise rate** card: Compares user responses to simulations against historical data within Microsoft 365. This analysis enables you to monitor threat readiness progress by documenting response rates to repeated simulated attacks.

While insights provide quick insight into actions that you need to do, reports offer snapshot insights into campaign progress, which helps track user training progress and aids in identifying areas that require improvement. While it's beneficial to first consult the insights for immediate, high-impact data, reports accessed via the Attack Simulation report page provide a comprehensive view. Clicking on any of the reports provided by the insights cards takes you to the **Attack Simulation report** page, which is divided into four distinct sections – **Training Efficacy**, **User Coverage**, **Training completion**, and **Repeat offenders**. These align with the different categories provided by the insight cards. We can see this page in the following figure.

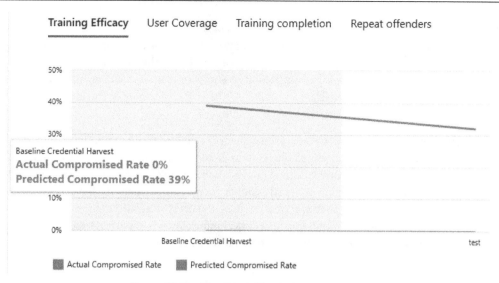

Figure 12.45 – The Attack Simulation report page

The default section on this page is the **Training Efficacy** tab, which mirrors the data found in the **Behavior impact on compromise rate** card and provides further context. A notable feature is a chart depicting both the actual and predicted compromised rates, as shown in the previous figure. Below this graph, a detailed table lists each simulation by name, the technique employed, tactics, the predicted and actual compromised rates, the total number of users targeted, and the number of users who interacted with the simulation. An **Export report** option is available to download this information in the CSV format for offline analysis.

In the **User Coverage** tab, a visual bar graph simplifies the difference between simulated and non-simulated users. Further down, a comprehensive table shows the specifics for each user, such as the username, the email address, the number of simulations included, the date and outcome of the last simulation, the number of simulations clicked, and instances of user compromise. This information, too, can be exported to a CSV file by clicking the **Export report** button. We can see this tab in the following figure.

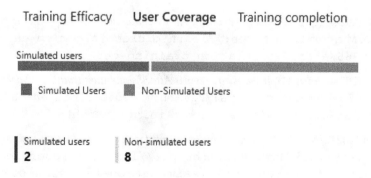

Figure 12.46 – The User Coverage tab

The **Training completion** tab highlights the progression of simulations – those completed, ongoing, or unfinished. Below the bar graph, a detailed table presents data for each participant, including the username, the email address, the count of simulations partaken, the date of the last simulation and its result, the most recent training completed, its completion date, and a comprehensive list of all the trainings the user has engaged in. This data is, likewise, exportable in the CSV format for further examination.

Finally, the **Repeat offenders** tab focuses on users repeatedly compromised in multiple simulations, spotlighting individuals needing intensified observation or additional training. A bar graph contrasts the count of repeat offenders against the general pool of simulated users. Accompanying this, a table lists each repeat offender, detailing the types and counts of engaged simulations, the email address, the last repeat count, the total number of repeat offenses, the latest simulation name and outcome, and the last assigned and completed training. This critical information is exportable, facilitating team discussions and strategic planning to enhance cybersecurity training initiatives.

Summary

In this chapter, we discussed the importance of providing effective security training to end users, who serve as the first line of defense against cyber threats. Traditional training methods often fail to engage employees and prepare them for real-world attacks. To overcome this challenge, we introduced attack simulation training, which immerses users in realistic scenarios, such as phishing emails and social engineering attempts, to improve their ability to recognize and respond to threats. Guidance was also provided on implementing attack simulation training, including available tools, platforms, and techniques to create custom simulations. Methods were explored to automate various aspects of the training process, and emphasis was placed on the importance of collecting and analyzing data from the simulations to identify areas for improvement. The topics covered in this chapter should help any organization adopt an innovative approach to security training, which can lead to a significant improvement in employees' awareness and a major reduction in the risk of a successful cyberattack.

References

- Microsoft (2024a, April 24). *Get started using Attack simulation training – Microsoft Defender for Office 365*. Microsoft Learn: `https://learn.microsoft.com/en-us/defender-office-365/attack-simulation-training-get-started`

- Microsoft (2024b, April 24). *Insights and reports Attack simulation training – Microsoft Defender for Office 365*. Microsoft Learn: `https://learn.microsoft.com/en-us/defender-office-365/attack-simulation-training-insights`

- Microsoft (2024c, May 2). *Simulation automations for Attack simulation training – Microsoft Defender for Office 365*. Microsoft Learn: `https://learn.microsoft.com/en-us/defender-office-365/attack-simulation-training-simulation-automations`

Index

Z

packtpub.com

Subscribe to our online digital library for full access to over 7,000 books and videos, as well as industry leading tools to help you plan your personal development and advance your career. For more information, please visit our website.

Why subscribe?

- Spend less time learning and more time coding with practical eBooks and Videos from over 4,000 industry professionals

- Improve your learning with Skill Plans built especially for you

- Get a free eBook or video every month

- Fully searchable for easy access to vital information

- Copy and paste, print, and bookmark content

Did you know that Packt offers eBook versions of every book published, with PDF and ePub files available? You can upgrade to the eBook version at packtpub.com and as a print book customer, you are entitled to a discount on the eBook copy. Get in touch with us at customercare@packtpub.com for more details.

At www.packtpub.com, you can also read a collection of free technical articles, sign up for a range of free newsletters, and receive exclusive discounts and offers on Packt books and eBooks.

Other Books You May Enjoy

If you enjoyed this book, you may be interested in these other books by Packt:

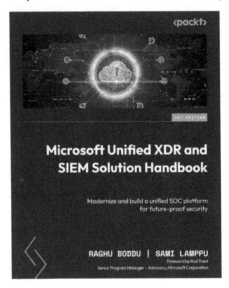

Microsoft Unified XDR and SIEM Solution Handbook

Raghu Boddu, Sami Lamppu

ISBN: 978-1-83508-685-8

- Optimize your security posture by mastering Microsoft's robust and unified solution
- Understand the synergy between Microsoft Defender's integrated tools and Sentinel SIEM and SOAR
- Explore practical use cases and case studies to improve your security posture
- See how Microsoft's XDR and SIEM proactively disrupt attacks, with examples
- Implement XDR and SIEM, incorporating assessments and best practices
- Discover the benefits of managed XDR and SOC services for enhanced protection

Mastering Microsoft 365 Defender

Ru Campbell, Viktor Hedberg

ISBN: 978-1-80324-170-8

- Understand the Threat Landscape for enterprises
- Effectively implement end-point security
- Manage identity and access management using Microsoft 365 defender
- Protect the productivity suite with Microsoft Defender for Office 365
- Hunting for threats using Microsoft 365 Defender

Packt is searching for authors like you

If you're interested in becoming an author for Packt, please visit `authors.packtpub.com` and apply today. We have worked with thousands of developers and tech professionals, just like you, to help them share their insight with the global tech community. You can make a general application, apply for a specific hot topic that we are recruiting an author for, or submit your own idea.

Share Your Thoughts

Now you've finished *Mastering Microsoft Defender for Office 365*, we'd love to hear your thoughts! Scan the QR code below to go straight to the Amazon review page for this book and share your feedback or leave a review on the site that you purchased it from.

https://packt.link/r/1835468284

Your review is important to us and the tech community and will help us make sure we're delivering excellent quality content.

Download a free PDF copy of this book

Thanks for purchasing this book!

Do you like to read on the go but are unable to carry your print books everywhere?

Is your eBook purchase not compatible with the device of your choice?

Don't worry, now with every Packt book you get a DRM-free PDF version of that book at no cost.

Read anywhere, any place, on any device. Search, copy, and paste code from your favorite technical books directly into your application.

The perks don't stop there, you can get exclusive access to discounts, newsletters, and great free content in your inbox daily

Follow these simple steps to get the benefits:

1. Scan the QR code or visit the link below

https://packt.link/free-ebook/978-1-83546-828-9

2. Submit your proof of purchase
3. That's it! We'll send your free PDF and other benefits to your email directly

www.ingramcontent.com/pod-product-compliance
Lightning Source LLC
Chambersburg PA
CBHW060649060326
40690CB00020B/4567